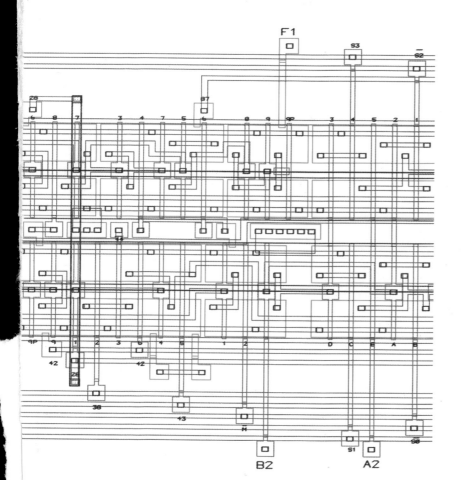

F1

D1370992

B2   A2

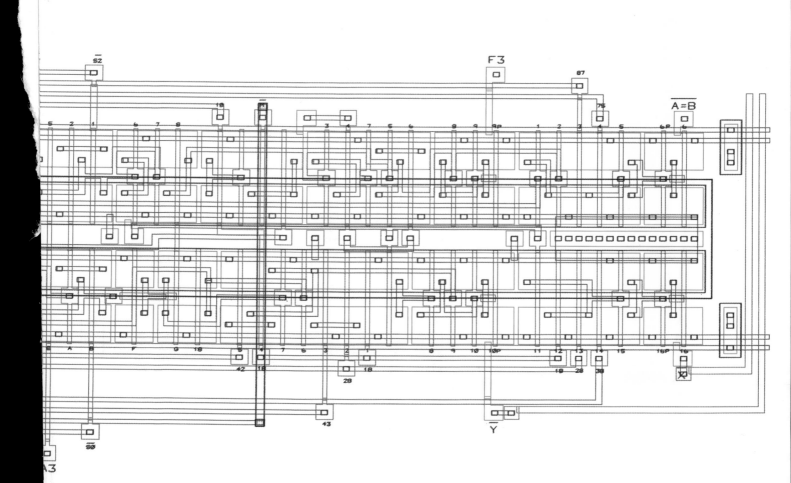

F3

87

$\overline{A=B}$

$\overline{Y}$

DIGITAL MOS
INTEGRATED
CIRCUITS

PRENTICE-HALL SERIES IN ELECTRICAL
AND COMPUTER ENGINEERING

Leon O. Chua, *Series editor*

BERGEN   Power Systems Analysis
CHUA AND LIN   Computer-Aided Analysis of Electronic Circuits: Algorithms
and Computational Techniques
GHAUSI AND LAKER   Modern Filter Design: Active RC and Switched
Capacitor
LAM   Analog and Digital Filters: Design and Realization
SCHAUMANN, GHAUSI, AND LAKER   Design of Analog Filters
WANG, N.,   Digital MOS Integrated Circuits: Design for Applications
WANG, S.,   Fundamentals of Semiconductor Theory and Device Physics

DIGITAL MOS INTEGRATED CIRCUITS

INTEGRATED CIRCUITS

Design for Applications

Niantsu Wang

Prentice Hall, Englewood Cliffs, New Jersey 07632

Library of Congress Cataloging-in-Publication Data

Wang, Niantsu.
    Digital MOS integrated circuits : design for application / Niantsu
Wang.
        p.   cm.
    Includes bibliographical references and index.
    ISBN 0-13-213109-9
    1. Digital integrated circuits. 2. Metal oxide semiconductors.
I. Title.
TK7874.W364   1989
621.381'73--dc19   88-8031
                                                                     CIP

Editorial/production supervision
  and interior design: Elaine Lynch
Cover design: Wanda Lubelska Design
Manufacturing buyer: Robert Anderson

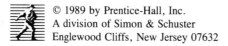 © 1989 by Prentice-Hall, Inc.
A division of Simon & Schuster
Englewood Cliffs, New Jersey 07632

The publisher offers discounts on this book when ordered
in bulk quantities. For information, write:

Special Sales/College Marketing
Prentice-Hall, Inc.
College Technical and Reference Division
Englewood Cliffs, NJ 07632

Printed in the United States of America

10 9 8 7 6 5 4 3 2 1

ISBN 0-13-213109-9

PRENTICE-HALL INTERNATIONAL (UK) LIMITED, *London*
PRENTICE-HALL OF AUSTRALIA PTY. LIMITED, *Sydney*
PRENTICE-HALL CANADA INC., *Toronto*
PRENTICE-HALL HISPANOAMERICANA, S.A., *Mexico*
PRENTICE-HALL OF INDIA PRIVATE LIMITED, *New Delhi*
PRENTICE-HALL OF JAPAN, INC., *Tokyo*
SIMON & SCHUSTER ASIA PTE. LTD., *Singapore*
EDITORA PRENTICE-HALL DO BRASIL, LTDA., *Rio de Janeiro*

To the memory of my dear mother

Hsin-Hsiang Fang

Grand Justice
Republic of China

CONTENTS

CONVENTIONS   xiii

PREFACE   xvii

CHAPTER 1   **CHARACTERISTICS AND OPERATION OF MOSFETS**   1

                  INTRODUCTION 1

                  1.   *PROPERTIES OF MOS SYSTEMS 1*

                  2.   *PROPERTIES OF MOS DEVICES 8*

                       2.1   Characteristics of Enhancement Mode Devices, 8
                       2.2   Characteristics of Depletion Mode Devices, 14

                  3.   *DEVICE CAPACITANCES 19*

                  4.   *THE TRANSCONDUCTANCE 22*

                  5.   *FREQUENCY RESPONSES 24*

                  6.   *TEMPERATURE EFFECTS 26*

                  7.   *A SIMPLE CIRCUIT MODEL FOR MOS DEVICES 27*

                       REFERENCES 30

                       APPENDIX: CIRCUIT MODELS FOR MOS DEVICES 30

CHAPTER 2   **MOS TECHNOLOGY REVIEW**   35

                  INTRODUCTION 35

                  1.   *BASIC PROCESS FLOW 35*

2.   PROCESS AND DEVICE DESIGN CONSTRAINTS   40

    2.1   Threshold Voltage, 41

        2.1.1   Substrate sensitivity, 41
        2.1.2   Punchthrough effect, 41
        2.1.3   Short-channel effect, 41
        2.1.4   Narrow-channel effect, 41
        2.1.5   Hot-carrier effect, 42
        2.1.6   Threshold voltage adjustment: channel tailoring, 42

    2.2   Subthreshold Current, 43
    2.3   Velocity Saturation Effect, 44
    2.4   Device Parasitic Capacitances and Resistances, 45
    2.5   Parasitic Bipolar Transistors: Snapback
          Breakdown and CMOS Latchup, 46

3.   SCALING THEORY   49

4.   ADVANCES IN PROCESS AND DEVICE DESIGN   50

    4.1   LDD Structure and Salicide Process, 50
    4.2   CMOS Processes, 52

        4.2.1   Trench isolations, 52
        4.2.2   Latchup control, 53

5.   MOS DYNAMIC RAM CELLS   54

6.   CONCLUSION   57

     REFERENCES   57

CHAPTER 3   DIGITAL INVERTERS—DC ANALYSIS   60

                 INTRODUCTION   60

1.   THE TRANSFER CHARACTERISTIC   62

2.   THE MAXIMUM DOWNLEVEL DESIGN   64

3.   DC STABILITY DESIGN   67

4.   THE NOISE MARGINS   70

5.   COMPARISON: NMOS VERSUS CMOS   73

6.   DEVICE TRACKING AND TEMPERATURE EFFECTS   74

7.   CONCLUSION   76

     REFERENCES   77

     PROBLEMS   77

     APPENDIX: PROGRAM OPTBETA   80

CHAPTER 4   DIGITAL INVERTERS—TRANSIENT ANALYSIS   88

                 INTRODUCTION   88

1.   CHARGING UP A CAPACITOR   89

    1.1   Constant Pullups, 89
    1.2   Depletion Mode Pullups, 97
    1.3   Bootstrap Pullups, 100

2.   DISCHARGING A CAPACITOR   107

    2.1   NMOS Pulldowns, 107
    2.2   PMOS Pullups, 109

3. DESIGN PRACTICE: THE STATISTICAL ANALYSIS 112

4. CONCEPT OF CIRCUIT DELAY 114

5. CALCULATION OF DELAYS 115

6. DELAY CALCULATORS 123

7. DEVICE CAPACITANCES AND THE MILLER EFFECT 131

8. CONCLUSION 136

PROBLEMS 136

APPENDICES:

1. DELAY CALCULATOR FOR CONSTANT PULLUPS 137

2. DELAY CALCULATOR FOR DEPLETION MODE PULLUPS 139

CHAPTER 5   RECEIVERS AND DRIVERS   142

INTRODUCTION 142

1. OFFCHIP RECEIVERS 142

1.1 Differential Receivers, 142

1.1.1 Static differential receivers, 143
1.1.2 Dynamic differential receivers, 148

1.2 Single-Ended Receivers, 149
1.3 Noise Margins of a Receiver, 155

2. HIGH-POWER DRIVERS 156

2.1 Differential Drivers, 156
2.2 Single-Ended Drivers, 157

2.2.1 Basic concept of pushpull drivers, 157
2.2.2 Pushpull drivers with depletion mode pullups, 161
2.2.3 Calculation of delays of pushpull drivers, 163
2.2.4 Dynamic drivers—The principle of invert and delay, 166
2.2.5 CMOS drivers, 175

3. OFFCHIP DRIVERS 179

3.1 TTL Compatibility, 179
3.2 Tristate Drivers, 179
3.3 Short-Circuit Protection, 181

4. MOS LINE DRIVERS 182

4.1 Analysis of Ideal Transmission Lines, 182
4.2 Analysis of MOS Driver/Receiver Pairs, 186
4.3 Termination of Transmission Lines, 188

5. SYSTEM NOISES 192

5.1 Crosstalk, 192
5.2 Simultaneous Switching of OCDs, 194

6. CONCLUSION 194

REFERENCES 195

APPENDICES:

1. FORTRAN LISTING OF SUBROUTINE RPUN 195

        2. IDEAL TRANSMISSION LINE THEORY 197

        3. SUBSTRATE BIAS GENERATION 206

CHAPTER 6   **MOS MEMORY CHIP DESIGNS**   **209**

       INTRODUCTION 209

    1.   *STATIC RAMS 210*

       1.1 Basic SRAM Architecture, 210

       1.2 Circuit Implementations, 215

       1.3 Advances in SRAM Architecture and Circuits, 225

           1.3.1 Power reduction, 226

           1.3.2 Performance enhancement, 226

    2.   *STATIC VERSUS DYNAMIC: 4-DEVICE CELLS 231*

       2.1 Static 4-Device Cells, 231

       2.2 Dynamic 4-Device Cells, 234

    3.   *DYNAMIC RAMS 237*

       3.1 Basic DRAM Architecture, 240

       3.2 Circuit Implementations, 242

       3.3 Dynamic Sense Amplifiers, 250

           3.3.1 Basic operations, 250

           3.3.2 Noise analysis, 257

           3.3.3 Variations in design, 264

           3.3.4 Bitline configurations, 267

       3.4 Advances in DRAM Architecture and Circuits, 268

           3.4.1 Chip modes, 268

           3.4.2 Refresh schemes, 269

           3.4.3 Redundancy, 270

    4.   *CONCLUSION 270*

       REFERENCES 271

CHAPTER 7   **MOS LOGIC CHIP DESIGNS**   **273**

       INTRODUCTION 273

    1.   *MASTERIMAGE DESIGN SYSTEM 274*

       1.1 Design Systems, 274

       1.2 Design Procedures, 274

    2.   *CHIP IMAGE 276*

    3.   *CIRCUIT LIBRARY 279*

       3.1 Static Books, 279

           3.1.1 Multiinput NOR: OIs, 279

           3.1.2 Multiinput NAND: AIs, 279

           3.1.3 Two-level logic functional blocks: OAIs and AOIs, 280

           3.1.4 Exclusive OR; XORs and XNORs, 284

           3.1.5 Parity generators, 286

           3.1.6 Decoders, 286

           3.1.7 Multiplexers, 286

           3.1.8 Rotators and barrel shifters, 288

      3.1.9   Latches, 288
      3.1.10 Shift register latch; SRLs, 293
      3.1.11 Clock generators and drivers, 295
      3.1.12 Flip-flops, 297
      3.1.13 Counters, 298
      3.1.14 Edge-triggered latches, 300

3.2   Dynamic Logic Circuits, 302

      3.2.1   Dynamic shifter registers: DSRs, 302
      3.2.2   Dynamic logic, 304
      3.2.3   Domino logic, 309
      3.2.4   Cascode logic, 311

3.3   Large Macros, 314

      3.3.1   Adders, 314
      3.3.2   Arithmetic/logic unit: ALU, 320
      3.3.3   Multipliers, 322
      3.3.4   Programmable logic arrays: PLAs, 328
      3.3.5   Read only memory: ROM/ROS, 333
      3.3.6   General-purpose register: GPR, 334

4.   *DESIGN FOR TESTABILITY*   *334*

4.1   Stuck-At Fault Models, 335
4.2   CMOS Faults, 337
4.3   Generation of Test Patterns, 338
4.4   Level-Sensitive Scan Design, 340
4.5   Selftest and Pseudorandom Test Pattern
       Generation, 345

5.   *TIMING VERIFICATION*   *347*

5.1   Delay Calculation, 347
5.2   Timing Analysis, 347

6.   *CONCLUSION*   *349*

     REFERENCES   *350*

     APPENDIX: LINEAR FEEDBACK SHIFT REGISTERS   *351*

     INDEX   *359*

CONVENTIONS

1.   Units

1.1   Voltages and Currents

Unless specified otherwise, voltages and currents are measured in volts and milliamperes.

$$[v] = V \quad \text{and} \quad [i] = mA$$

1.2   Electrical Elements

Electrical elements R, L, and C are specified by

$$[R] = k\Omega \quad (\text{kilo-ohm} = 10^3 \, \Omega)$$
$$[L] = nH \quad (\text{nano-henry}) = 10^{-9} \, H$$

and

$$[C] = pF \quad (\text{pico-farad} = 10^{-12} \, F)$$

1.3   Time

Unit for time t is given by

$$[t] = \frac{[C][v]}{[i]} = \frac{[L][i]}{[v]} = nS \, (\text{nanosecond} = 10^{-9} \, S)$$

Physical quantities without explicit units should be interpreted as, or derived from, the preceding expression.

2.   MOS Device Symbols

Various symbols are used for different types of devices with different threshold voltages.

2.1   N-Channel Devices

2.1.1   Enhancement Mode Devices

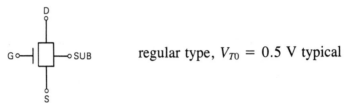

regular type, $V_{T0} = 0.5$ V typical

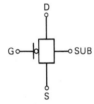

low V_T type, $V_{T0} = 0.15$ V typical

2.1.2   Depletion Mode Devices

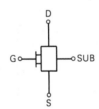

regular type, $V_{T0} = -1.3$ V typical

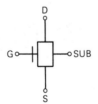

weak depletion type, $V_{T0} = -0.5$ V typical

2.2   P-Channel Devices

2.2.1   Enhancement Mode Devices

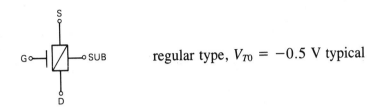

regular type, $V_{T0} = -0.5$ V typical

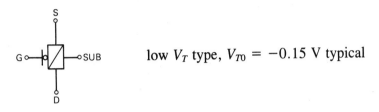

low V_T type, $V_{T0} = -0.15$ V typical

2.2.2   Depletion Mode Devices

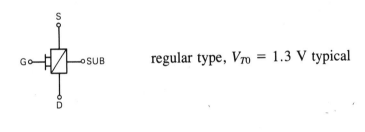

regular type, $V_{T0} = 1.3$ V typical

weak depletion type, $V_{T0} = 0.5$ V typical

In case the substrate is connected to a constant bias voltage, the connection SUB is omitted for simplicity.

3.   Circuit Schematic Conventions

Devices are referred to by numbers or characters written in their boxes. Power supplies are represented by single bars labeled V_H, and ground is represented by two or more bars labeled GD. When both n-channel and p-channel devices are present, the p-channel devices are sometimes labeled with a primed number or character without a diagonal in the box. For example, a simple CMOS inverter can be represented either by

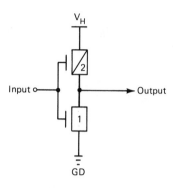

or by

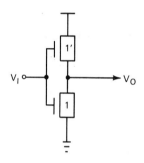

Unless specified otherwise, $V_H = 3$ V typical in this book.

PREFACE

The MOS technology has been advancing at an astounding rate in the past decade. Memory chip density quadrupled and performance doubled with each new product. Logic chip function, at the same time, grew from simple 4-bit microcontrollers to full-blown 32-bit micromainframes. While the basic structure of the MOS devices remains the same, processing technology and circuit designs are constantly changing. The process becomes extremely sophisticated, and new, innovative circuits are invented every day.

This book is an account of these achievements—from a circuit's vantage point. Various circuit techniques and ideas, which have been responsible for the tremendous success of the MOS IC industry, are introduced to the reader as different products are discussed. It is my sincere hope that this book, in addition to being a technical text, serves as a tribute to all IC designers—from whose innovative minds has sprung the whole VLSI industry.

Presentation of the material in this book follows a conventional approach. Chapter 1 is a brief but complete review of the MOS device theory. Since all device equations are derived in this chapter, the reader is encouraged to go through the whole chapter to become familiar with the basic device concepts and notations used throughout this book. Chapter 2 is an introduction to the MOS process. Emphasis here is on the interrelationship between processes and device designs. In Chapters 3 and 4 we discuss the basic operations of MOS digital inverters. Chapter 3 is concerned with the DC design, where the concepts of noise margins and DC stability are introduced. Chapter 4 is a detailed discussion of transient responses of different inverter configurations. The concept of circuit delays is defined, and delay calculators are derived that are of paramount importance to the development of VLSI chip design tools. With these chapters as background, we can now design high-power drivers and receivers. In Chapter 5 the reader will see the application of delay calculators to the design of practical drivers and receivers. Chapters 6 and 7 discuss products. In Chapter 6 two small but complete RAMs are analyzed that give the reader the flavor of a real product design. In Chapter 7 a simple but powerful semicustom logic-design approach is discussed that effectively summarizes today's trends in MOS logic chip design methodologies. Since this book is mainly concerned with MOS circuits, no specific design tools are mentioned, but the design approaches are clearly oriented toward a structured methodology. I hope this book will be useful not only to circuit designers, but also to the programmers who develop design tools as well.

Circuit design is both a science and an art. As such, it is easy to learn but difficult to master. The most effective way to learn design is through examples. By studying some nontrivial practical designs we can better understand and appreciate the problems and their solutions. Examples in this book, however, serve only as pedagogical tools. While most of them can be readily applied to practical situations, the main purpose of the examples is to help the reader develop skills in designing MOS circuits. For example, while many readers may not think that they will ever be involved in a RAM design (not a terribly valid assumption, as my experiences can tell), the generation of timing chain and the analysis of a balanced latch in RAM chip designs, as we shall discuss in Chapter 6, are certainly important concepts that any professional MOS circuit designer should master. In this age of system integration—where digital logic, memory, and analog interfaces are all processed in a single chip—the most effective weapon a circuit designer can possess, to defend himself from becoming obsolete, is his analytical ability to grasp the main theme of a new problem and then apply his design skills to solve it. Thus the purpose of this book is more of a guide, to help the reader find his own way in the forest of technology, than a cookbook full of recipes.

During the writing of this book I have received much help and encouragement, either directly or indirectly, from my colleagues. In addition to the management of IBM, I am much indebted to Steve Doran, Larry Heller, Paul Heudorfer, Paul Hwang, Jack Gerschbach, Hsing San Lee, Sol Lewin, Bob Mao, and Moo Wen—all with IBM Corporation. I am also deeply indebted to Haik Marcar of the Technical Communication Department of IBM, San Jose for his excellent editing work. His enthusiasm and great sense of humor have certainly made the task of writing this book much easier.

Finally, I would like to express my deep gratitude to my family. This book would not have been possible without their support. The encouragement from my parents, the understanding of my wife Alice, and the cooperation of my children, all were indispensible in making this book a reality. To my children, Charlie and Shirley, Daddy really should have spent more time with you going fishing and camping.

Niantsu Wang
San Jose, California

DIGITAL MOS
INTEGRATED
CIRCUITS

CHARACTERISTICS AND OPERATION OF MOSFETS

INTRODUCTION

Electrical circuits are composed of physical devices. And the behavior of a circuit is affected by these devices. A circuit designer has to work closely with both the device designers and the system engineers, maintaining communication between the component manufacturers and the system architects. Since a technology's measurable merit—power, density, performance, reliability, and so on—is ultimately tested on the circuit level, it is a circuit designer's responsibility to optimize the technology for a particular application. Given this strong relationship between devices and circuits, to appreciate both the flexibility and limitations of the technology, it is essential that a circuit designer have a solid background in basic device theory.

This chapter is a brief review of MOS device theories relevant to circuit applications. It starts with an idealized MOS junction theory, formulating the concept of threshold voltages, deriving IV characteristics and device capacitances, and then considering practical limitations on device performance, such as transconductance, frequency responses, and temperature effects.

To keep it simple, only n-channel devices are considered here, with the understanding that p-channel devices can be treated in the same fashion with some modification.

Finally, at the end of the chapter, a simple circuit model of an MOS device is derived for use throughout this book.

1. PROPERTIES OF MOS SYSTEMS

Consider the MOS structure shown in Figure 1, where a voltage source, V_{GB}, is applied from the metal or polysilicon *gate, M*, to the silicon *bulk, S*. Because electrons in different materials reside in different energy levels, the mobile charge carriers in each material tend to move into each other. As a result, built-in potentials are

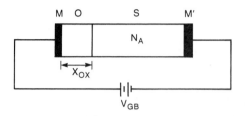

Figure 1. An MOS system with bias voltage V_{GB}.

developed across metallurgical junctions. It is these internal potential variations that cause different electrical properties in the device. Since the internal potentials are affected by the boundary conditions at device terminals, various electric properties under different operating conditions can be achieved by varying the bias voltage, V_{GB}.

The simplest method of analyzing internal potential variations in an MOS structure is through the construction of an energy band diagram. Before materials are brought into contact with each other, the energy band of each material can be plotted independently relative to a common reference. As is shown in Figure 2, this common reference, E_o, called *the vacuum level,* is the energy of an electron that has just been freed from the influence of the material. Since the materials do not interact, E_o is at a constant level independent of positions.

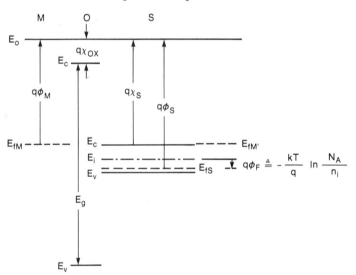

Figure 2. The energy band diagrams of each material before being brought into contact.

The energy difference, $E_o - E_f$, where E_f is the Fermi level of the material, is called the *work function.* It is usually denoted by $q\phi$ in eV. For *metal* and some other conductors—because the Fermi level is immersed in the conduction band— $q\phi_M \triangleq E_o - E_{fM}$ is also the energy difference between a "just-free" electron and the conduction electrons in the material. For *semiconductors,* however, the minimum amount of energy possessed by an electron in escaping from the surface $E_o - E_c$ is smaller than its work function because the Fermi level E_{fs} always lies below E_c. This quantity $q\chi_S \triangleq E_o - E_c$ is called *the electron affinity* of the semiconductor. It is a property of the semiconductor's crystal lattice structure and, hence, is independent of the doping concentrations.

Once the MOS system is formed, built-in potentials are established across the junctions corresponding to different values of V_{GB}. The following is a brief discussion of the cases presented in Figure 3.

(1) $V_{GB} = 0$

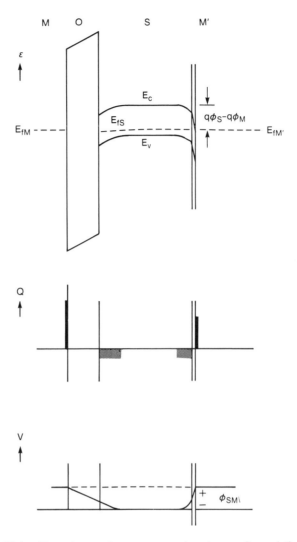

Figure 3(a). The electronic energy ε, the charge Q, and the electrostatic potential V, of an MOS system with $V_{GB} = 0$.

In this case the whole system is said to be in thermal equilibrium. Since the Fermi level determines the probability of occupancy of an allowed electronic energy state, E, which should be the same everywhere in a system that is in thermal equilibrium, the Fermi levels E_{fM}, E_{fs}, and $E_{fM'}$ must all align. As a result, the conduction band edge, E_c, bends down at the Si-SiO₂ interface. The bending causes this region to be depleted of mobile carriers and exposes the negatively charged acceptor ions. The MOS junction is therefore charged up with the metal positive relative to the silicon bulk. At the other end of the region, however, the silicon is in direct contact with metal. Ideally, the band bending of E_c would be the work function difference, $q\phi_S - q\phi_M$. The Fermi level E_{fs} would then be very close to E_c, and the silicon would become n-type near the interface. In reality, however, because of the surface states in the interface, E_{fs} is pinned at about one-third of the energy gap above E_v. A depletion region is formed in the silicon as E_c bends down, with additional voltage drop across the thin layer of the interface. The total voltage drop across SM' junction is equal to $\phi_{MS} = \phi_M - \phi_S$. Since the voltage drop across the MOS junction is $\phi_{SM} = \phi_S - \phi_M$, KVL (Kirchhoff's Voltage Law) is satisfied with $V_{GB} = 0$. See Figure 3(a).

(2) $V_{GB} \neq 0$

Now, let $V_{GB} \neq 0$, so the whole system is no longer in thermal equilibrium. Since the substrate current is negligible over a wide range of V_{GB}, the Fermi levels E_{fS} and $E_{fM'}$ are always aligned, maintaining the constant band bending. The voltage drop across the SM' junction is therefore constant at ϕ_{MS}.

The voltage across the MOS junction, however, varies as a function of V_{GB}. In particular, the voltage across the MOS junction equals zero when $V_{GB} = \phi_{MS}$. If there is no charge either within the oxide or betweeen the Si-SiO$_2$ interface the electric field in the oxide and silicon will be zero. At this point, the MOS system is characterized by flat energy bands maintaining a neutral charge. This situation is illustrated in Figure 3(b).

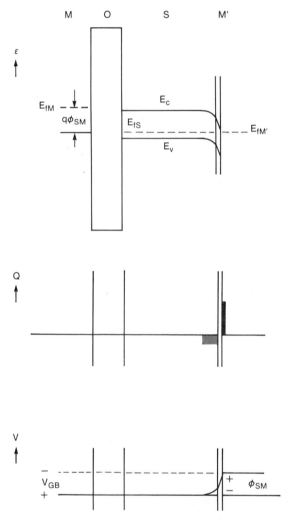

Figure 3(b).   The flat band voltage of an idealized MOS system.

In reality, however, there are always positive charges in both the oxide and the Si-SiO$_2$ interface. As shown in Figure 3(c), there is a positive charge distribution, $\rho(x)$, in the oxide and some fixed positive charge, Q_{ss}, due to surface states in the Si-SiO$_2$ interface.* The presence of these charges makes it impossible to maintain a constant zero field simultaneously in the oxide and the silicon. The applied voltage, V_{GB} at which the silicon bulk exhibits a flat energy band at the MOS junction and a neutral charge, is called the *flat band voltage,* denoted by V_{FB}. Under the flat band condition the field in the silicon is zero from Si-SiO$_2$ interface to the bulk. The field within the oxide, however, is generally negative.

---

* $\rho(x)$ and Q_{ss} are densities per unit area.

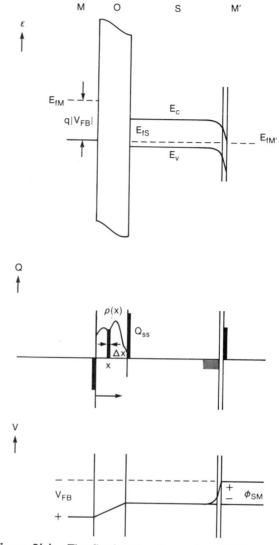

Figure 3(c). The flat band voltage of an MOS system.

The flat band voltage, V_{FB}, is an important design parameter of MOS devices. To understand V_{FB}, consider a thin layer of charge, $\rho(x)\Delta x$, at coordinate x in the oxide as shown in Figure 3(c). To prevent it from inducing an image charge on the silicon destroying the charge neutrality, a negative voltage, $\Delta V_1(x)$, must be applied from the gate to the bulk to induce an equal amount of charge in the metal. According to Gauss law,

$$\Delta V_1(x) = -\frac{\rho(x)\Delta x}{\varepsilon_{ox}/x} = -\frac{x}{\varepsilon_{ox}}\rho(x)\Delta x$$

Integrating the above expression from 0 to x_{ox}, the total contribution of the oxide charge to V_{FB} is given by:

$$V_1 = -\frac{1}{\varepsilon_{ox}}\int_0^{x_{ox}} x\rho(x)\,dx$$

$$= -\frac{1}{C_o'}\int_0^{x_{ox}} \frac{x}{x_{ox}}\rho(x)\,dx$$

where $C_o' \triangleq \varepsilon_{ox}/x_{ox}$. Similarly, the contribution of Q_{ss} is given by:

$$V_2 = -\frac{Q_{ss}}{C_o'}$$

The flat band voltage, V_{FB}, is therefore given by:

$$V_{FB} = \phi_{MS} + V_1 + V_2$$

$$= \phi_{MS} - \frac{1}{C_o'} \int_0^{x_{ox}} \frac{x}{x_{ox}} \rho(x)\, dx - \frac{Q_{ss}}{C_o'}$$

With V_{FB} as a reference, we can now discuss three interesting operating conditions: (a) $V_{GB} < V_{FB}$, (b) $V_{GB} > V_{FB}$, and (c) $V_{GB} \gg V_{FB}$.

(a) $V_{GB} < V_{FB}$, see Figure 3(d)

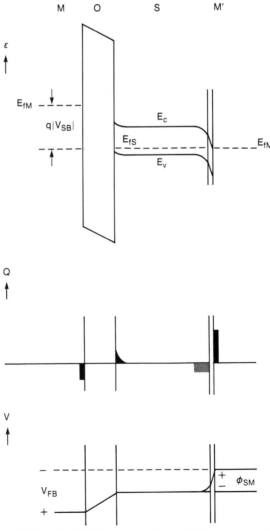

Figure 3(d).   An MOS system biased into accumulation.

In this case the energy band bends upward at the Si-SiO₂ interface. Holes accumulate and the silicon is charged positive relative to the metal.

(b) $V_{GB} > V_{FB}$, see Figure 3(a)

In this case the energy band bends downward at the Si-SiO₂ interface. Holes are depleted from the surface of the silicon. Since $V_{FB} < 0$, the condition of thermal equilibrium $V_{GB} = 0$ is significant in this case.

(c) $V_{GB} \gg V_{FB}$, see Figure 3(e)

As the band bends more and more with V_{GB}, the Si-SiO₂ interface eventually takes on n-type characteristics. Electrons are attracted to the surface, and the surface area is then said to be *inverted*. If V_{GB} is so large that the electron concentration in the surface exceeds the hole concentration in the bulk, the surface is said to be *strongly inverted*. Once strong inversion sets in, band bending at the interface remains rela-

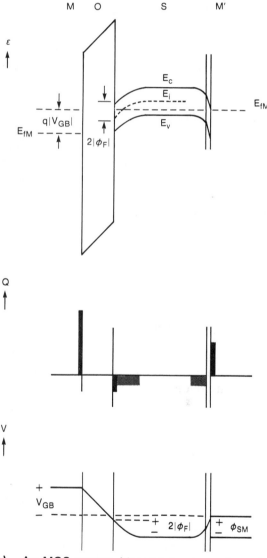

Figure 3(e).   An MOS system biased into strong inversion (oxide charges and Q_{ss} neglected).

tively constant; $|E_i - E_{fS}|_s \cong |E_i - E_{fS}|_b$, where s stands for the surface and b stands for the bulk. Any further increase in V_{GB} will attract a large number of electrons into the surface without appreciably affecting the band bending because the concentration of electrons increases exponentially with $|E_i - E_{fS}|_s$. Thus, *the surface potential, V,* defined as the voltage drop from the surface to the bulk, remains constant at

$$V = 2|\phi_F|$$

where $q\phi_F \triangleq E_{fS} - E_i$ in the bulk.

EXERCISES

Refer to Figure 3(e)

1. Redraw Figure 3(e) to include the effect of the oxide charge and Q_{ss}.
2. Let E_x be the uniform field across the oxide due to charges in the depletion region and free electrons in the Si-SiO₂ interface. Prove that under strong inversion,

$$V_{GB} = E_x x_{ox} + 2|\phi_F| + V_{FB}$$

The problem is equivalent to analyzing an idealized MOS junction (i.e., one with no oxide charge and Q_{ss}) with an applied voltage: $V_{GB} - V_{FB}$.

2. PROPERTIES OF MOS DEVICES

This section will cover the basic principles underlying the operation of MOS devices. Both enhancement mode and depletion mode devices will be considered here. To keep the results mathematically tractable, simple "square laws" will be derived assuming uniform substrate doping. Practical, in-use devices do not follow the square law exactly; however, (as shown in Section 7) most of the discrepancies pertinent to circuit design can be taken care of by properly defining the threshold voltages and the substrate sensitivity.

2.1 Characteristics of Enhancement Mode Devices

(1) The Threshold Voltage: Consider the MOS structure shown in Figure 4(a). The device operates under the condition of *strong inversion*. Let V_G be the positive voltage applied to the gate and V_B be the constant bias applied to the substrate. The voltage, V_S, applied to the n+ diffusion of the device fixes the surface potential of the inversion layer underneath the gate. From the energy band diagram, Figure 4(b), it is clear that the surface potential V, relative to the bulk, is given by:

$$V = V_S - V_B + 2|\phi_F| \tag{1}$$

where $\phi_F \triangleq (E_f - E_i)/q$ is calculated in the silicon bulk. For uniform doping, $\phi_F = -kT/q \ln(N_A/n_i)$. Now, applying KVL from the gate to the substrate,

$$V_G - V_B = E_x x_{ox} + V + V_{FB} \tag{2}$$

where E_x denotes the uniform electric field across the gate oxide and V_{FB} is the flat band voltage. Equations (1) and (2) are the fundamental equations describing the operation of MOS capacitors.

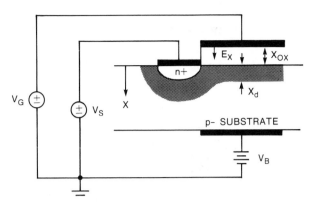

Figure 4(a). An MOS structure biased in strong inversion. The dark areas are metal contacts, and the shaded area designates depletion region.

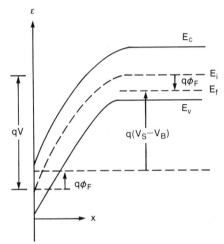

Figure 4(b). The energy band diagram for above device.

Under the condition of strong inversion, free electrons exist in the inversion layer. Let Q_n' be the free electron charge density per unit area; then, according to Gauss law,

$$E_x = -\frac{Q_n' - qN_Ax_d}{\varepsilon_{ox}} \tag{3}$$

where N_A is the uniform doping concentration of acceptors underneath the gate and x_d is the width of the depletion region. Assuming a negligible width for the inversion layer, the solution of Poisson's equation in the depletion region

$$\frac{d^2\psi}{dx^2} = -\frac{1}{\varepsilon_s}qN_A$$

with boundary conditions $\psi(x_d) = 0$ and $d\psi(x_d)/dx = 0$ gives (notice that $V = \psi(0) - \psi(x_d)$)

$$x_d = \sqrt{\frac{2\varepsilon_s V}{qN_A}} \tag{4}$$

Substituting (3) and (4) into (2),

$$V_G - V_B = -\frac{x_{ox}}{\varepsilon_{ox}}(Q'_n - qN_A x_d) + V + V_{FB}$$

$$= -\frac{Q'_n}{C'_o} + \left(1 + \frac{2C'_s}{C'_o}\right)V + V_{FB}$$

where $C'_o = \varepsilon_{ox}/x_{ox}$ is the capacitance per unit area of the gate oxide and $C'_s \triangleq \varepsilon_s/x_d = \sqrt{\varepsilon_s q N_A/2V}$. Rearranging the terms in the above equation,

$$V = \frac{C'_o}{C'_o + 2C'_s}(V_G - V_B - V_{FB}) + \frac{Q'_n}{C'_o + 2C'_s} \qquad (5)$$

Since C'_s is usually of the same or smaller order of magnitude as C'_o, a constant, Equation (5) states that the surface potential, V, is almost an affine function of the free electron charge density, Q'_n, for a given gate voltage V_G.

From the point of view of circuit design, however, it is more convenient to express the free electron charge density, Q'_n, in terms of the "gatedrive" $V_{GS} = V_G - V_S$. Substitution of (1) into (5) yields

$$Q'_n = -C'_o(V_{GS} - 2|\phi_F| - V_{FB})$$
$$+ \sqrt{2 \, \varepsilon_s q N_A(V_{SB} + 2|\phi_F|)} \triangleq -C'_o(V_{GS} - V_T) \qquad (6)$$

where $V_T \triangleq V_{FB} + 2|\phi_F| + \sqrt{2 \, \varepsilon_s q N_A(V_{SB} + 2|\phi_F|)}/C'_o$ is called the *threshold voltage* of the device and $V_{SB} = V_S - V_B$. Notice that V_T is a function of the source to substrate voltage, V_{SB}, and hence depends on V_S once V_B is fixed.

Strictly speaking, the above analysis is valid only if the device operates under the condition of strong inversion when $V_{GS} \geq V_T$. To express this condition explicitly,

$$Q'_n = -C'_o(V_{GS} - V_T) \, \text{sgn} \, (V_{GS} - V_T) \qquad (7)$$

where the signum function sgn (•) is defined by

$$\text{sgn} \, (x) = \begin{cases} 1, & \text{if } x > 0 \\ 0, & \text{otherwise} \end{cases}$$

(2) The IV Characteristics: Consider the MOS device shown in Figure 5. Two n+ diffusions are now created on both sides of the gate, Ⓖ, so current can flow from one side to another. The gate voltage, V_G, controls the current. The Si-SiO₂ interface must be biased into strong inversion to form a channel in which carriers may pass. Carriers always flow in one direction, from the source, Ⓢ, to the drain, Ⓓ. Since the carriers are electrons in n-channel devices, $V_S \leq V_D$ and $I_{DS} \geq 0$. For p-channel devices in which the carriers are holes, $V_S \geq V_D$ and $I_{DS} \leq 0$.

The distance from source to drain is called *the channel length*, designated by L. The dimension perpendicular to current flow is called *the channel width, W*. As we shall see shortly, I_{DS} can be controlled by the ratio W/L.

Now, consider the major mechanism of current flow in an MOS device: the drift current caused by an extended field from source to drain.

Assume $V_D > V_S$ so that there is a current flowing from drain to source. Assigning the lateral direction the z-axis, both the surface potential V and the free

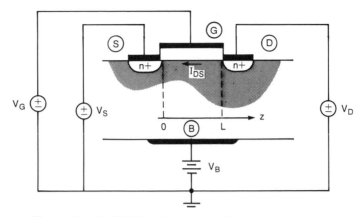

Figure 5.   An MOS device operated in strong inversion.

electron charge density Q'_n are now functions of z. Differentiating Equation (5) with respect to z (notice that $C'_s = \hat{C}'_s(z)$ is also a function of z) yields:

$$(C'_o + C'_s)\frac{\partial V}{\partial z} = C'_o\frac{\partial V_G}{\partial z} + \frac{\partial Q'_n}{\partial z}$$

The above equation states that the gradient of the surface potential, $\partial V/\partial z$, is related to both the concentration gradient of free electrons via $\partial Q'_n/\partial z$ *and the "fringe field"* $\partial V_G/\partial z$ due to variations of V_G along the channel. If V_G is uniform along the channel making $\partial V_G/\partial z = 0$, the surface field $-\partial V/\partial z$ is proportional to the concentration gradient, $\partial Q'_n/\partial z$, by the relation

$$E_z = -\frac{\partial V}{\partial z} = -\frac{1}{C'_o + C'_s}\frac{\partial Q'_n}{\partial z} \qquad (8)$$

The drift current density per unit channel width is given by

$$J = <v(z)> Q'_n(z)$$

where $<v(z)>$ is the average velocity of the electrons. To a first-order approximation, $<v(z)> = \mu E_z(z)$ where

$$\mu = \frac{\mu_o}{1 + |\mu_o E_z(z)|/v_s}$$

and μ_o and v_s are the low field surface mobility and the thermal velocity of the electrons, respectively. Notice that $|<v(z)>| \rightarrow v_s$ asymptotically as $|E_z(z)| \rightarrow \infty$; the velocity of the electrons saturates under strong field. Now, substituting Equation (8) into the above equations, we obtain

$$J = \frac{-[\mu_o/(C'_o + C'_s)](\partial Q'_n/\partial z)}{1 + [\mu_o/v_s(C'_o + C'_s)]|\partial Q'_n/\partial z|}Q'_n$$

$\Rightarrow$
$$J\left[1 + \frac{\mu_o}{v_s(C'_o + C'_s)}\left|\frac{\partial Q'_n}{\partial z}\right|\right]$$

$$= -\frac{\mu_o}{C'_o + C'_s}Q'_n\frac{\partial Q'_n}{\partial z}$$

Integrating the above equation from source to drain with respect to z, recognizing the fact that J is a constant independent of z, yields

$$J\int_0^L \left[1 + \frac{\mu_o}{v_s(C'_o + C'_s)}\frac{\partial Q'_n}{\partial z}\right]dz = -\int_0^L \frac{\mu_o}{C'_o + C'_s}Q'_n\frac{\partial Q'_n}{\partial z}\,dz$$

Replacing $(\partial Q'_n/\partial z)\,dz$ by dQ'_n, we have

$$J\left[L + \int_{-C'_o(V_{GS}-V_{TS})\,\mathrm{sgn}\,(V_{GS}-V_{TS})}^{-C'_o(V_{GD}-V_{TD})\,\mathrm{sgn}\,(V_{GD}-V_{TD})} \frac{\mu_o}{v_s(C'_o + C'_s)}\,dQ'_n\right]$$

$$= -\int_{-C'_o(V_{GS}-V_{TS})\,\mathrm{sgn}\,(V_{GS}-V_{TS})}^{-C'_o(V_{GD}-V_{TD})\,\mathrm{sgn}\,(V_{GD}-V_{TD})} \frac{\mu_o}{C'_o + C'_s}Q'_n\,dQ'_n$$

where $|\partial Q_n/\partial z| = \partial Q'_n/\partial z > 0$ because $V_D > V_S$. The limits of integration are obtained from Equation (6). Since C'_s is usually of a smaller order than C'_o, it can be assumed a constant. Now we have

$$J\left\{L + \frac{\gamma}{C'_o v_s}[(V_{GS} - V_{TS})\,\mathrm{sgn}\,(V_{GS} - V_{TS}) - (V_{GD} - V_{TD})\,\mathrm{sgn}\,(V_{GD} - V_{TD})]\right\}$$

$$= \frac{\gamma}{2}[(V_{GS} - V_{TS})^2\,\mathrm{sgn}\,(V_{GS} - V_{TS}) - (V_{GD} - V_{TD})^2\,\mathrm{sgn}\,(V_{GD} - V_{TD})]$$

where $\gamma \triangleq \mu_o C'^2_o/(C'_o + C'_s)$ is sometimes erroneously called the transconductance of the device. The total current flowing through the device is then equal to

$$I_{DS} = WJ \tag{9}$$

$$= \frac{(\gamma/2)\,W[(V_{GS} - V_{TS})^2\,\mathrm{sgn}\,(V_{GS} - V_{TS}) - (V_{GD} - V_{TD})^2\,\mathrm{sgn}\,(V_{GD} - V_{TD})]}{L + \theta|(V_{GS} - V_{TS})\,\mathrm{sgn}\,(V_{GS} - V_{TS}) - (V_{GD} - V_{TD})\,\mathrm{sgn}\,(V_{GD} - V_{TD})|}$$

where $\theta \triangleq \gamma/C'_o v_s = \mu_o C'_o/v_s(C'_o + C'_s) \cong \mu_o/v_s$ in case $C'_s \ll C'_o$.

Equation (9) describes the so-called "square law" characteristics of MOS devices. A typical plot of (9) is given in Figure 6, where the currents I_{DS} are plotted as functions of V_D with V_G as parameters. The source ⓈＳ is grounded so $V_S = 0$.

As Figure 6 shows, I_{DS} increases almost linearly with V_D for $V_D \ll V_P$. The current then passes a maximum at $V_D = V_P$ and starts to decline for $V_D > V_P$. In reality, of course, I_{DS} stays at the maximum value and saturates when $V_D \geq V_P$. The voltage, V_P, at which the device current saturates is called *the pinchoff voltage*. Notice that $V_P = \hat{V}_P(V_G)$ is a function of V_G.

To calculate V_P for the general case, let

$$X_S \triangleq V_{GS} - V_{TS}$$

and

$$X_D \triangleq V_{GD} - V_{TD}$$

and we have

$$J = \frac{(\gamma/2)(X_S^2 - X_D^2)}{L + \theta(X_S - X_D)}$$

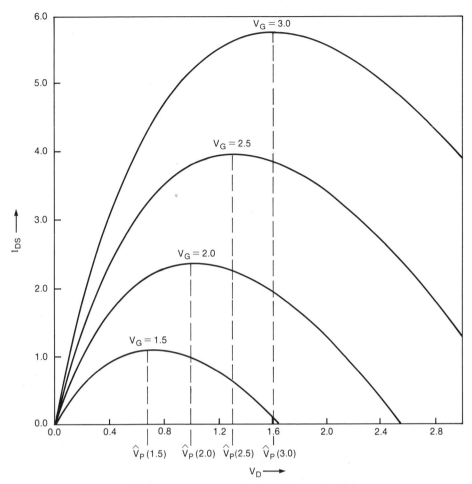

Figure 6. A typical family of Equation (9).

By setting

$$\frac{\partial J}{\partial X_D} = \frac{\gamma\theta}{2\Delta^2}\left[X_D^2 - 2\left(X_S + \frac{L}{\theta}\right)X_D + X_S^2\right] = 0$$

where $\Delta \triangleq L + \theta(X_S - X_D)$, we obtain

$$X_D = X_S + \left[\frac{L}{\theta} - \sqrt{\left(\frac{L}{\theta}\right)^2 + 2X_S\frac{L}{\theta}}\right] \tag{10a}$$

or

$$V_P = V_S - (V_{TD} - V_{TS}) - \left[\frac{L}{\theta} - \sqrt{\left(\frac{L}{\theta}\right)^2 + 2X_S\frac{L}{\theta}}\right] \tag{10b}$$

The factor L/θ in Equation (10) is an important parameter for measuring velocity saturation. In the limiting case for $L \gg \theta$, it is easily shown that

$$V_P \cong V_G - V_{TD}$$

and the current saturates when the channel at the drain is completely depleted. The total current through the device then becomes

$$I_{DS} = \frac{\gamma}{2}\left(\frac{W}{L}\right)[(V_{GS} - V_{TS})^2 \text{ sgn } (V_{GS} - V_{TS}) - (V_{GD} - V_{TD})^2 \text{ sgn } (V_{GD} - V_{TD})]$$

which is directly proportional to the channel width to length ratio: W/L. For many practical devices, however, L/θ can be of comparable order of magnitude to X_S in Equation (10b). The current as described here degrades appreciably due to velocity saturation. Now, defining the effective channel length, L_e, as

$$L_e \triangleq L\left[1 + \left(\frac{\theta}{L}\right)\left|(V_{GS} - V_{TS}) \text{ sgn } (V_{GS} - V_{TS}) - (V_{GD} - V_{TD}) \text{ sgn } (V_{GD} - V_{TD})\right|\right]$$

it becomes clear that the effect of velocity saturation is equivalent to lengthening the channel by a factor roughly proportional to V_{DS}. The proportionality constant $\lambda \triangleq \theta/L$ measures the channel-length dilation. Notice that λ is inversely proportional to the physical channel length L.

2.2 Characteristics of Depletion Mode Devices

The threshold voltage of an n-channel enhancement mode device is positive. Such devices can be turned off easily by grounding the gate voltage. This is certainly a desirable feature in logic circuits, as we shall see in later chapters. There are many cases, however, in which the device is used as a pullup to charge a capacitor connected to its source. Since the device current approaches zero as $V_G - V_S \rightarrow V_T$, the highest potential V_S can ever reach (for an n-channel enhancement mode device) is $V_G - V_T < V_G$. In this case the logic level degrades. To maintain uniform logic levels a device with negative threshold voltage is needed; the depletion mode device is a case in point.

The operation of depletion mode devices is much more complicated than the operation of enhancement mode devices. In many circuit applications, however, care in the fabrication process ensures that the device behaves just like an enhancement mode device with a negative threshold voltage. This is accomplished by a rather shallow channel implant with a light dose of donor ions.

Figure 7 displays a typical n-channel depletion mode structure. The structure is similar to that of an enhancement mode device except that here donor ions are introduced into the silicon forming a thin n-type layer underneath the gate. As we shall see later, different negative threshold voltages can be affected by varying the dopant concentration and the thickness of this layer. In practice the gate, P_N^+, is usually made of heavily doped n-type polysilicon whose property is very similar to metal. For simplicity, assume the n-type layer is uniformly doped with density N_D for a depth x_I. The total "dose" per unit area is therefore $Q_I' = qN_Dx_I$.

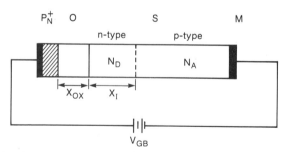

Figure 7. An n-channel depletion mode structure. The shaded area denotes the polysilicon gate.

The following is a brief discussion of the different operating conditions affected by the bias V_{GB}:

(a) $V_{GB} = 0$

As shown in Figure 8, the whole system under examination here is in thermal equilibrium, and all Fermi levels align. The energy band in the n-type layer near the Si-SiO$_2$ interface bends down, indicating an accumulation of electrons. In this figure $\phi_{MS} \triangleq \phi_M - \phi_S$ and $\phi_{P_N^+ M} \triangleq \phi_{P_N^+} - \phi_M$ are the work function differences between the metal and silicon and the n+ doped polysilicon and metal, respectively, and ϕ_B is the built-in voltage across the pn junction at equilibrium

$$\phi_B = \frac{kT}{q} \ln \frac{N_D N_A}{n_i^2}$$

(b) $V_{GB} \neq 0$

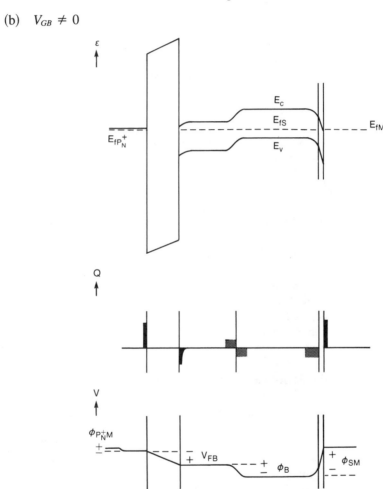

Figure 8.   A depletion mode structure in thermal equilibrium.

The flat band voltage V_{FB} is defined as the applied voltage V_{GB} where the energy band near the Si-SiO$_2$ interface is flat with no charge accumulation near the silicon surface. The energy band diagram makes it clear that the V_{FB} for an idealized MOS system is given by

$$V_{FB} = \phi_{MS} + \phi_{P_N^+ M} + \phi_B$$
$$= \phi_{P_N^+ S} + \phi_B$$

Taking the oxide charge and Q_{ss} into account as we did in Section 1, we have

$$V_{FB} = \phi_{P\mathring{N}S} + \phi_B - \frac{1}{C_o'} \int_0^{x_{ox}} \frac{x}{x_{ox}} \rho(x) \, dx - \frac{Q_{ss}}{C_o'}$$

In Exercise 2 of Section 1, the voltage drop from gate to the bulk is given by $V_{GB} - V_{FB}$, including the effects of oxide charge and Q_{ss}. With depletion mode devices, however, this statement is not true because of the voltage across the pn junction. It is, therefore, more convenient to define the following relationship as a reference:

$$V_{fb} \triangleq V_{FB} - \phi_B$$

$$= \phi_{P\mathring{N}S} - \frac{1}{C_o'} \int_0^{x_{ox}} \frac{x}{x_{ox}} \rho(x) \, dx - \frac{Q_{ss}}{C_o'}$$

The voltage drop from the gate to the bulk is then given by $V_{GB} - V_{fb}$, including the drop across the pn junction. With V_{fb} as a reference, the discussion will turn to three interesting operating conditions: (a) $V_{GB} > V_{fb}$, (b) $V_{GB} < V_{fb}$, and (c) $V_{GB} \ll V_{fb}$.

(a) $V_{GB} > V_{fb}$

Since $V_{fb} < 0$, the condition $V_{GB} = 0$ belongs here. The surface is accumulated with electrons.

(b) $V_{GB} < V_{fb}$

In this case the energy band bends upward and the Si-SiO$_2$ interface is depleted.

(c) $V_{GB} \ll V_{fb}$

As V_{GB} decreases further below V_{fb}, the depletion region in the surface widens. Eventually there are two possibilities: the depletion region in the surface merges with the depletion region across the pn junction and the channel disappears, or the interface converts into p-type before the two depletion regions ever meet, resulting in a dual channel situation—one channel formed by the inversion layer and the other formed by part of the original channel between the two depletion regions. In this last case the device cannot be turned off.

In our applications where the channel implant is shallow and light, the first case—one channel formed between two depletion layers—is more common than the second.

In its normal operation as a pullup, $V_{GS} \gtrsim 0$. With a heavily doped n-type polysilicon gate, the device always operates with some electron accumulation in the channel. With this assumption the total mobile charge in the channel can be calculated, defining the threshold voltage of the depletion mode device.

(1) The Threshold Voltage: Consider the depletion mode device shown in Figure 9 where the surface potential in the channel is now controlled by the source voltage V_S. Let V be the surface potential at the Si-SiO$_2$ interface relative to the p-type silicon bulk. From the energy band diagram;

$$V = V_S - V_B + \phi_B$$
$$= V_{SB} + \phi_B \tag{11}$$

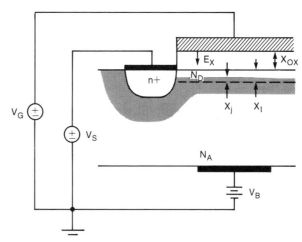

Figure 9(a). A biased MOS structure with depletion implant.

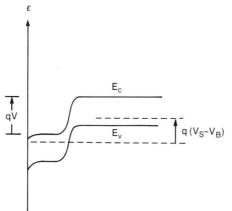

Figure 9(b). Energy band diagram for above device.

Let Q_s' be the charge density per unit area due to electron accumulation, then

$$Q_s' = -C_o'(V_{GB} - V_{fb} - V)$$
$$= -C_o'(V_{GS} - V_{FB})$$

Since there is a pn junction formed in the bottom of the n-type layer, the charge density per unit area Q_j' because of the depletion of electrons is given by

$$Q_j' = -qN_Dx_j$$
$$= -\sqrt{2\varepsilon_s qN_E(V_{SB} + \phi_B)}$$

where x_j is the width of the depletion region *inside* the n-type layer and $N_E \triangleq N_DN_A/(N_D + N_A)$. Taking these charges into account, the total electronic charge per unit area Q_n' in the channel is given by

$$Q_n' = -Q_I' + Q_s' - Q_j'$$
$$= -Q_I' - C_o'(V_{GS} - V_{FB}) + \sqrt{2\varepsilon_s qN_E(V_{SB} + \phi_B)} \qquad (12)$$
$$\triangleq -C_o'(V_{GS} - V_T)$$

where

$$V_T \triangleq V_{FB} + \frac{1}{C'_o}\sqrt{2\varepsilon_s q N_E (V_{SB} + \phi_B)} - \frac{Q'_I}{C'_o} \tag{13}$$

Equation (13) defines the *threshold voltage*. Compares with Equation (6) we see that the threshold voltage of a depletion mode device can be considered a negative shift of the threshold voltage of an enhancement mode device in an amount determined by the channel implant of the n-type layer. As a matter of fact, the name "threshold voltage" applied to depletion mode devices comes about more from an analogy to its enhancement counterpart than any accurate understanding of the term.

(2) The IV Characteristics: The IV characteristics of depletion mode devices are quite similar to those of enhancement mode devices. Figure 10 displays a complete depletion mode device with $V_D > V_S$. To calculate the drift current caused by surface potential variation along the channel, substitute Equation (11) into (12) and (13):

$$Q'_n = Q'_I - C'_o(V_{GB} - V_{fb} - V) + \sqrt{2\varepsilon_s q N_E V} \tag{14}$$

Differentiating Equation (14) with respect to z, we obtain

$$\frac{\partial V}{\partial z} = \frac{1}{C'_o + C'_s}\frac{\partial Q'_n}{\partial z} \tag{15}$$

where $C'_s \triangleq \sqrt{\varepsilon_s q N_E / 2V}$. The identical forms of Equations (6) and (12) and Equations (8) and (15) suggest that Equations (9) and (10) will be equally applicable to depletion mode devices with proper threshold voltage substitutions. This is more than a coincidence because, as we mentioned earlier, the depletion mode device is intended to be an "enhancement" device with a negative threshold voltage.

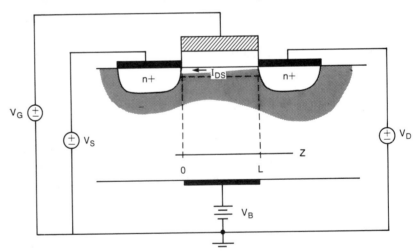

Figure 10. A depletion mode MOS device.

A minor difference between the two modes worth noting is that while the γ in Equation (9) is proportional to the surface mobility μ_o in an enhancement mode device, the γ of a depletion mode device is a rather complicated function of both the surface and bulk mobilities. Depending on different process assumptions, this "effective mobility" may be better or worse than the relatively constant surface mobility in the enhancement mode devices.

In this section only the simplest operation of the depletion mode device relevant to circuit design has been considered. Since a complete treatment of this device would take us off the track, interested readers are left to themselves to work through the references.

3. DEVICE CAPACITANCES

The performance of MOS circuits are limited by the various capacitances associated with the interconnection wiring and the devices themselves. An MOS device, in the on mode, manifests itself as a capacitive load to its driver. To calculate these device capacitances, refer back to Figure 5. The total charge Q_n stored in the device is given by

$$Q_n = W \int_0^L Q_n'(z) \, dz \qquad (16)$$

where Q_n' is the electronic charge density in the channel. Now consider an elemental section of the channel of length dz—the voltage across such an element is

$$dV = I_{DS} dR = \frac{I_{DS}}{\mu Q_n'} \frac{dz}{W}$$

where

$$\mu = \frac{\mu_o}{1 + [(\mu_o/v_s)(dV/dz)]}$$

$$\Rightarrow \qquad dz = \left(\frac{\mu_o W Q_n'}{I_{DS}} - \frac{\mu_o}{v_s} \right) dV \qquad (17)$$

Substitution of Equation (17) into (16) yields

$$Q_n = \frac{\mu_o W}{C_o' + C_s'} \frac{W}{I_{DS}} \left(\int Q_n'^2 \, dQ_n' - \frac{1}{v_s} \int Q_n' \, dQ_n' \right) \qquad (18)$$

In Equation (18), again, $C_o' + C_s'$ is regarded as a constant. We are now ready to calculate the device capacitances according to operating conditions.

(a)   Device operated in linear region

In this case $Q_n' = -C_o'(V_{GS} - V_{TS}) = -C_o' X_S$ and $Q_n' = -C_o'(V_{GD} - V_{TD}) = -C_o' X_D$ at the source and the drain, respectively. Carrying out the integration and substituting Equation (9) for I_{DS}, we obtain

$$Q_n = -\frac{2}{3} C_o' WL \frac{X_S^2 + X_S X_D + X_D^2}{X_S + X_D} \qquad (19)$$

Since $Q_n = \hat{Q}_n(V_S, V_D)$, there are two capacitors associated with the device: C_S, accounting for the charge due to V_S, and C_D, the charge due to V_D. Now, by definition

$$C_S \triangleq \frac{\partial Q_n}{\partial V_S} = \frac{\partial Q_n}{\partial X_S} \frac{dX_S}{dV_S}$$

Differentiating Equation (19) with respect to X_S,

$$\frac{\partial Q_n}{\partial X_S} = -\frac{2}{3} C_o' WL \frac{X_S^2 + 2X_S X_D}{(X_S + X_D)^2}$$

In case of uniform doping, $V_{TS} = V_{FB} + 2|\phi_F| + \sqrt{2\varepsilon_s q N_A (V_S - V_B + 2|\phi_F|)}/C_o'$, and we have

$$\frac{dX_S}{dV_S} = -\left(1 + \frac{C_s'}{C_o'}\right)$$

Therefore,

$$C_S = \frac{2}{3} C_o' WL \left(1 + \frac{C_s'}{C_o'}\right) \frac{X_S^2 + 2X_S X_D}{(X_S + X_D)^2}$$

$$\triangleq C_{gs} + C_{bs}$$

where

$$C_{gs} \triangleq \frac{2}{3} C_o' WL \frac{X_S^2 + 2X_S X_D}{(X_S + X_D)^2} \tag{20}$$

represents a capacitor connected from the gate to the source, and

$$C_{bs} \triangleq \frac{2}{3} C_s' WL \frac{X_S^2 + 2X_S X_D}{(X_S + X_D)^2} \tag{21}$$

represents a capacitor connected from the substrate to the source. The total capacitance associated with V_S is the parallel combination of the two, as shown in Figure 11.

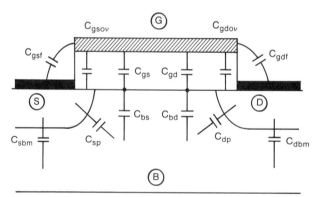

Figure 11.   Definitions of device and parasitic capacitances.

Similarly, the capacitance associated with the drain is given by

$$C_D = C_{gd} + C_{bd}$$

where

$$C_{gd} \triangleq \frac{2}{3} C_o' WL \frac{X_D^2 + 2X_S X_D}{(X_S + X_D)^2}$$

and

$$C_{bd} \triangleq \frac{2}{3} C'_s WL \frac{X_D^2 + 2X_S X_D}{(X_S + X_D)^2}$$

represents the capacitors connnected between gate and drain and substrate and drain, respectively.

(b)   Device operated in saturation region

In this case I_{DS} is independent of V_D and $C_D = 0$. The capacitance associated with the source can be evaluated from Equations (20) and (21) by setting $X_D = 0$:

$$C_S = C_{gs} + C_{bs}$$

where

$$C_{gs} = \frac{2}{3} C'_o WL \quad \text{and} \quad C_{bs} = \frac{2}{3} C'_s WL$$

A typical plot of the gate capacitances is shown in Figure 12 as functions of V_D. As shown in the figure, $C_{gs} = C_{gd} = (1/2)C'_o WL$ at $V_D = V_S = 0$. As V_D increases, C_{gs} increases and C_{gd} decreases. Finally, $G_{gs} = (2/3)C'_o WL$, $C_{gd} = 0$ when the device enters the saturation region.

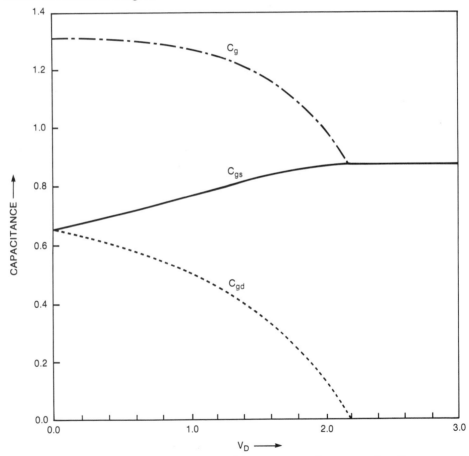

Figure 12.   Typical variations of the gate capacitances as functions of V_D with $V_G = 3V$ and $V_S = 0V$. $C_g = C_{gs} + C_{gd}$.

In addition to these "device capacitances," there are what are called "parasitic capacitances" surrounding each device. For example, at the source Ⓢ, C_{sbm} extends from the bottom of the source diffusion to substrate, and C_{sp} extends from the perimeter of the source diffusion to the channel area and the substrate. Finally, there is C_{gsov} due to the overlap between the gate and the lateral diffusion of the source. Furthermore, because of the finite thickness of the gate material, C_{gsf} fringes from the edge of the gate to some adjacent conductor. All these capacitances can affect device speed appreciably in MOS circuits.

EXERCISES

1. Let $V_{TS} \cong V_{TD} \triangleq V_T$, and prove that

$$\frac{X_S^2 + 2X_S X_D}{(X_S + X_D)^2} = \frac{[3(V_{GS} - V_T) - 2V_{DS}](V_{GS} - V_T)}{[2(V_{GS} - V_T) - V_{DS}]^2}$$

and

$$\frac{X_D^2 + 2X_S X_D}{(X_S + X_D)^2} = \frac{[3(V_{GS} - V_T) - 2V_{DS}](V_{GS} - V_T - V_{DS})}{[2(V_{GS} - V_T) - V_{DS}]^2}$$

2. Prove that the so-called channel capacitance

$$C_g \triangleq C_{gd} + C_{gs}$$

attains its maximum value $C_o' WL$ when $V_S = V_D$.

3. Describe the nature of each capacitor shown in Figure 11.

4. The capacitances discussed so far are the capacitances visible to the external source when the device is on. What are the capacitances visible to the external source when the device is off?

4. THE TRANSCONDUCTANCE

Transconductance is one of the most important device parameters in MOS linear circuit design. Consider the MOS device shown in Figure 13(a) with DC bias voltages V_{GS} and V_{DS}. The transconductance, g_m, defined at (V_{GS}, V_{DS}) is given by:

$$g_m \triangleq \left. \frac{\partial i_{DS}}{\partial v_{GS}} \right|_{V_{DS}} = \text{const.}$$

From a small signal point of view,

$$i_{ds} \cong \frac{\partial i_{DS}}{\partial v_{GS}} v_{gs}$$

Hence $i_{ds} \cong g_m v_i$.

For devices operating in saturation region,

$$i_{DS} = \frac{\gamma W}{2L} \frac{X_S^2 - X_P^2}{1 + \lambda |X_S - X_P|} \tag{22}$$

where X_P is the pinchoff point defined in Equation (10b) and the transconductance

$$g_m = \frac{\gamma W}{2L} \frac{2[X_S - (1 - \eta)X_P] + \lambda(2 - \eta)(X_S - X_P)^2}{[1 + \lambda |X_S - X_P|]^2} \tag{23}$$

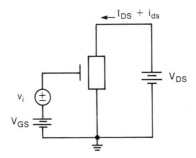

Figure 13(a). An MOS device under DC bias.

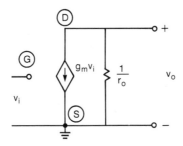

Figure 13(b). A small-signal equivalent circuit for above device.

where $\eta \triangleq 1/\sqrt{1 + 2\lambda X_S}$. Notice that $X_P = \hat{X}_P(X_S)$ was defined in Equation 10(a). For long-channel devices where $\lambda \rightarrow 0$, $X_P \rightarrow 0$, and $\eta \rightarrow 1$, Equation (23) becomes

$$g_m = \gamma \frac{W}{L} X_S = \gamma \frac{W}{L}(V_{GS} - V_{TS})$$

For devices operating in a linear region,

$$i_{DS} = \frac{\gamma}{2} \frac{W}{L} \frac{X_S^2 - X_D^2}{1 + \lambda|X_S - X_D|} \tag{24}$$

and the transconductance is

$$g_m = \gamma \frac{W}{L} \frac{X_S - X_D}{1 + \lambda|X_S - X_D|} \tag{25}$$

In case $V_{TS} \cong V_{TD} \triangleq V_T$, Equation (25) becomes

$$g_m \cong \gamma \frac{W}{L} \frac{V_{DS}}{1 + \lambda|V_{DS}|}$$

Another important parameter in linear circuit design is the output resistance, r_o, defined as

$$\frac{1}{r_o} \triangleq \frac{\partial i_{DS}}{\partial v_{DS}} \Big| V_{GS} = \text{const.}$$

When biased in linear region,

$$\frac{1}{r_o} = \frac{\gamma W}{2L} \frac{2(1 + \lambda|X_S - X_D|) X_D - \lambda(X_S^2 - X_D^2)}{(1 + \lambda|X_S - X_D|)^2}$$

For long-channel devices with $\lambda = 0$,

$$\frac{1}{r_o} \cong \gamma \frac{W}{L}(V_{GD} - V_{TD})$$

When biased in the saturation region, $1/r_o = 0$ and the device behaves like an ideal voltage-controlled current source. In reality, however, r_o is large but finite due to channel length modulation (see Chapter 2). A small-signal equivalent circuit of the device in Figure 13(a) is shown in Figure 13(b).

5. FREQUENCY RESPONSES

Since it takes a finite amount of time for an electron to travel from source to drain, there is a fundamental limit—and hence, a finite frequency response—to the speed of a device. *The channel transit time τ* is defined as the average time an electron takes to move from the source to the drain:

$$\tau \triangleq \frac{-Q_n}{I_{DS}}$$

Substituting Equations (19) and (24) into the above equation yields

$$\tau = \frac{4}{3} \frac{C_o' L^2}{\gamma} \frac{1 + \lambda|X_S - X_D|}{(X_S - X_D)[1 + X_S X_D/(X_S^2 + X_S X_D + X_D^2)]}$$

Since $X_S X_D/(X_S^2 + X_S X_D + X_D^2) \le 1/3$, in case $V_{TS} \cong V_{TD}$ and $\gamma \cong \mu_o C_o'$,

$$\tau \cong \frac{4}{3} \frac{L^2}{\mu_o} \left(\frac{1}{V_{DS}} + \lambda\right)$$

Many system designers believe the channel transit time τ can be viewed as the basic time unit of an integrated circuit system. This is the "minimum delay" for a device to pass one signal to another. Notice, however, that the channel transit time is a small-signal parameter. In reality, the delay of even the simplest digital circuit is usually on an order of magnitude greater than τ.

For small-signal analysis the speed of a device is more conveniently described in frequency domain. Since an MOS device is a voltage-controlled current source, consider the short-circuit current gain of the device shown in Figure 14. Let $\hat{x}(s)$ be the Laplace transform of a variable $x(t)$. At the input,

$$\hat{i}_i(s) = s(C_{gs} + C_{gd})\hat{v}'(s)$$
$$\triangleq sC_g\hat{v}'(s)$$

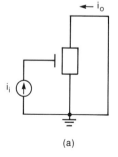

(a)

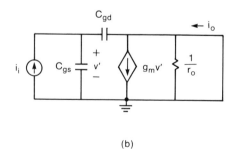

(b)

Figure 14.

At the output, neglecting the current through C_{gd} and r_o, we obtain

$$\hat{i}_o(s) \cong g_m \hat{v}'(s)$$

$$= \frac{g_m}{sC_g} \hat{i}_i(s)$$

The current gain is

$$A(j\omega) \triangleq \frac{\hat{i}_o(j\omega)}{\hat{i}_i(j\omega)}$$

$$= \frac{g_m}{j\omega C_g}$$

The transition frequency ω_T is defined at $|A(j\omega)|_{\omega=\omega_T} = 1$. From the above relationship, it is clear that

$$\omega_T = \frac{g_m}{C_g}$$

The transition frequency is an intrinsic property of a device. It depends only on the channel length L because both g_m and C_g are proportional to channel width W.

When a device is used in a complex circuit, its frequency characteristic interacts with the rest of the system, resulting in a slower overall frequency response. The system frequency response is described by the so-called *cut-off frequency*, ω_C. As a simple example, consider the common source configuration shown in Figure 15, where the resistor r_g accounts for the internal resistance of the current source (i.e., a Norton equivalent circuit). The current gain,

$$A(j\omega) \triangleq \frac{\hat{i}_o(j\omega)}{\hat{i}_i(j\omega)}$$

$$\cong \frac{g_m r_g}{1 + (g_m r_g) j\omega \dfrac{C_g}{g_m}}$$

$$= \frac{A_o}{1 + j\omega A_o \dfrac{C_g}{g_m}}$$

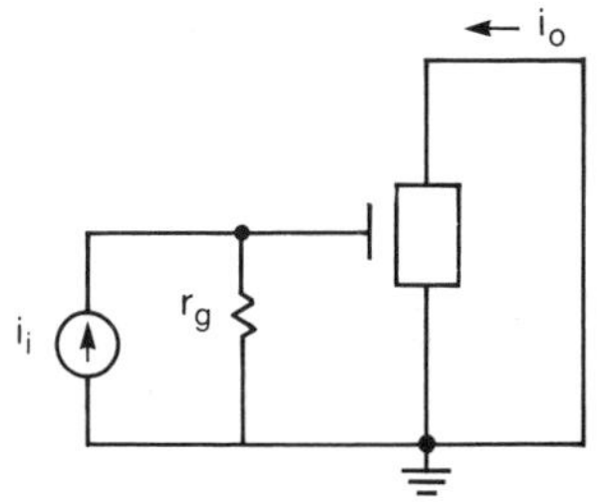

(a)

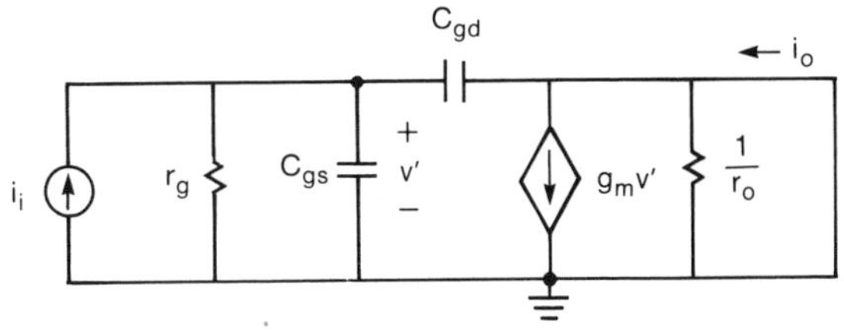

(b)

Figure 15.

where $A_o = A(j0) = g_m r_g$ is the low-frequency current gain. The cut-off frequency ω_C is defined as the frequency at which

$$|A(j\omega_C)| = \frac{A_o}{\sqrt{2}}$$

$\Rightarrow$
$$\omega_C = \frac{g_m}{A_o C_g} = \frac{\omega_T}{A_o}$$

If we plot the "power" $|A(j\omega)|^2$ on log-log scale in decibels; i.e., $10 \log |A(j\omega)|^2 = 20 \log |A(j\omega)|$, the cut-off frequency ω_C occurs when the power drops by $\cong 3$ db. For this reason ω_C is also called the half-power or -3 db frequency. The frequency range $(0, \omega_C)$ is conveniently called the *bandwidth* of the circuit (see Figure 16). Notice that the gain-bandwidth product $A_o \omega_C = \omega_T$ is a constant determined by device parameters.

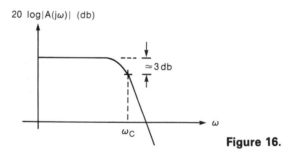

Figure 16.

EXERCISE

Relate the transition frequency ω_T to the channel transit time τ.

6. TEMPERATURE EFFECTS

Both threshold voltage and mobility are functions of temperature. For example, the threshold voltage of an n-channel enhancement mode device with a uniformly doped substrate is given by

$$V_T = V_{FB} + 2|\phi_F| + \frac{1}{C'_o}\sqrt{2\varepsilon_s q N_A(V_{SB} + 2|\phi_F|)}$$

where $\phi_F = (kT/q) \ln (N_A/n_i)$ and $n_i = KT^{3/2} e^{-E_{GO}/2kT}$. The temperature T is measured in °K. It is easily shown that

$$\frac{d}{dT} n_i = \frac{n_i}{T}\left(\frac{3}{2} + \frac{E_{GO}}{2kT}\right)$$

$$\cong \frac{n_i}{T}\frac{E_{GO}}{2kT}$$

and that

$$\frac{d}{dT}|\phi_F| \cong \frac{1}{T}\left(|\phi_F| - \frac{E_{GO}}{2q}\right)$$

Therefore,

$$\frac{d}{dT} V_T = \left(\frac{d}{dT} | \phi_F | \right) \left[2 + \frac{1}{C_o'} \sqrt{2 \varepsilon_s q N_A / (V_{SB} + 2 | \phi_F |)} \right]$$

$$\cong \frac{1}{T} \left(| \phi_F | - \frac{E_{GO}}{2q} \right) \left[2 + \frac{1}{C_o'} \sqrt{2 \varepsilon_s q N_A / (V_{SB} + 2 | \phi_F |)} \right]$$

As an example, let $N_A = 2 \times 10^{16}$ cm^{-3}, $n_i \cong 10^{10}$ cm^{-3}, and $E_{GO} \cong 1.21$ eV, then at $T = 300°$K, $| \phi_F | - E_{GO}/2q \cong -0.29$ and dV_T/dT is of the order of -2mV/°K. Notice that $dV_T/dT < 0$. The negative temperature coefficient of V_T tends to increase I_{DS} when temperature increases.

The increase in I_{DS} due to V_T, however, is overwhelmingly offset by the degradation of mobility μ_o:

$$\mu_o(T) = \mu_o(295°\text{K}) \left[\frac{295}{T} \right]^\alpha$$

where $\alpha \cong 1.5$ for electrons and $\alpha \cong 1$ for holes. As a result, an MOS device will slow down when temperature increases.

7. A SIMPLE CIRCUIT MODEL FOR MOS DEVICES

This section will concentrate on the construction of a simple circuit model for MOS devices. The model will be based on the first-order analysis we have carried out previously. To account for higher-order effects, many important variables such as threshold voltages and substrate sensitivity will be defined as parameters that can be changed by the user. The model is written with an IBM ASTAP model in mind because the simple syntax of ASTAP makes it almost self-explanatory. Coding the model in any other CAD programming language, however, should be straightforward as well.

Notice that the threshold voltage V_{TS} (resp. V_{TD}) is a nonlinear, monotonically increasing function of V_{SB} (resp. V_{DB}), as shown in Figure 17. For practical purposes it is convenient to approximate V_{TS} (resp. V_{TD}) with a straight line over the voltage swing of v_S (resp. v_D). Since in most cases v_S varies from 0, the ground voltage, to V_H, the power supply, we can say for n-channel devices

$$V_{TS} \cong V_{T0} + k_1 v_S$$

where V_{T0} is the threshold voltage with the source grounded; i.e.,

$$V_{T0} \triangleq \hat{V}_T(-V_B)$$

and k_1 is the substrate sensitivity defined by:

$$k_1 \triangleq \frac{\hat{V}_T(V_H - V_B) - V_{T0}}{V_H}$$

whereas for p-channel devices

$$V_{TS} \cong V_{T0} - k_1(V_H - v_S)$$

where V_{T0} is the threshold voltage when the source voltage is at V_H; i.e.,

$$V_{T0} = \hat{V}_T(V_H - V_B)$$

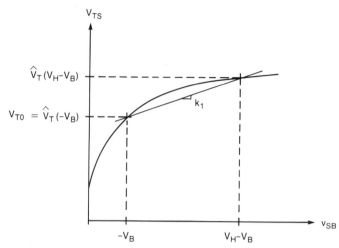

Figure 17(a).   Substrate sensitivity of n-channel devices.

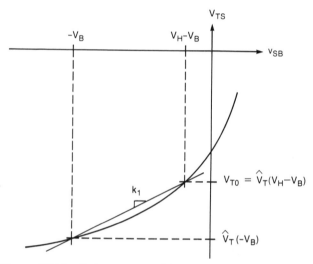

Figure 17(b).   Substrate sensitivity of p-channel devices.

and k_1 is defined by:

$$k_1 \triangleq \frac{V_{T0} - \hat{V}_T(-V_B)}{V_H}$$

The definitions of k_1 are shown in Figure 17. In practice, the values of k_1 do not have to be as rigidly defined as above. They can be modified so as to match the empirical data.

The same equations also apply to V_{TD} because an MOS device is physically bilateral.

Now from Equations (9) and (10) we can write

$$J_{DS} = \frac{(\gamma/2)W(X_S'^2 - X_D'^2)}{L + \theta|X_S' - X_D'|}$$

where

$$X_D' = \begin{cases} X_D, & \text{if } X_D' > X_{PD} \\ X_{PD}, & \text{if } X_D' \le X_{PD} \end{cases}$$

and

$$X_{PD} = X_S + (1 - \sqrt{1 + 2\lambda X_S})/\lambda$$

is the pinchoff point at the drain defined in Equation (10a). Similar expressions for X'_S can be obtained from the above equations by replacing the subscript S with D, and vice versa. Since in ASTAP current and voltage sources are designated by J and E, J_{DS} is used instead of I_{DS} in the model. Definitions of device capacitances C_{gd}, C_{gs}, C_{bd}, and C_{bs} in the model consist of both the device channel capacitance and the gate-diffusion or gate-substrate overlap capacitances. The channel capacitances vary with terminal voltages and become zero when the device is off, whereas the overlap capacitances remain constant, regardless of the operating conditions.

A circuit schematic of the complete model is shown in Figure 18.

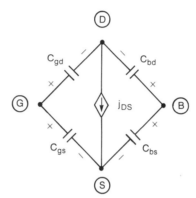

Figure 18. Device circuit model.

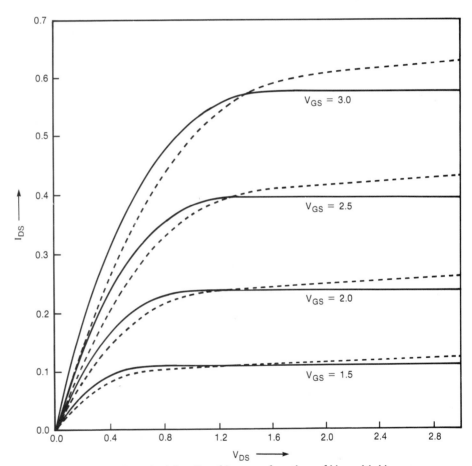

Figure 19. A typical family of I_{DS} as a function of V_{DS} with V_{GS} as a parameter. The solid curves are obtained from the model. The dashed curves represent actual data.

A typical plot of the drain characteristic I_{DS} versus V_{DS} with V_{GS} as a parameter is shown in Figure 19. The drain characteristics match very well with experimental results. A model listing is in the appendix.

REFERENCES

1. Sah, C. T. "Characteristics of the Metal-Oxide-Semiconductor Transistors." *IEEE Trans. of Electron Devices,* July 1964, pp. 324–345.

2. Grove, A. S. *Physics and Technology of Semiconductor Devices.* New York; John Wiley, 1967.

3. Shichman, H., and D. A. Hodges. "Modeling and Simulation of Insulated-Gate Field Effect Transistor Switching Circuit." *IEEE J. of Solid State Circuits,* Vol. SC-3, no. 3, September 1968, pp. 285–289.

4. Frohman-Bentchkowsky, D., and A. S. Grove. "Conductance of MOS Transistors in Saturation." *IEEE Trans. on Electron Devices,* Vol. ED-16, no. 1, January 1969, pp. 108–113.

5. Meyer, J. E. "MOS Models and Circuit Simulation," *RCA Review,* Vol. 32, March 1971, pp. 42–63.

6. Carr, W. N., and J. P. Mize. *MOS/LSI Design and Application.* New York: McGraw-Hill, 1972.

7. Muller, R. S., and T. I. Kamins. *Device Electronics for Integrated Circuits,* New York: John Wiley, 1977.

8. El-Mansy, Y. A., "Analysis and Characterization of the Depletion-Mode IGFET." *IEEE J. of Solid State Circuits,* Vol. SC-15, no. 3, June 1980, pp. 331–340.

9. Streetman, G. G. *Solid State Electronics,* 2nd ed. Englewood Cliffs, NJ: Prentice Hall, 1980.

10. Sze, S. M. *Physics of Semiconductor Devices,* 2nd ed., New York: John Wiley, 1981.

11. IBM Corporation. *Advanced Statistical Analysis Program (ASTAP).* Program Reference Manual, Program Number 5796-PBH. IBM Data Processing Division, White Plains, NY: 1984.

APPENDIX

CIRCUIT MODELS FOR MOS DEVICES

In this appendix are the ASTAP models for MOS devices. Each model has a model name and a list of circuit nodes to be connected to the outside world; for example,

```
MODEL FETN (G, D, SUB, S)
```

Here the model name is FETN (n-channel device). The nodes that are accessible to the user are G (gate), D (drain), SUB (substrate), and S (source), in that order. The model is invoked by a statement in the following format:

```
T1 = MODEL FETN (VI, VH, GD, 1) (PWM=10, PLM=2, PTEMP=85, PVSUB=0)
```

In the statement T1 is the device name. The gate, drain, and source of T1 are tied to nodes VI, VH, and ①, respectively. The substrate is grounded. In an ASTAP model any constant defined in the model can be changed by specifying its value in a change list enclosed in parentheses. In the statement, for example, the temperature is specified at 85° C. Parameter PVSUB is the substrate bias voltage. It is equal to 0 because the substrate is grounded in this case.

Physical and Electrical Units

Voltage	V (volts)
Current	mA (milliamperes)
Capacitance	pF (picofarads)
Length	μ (micron or micrometers)
Velocity	μ/sec (micron per second)

Constant Parameters

All parameter names begin with letter p or x.

```
PTEMP    = JUNCTION TEMPERATURE, DEFAULT AT 22 DEGREE C
PVH      = POWER SUPPLY VOLTAGE, DEFAULT AT 3.00
PVSUB    = SUBSTRATE BIAS VOLTAGE, DEFAULT AT -1.00 FOR N-CHANNEL
             AND 3.00 FOR P-CHANNEL DEVICES
XOX      = tox, THICKNESS OF THE GATE OXIDE, DEFAULT AT .025
PWM      = CHANNEL WIDTH (MASK LEVEL), DEFAULT AT 4.0
PLM      = CHANNEL LENGTH (MASK LEVEL), DEFALT AT 1.0
PWB      = MASK TO WAFER BIAS OF CHANNEL WIDTH, DEFAULT AT 0.00
PLB      = MASK TO WAFER BIAS OF CHANNEL LENGTH, DEFAULT AT 0.00
PDW      = TOLERANCE OF CHANNEL WIDTH, DEFAULT AT 0.00
PDL      = TOLERANCE OF CHANNEL LENGTH, DEFAULT AT 0.00
PGAMMA22= γ, CONDUCTION FACTOR IN MILLIMHO/V AT 22 DEGREE C, DEFAULT
             AT 0.082 FOR N-CHANNEL AND 0.032 FOR P-CHANNEL DEVICES
PVT0     = VT0, THRESHOLD VOLTAGE, DEFAULT AT 0.50 FOR N-CHANNEL
             AND -0.50 FOR P-CHANNEL DEVICES
PVS      = vs, SATURATION VELOCITY OF ELECTRONS AND HOLES, DEFAULT AT
             1.0D11
PK1      = k1, COEFFICIENT OF SUBSTRATE SENSITIVITY, DEFAULT AT 0.08
             FOR NMOS AND 0.20 FOR CMOS DEVICES
PGBOVLP  = OVERLAP BETWEEN THE GATE AND BULK (DIFFUSION)
             USED TO CALCULATE THE GATE-DRAIN OR GATE-SOURCE
             OVERLAP CAPACITANCE, DEFALT AT .05
XXXPUA   = XXX PER UNIT AREA
PCO      = GATE OXIDE CAPACITANCE
PCS      = CHANNEL-SUBSTRATE CAPACITANCE, DEFAULT AT 0.1 × PCO
0.5D-5   = SINCE THE VALUE OF AN ELEMENT MUST BE NONZERO IN ASTAP
             0.5D-5 IS USED AS THE SMALLEST VALUE OF A CAPACITANCE
```

For clarity the models of n-channel and p-channel devices are listed separately.

(1) Model of n-Channel Device

```
MODEL FETN (G, D, SUB, S)
THE THRESHOLD VOLTAGE VT = VT0 + K1 × VS = VT0 + K1 × (-VCBS+PVSUB)
WHERE VCBS IS THE VOLTAGE ACROSS CAPACITOR Cbs.


ELEMENTS

PTEMP   = 22
XOX     = .025
PVT0    = .500
```

```
PK1      = .08
PVS      = (1.0D11*DSQRT((273+PTEMP)/295))
PGAMMA22= .082


PVH    = 3.0
PVSUB  = -1.0


PWM  = 4.0
PLM  = 1.0
PWB  = 0.00
PLB  = 0.00
PDW  = 0.00
PDL  = 0.00
PW   = (PWM-PWB+PDW)
PL   = (PLM-PLB+PDL)
PCOPUA  = (3.9*8.854D-6/XOX)
PCSPUA  = (.1*PCOPUA)
PCOMAX  = (0.67*PCOPUA*PW*PL)
PCSMAX  = (0.67*PCSPUA*PW*PL)
PGBOVLP = .05
PCOVLP  = (PCOPUA*PGBOVLP*PW)
PGAMMA  = (PGAMMA22*(295/(273+PTEMP))**1.5)
PTHETA  = (1.0D9*PGAMMA/(PCOPUA*PVS))
PLAMBDA = (PTHETA/PL)


XD = (VCGD-(PVT0+PK1*(-VCBD+PSUB)))
XS = (VCGS-(PVT0+PK1*(-VCBS+PSUB)))
PD = (XD*PSGN(XD))
PS = (XS*PSGN(XS))


XPD     = (PS+(1-DSQRT(1+2*PLAMBDA*PS))/PLAMBDA)
PINCHD  = (PSGN(XPD-PD))
XPS     = (PD+(1-DSQRT(1+2*PLAMBDA*PD))/PLAMBDA)
PINCHS  = (PSGN(XPS-PS))


XDP     = (PD*(1-PINCHD) + XPD*PINCHD)
XSP     = (PS*(1-PINCHS) + XPS*PINCHS)
JDS, D-S = (PW*PGAMMA*(XSP**2-XDP**2)/(2*(PL+PTHETA*DABS(XSP-XDP))))


PCGD  = ((PD**2+2*PS*PD)/(PS+PD+0.5D-5)**2)
PCGS  = ((PS**2+2*PD*PS)/(PS+PD+0.5D-5)**2)
CGD, G-D    = (PCOVLP+PCOMAX*PCGD)
CGS, G-S    = (PCOVLP+PCOMAX*PCGS)
CBD, SUB-D  = (0.5D-5+PCSMAX*PCGD)
CBS, SUB-S  = (0.5D-5+PCSMAX*PCGS)



//SOURCE.DECKS DD *
     FUNCTION PSGN (X)
     IMPLICIT REAL*8 (A-H, 0-Z)
     IF (X .GT. 0.0) GO TO 10
     PSGN = 0.0
     RETURN
  10 PSGN = 1.0
     RETURN
     END
     FUNCTION PSGNC (X)
     IMPLICIT REAL*8 (A-H,0-Z)
     IF (X .GT. 0.0) GO TO 10
```

```
                         PSGNC = 1.0
                         RETURN
                  10 PSGNC = 0.0
                         RETURN
                         END
```

(2) Model of p-Channel Device

```
MODEL FETP (G, D, SUB, S)
THE THRESHOLD VOLTAGE VT = VT0 - K1 × (PVH-VS) = VT0 - K1 × (PVH+VCBS-
    PVSUB)
WHERE VCBS IS THE VOLTAGE ACROSS CAPACITOR Cbs.


   ELEMENTS

   PTEMP      = 22
   XOX        = .025
   PVT0       = -0.50
   PK1        = .20
   PVS        = (1.0D11*DSQRT((273.15+PTEMP)/295.15))
   PGAMMA22 = .032


   PVH    = 3.00
   PVSUB = 3.00


   PWM = 4.0
   PLM = 1.0
   PWB = 0.00
   PLB = 0.00
   PDW = 0.00
   PDL = 0.00
   PW   = (PWM-PWB+PDW)
   PL   = (PLM-PLB+PDL)


   PCOPUA   = (3.9*8.854D-6/XOX)
   PCSPUA   = (.1*PCOPUA)
   PCOMAX   = (0.67*PCOPUA*PW*PL)
   PCSMAX   = (0.67*PCSPUA*PW*PL)
   PGBOVLP = .05
   PCOVLP   = (PCOPUA*PGBOVLP*PW)
   PGAMMA   = (PGAMMA22*(295/(273+PTEMP)))
   PTHETA   = (1.0D9*PGAMMA/(PCOPUA*PVS))
   PLAMBDA = (PTHETA/PL)


   XD = (VCGD-(PVT0-PK1*(PVH+VCBD-PVSUB)))
   XS = (VCGS-(PVT0-PK1*(PVH+VCBS-PVSUB)))
   PD = (XD*PSGNC(XD))
   PS = (XS*PSGNC(XS))


   XPD      = (PS+(-1+DSQRT(1-2*PLAMBDA*PS))/PLAMBDA)
   PINCHD = (PSGNC(XPD-PD))
   XPS      = (PD+(-1+DSQRT(1-2*PLAMBDA*PD))/PLAMBDA)
   PINCHS = (PSGNC(XPS-PS))


   XDP      = (PD*(1-PINCHD) + XPD*PINCHD)
   XSP      = (PS*(1-PINCHS) + XPS*PINCHS)
   JDS, D-S = (PW*PGAMMA*(XDP**2-XSP**2)/(2*(PL+PTHETA*DABS(XDP-XSP))))
```

```
PCGD = ((PD**2+2*PS*PD)/(PS+PD+0.5D-5)**2)
PCGS = ((PS**2+2*PD*PS)/(PS+PD+0.5D-5)**2)
CGD, G-D = (PCOVLP+PCOMAX*PCGD)
CGS, G-S = (PCOVLP+PCOMAX*PCGS)
CBD, SUB-D = (.5D-5+PCSMAX*PCGD)
CBS, SUB-S = (.5D-5+PCSMAX*PCGS)


//SOURCE.DECKS DD *
      FUNCTION PSGN (X)
      IMPLICIT REAL*8 (A-H,O-Z)
      IF (X .GT. 0.0) GO TO 10
      PSGN = 0.0
      RETURN
   10 PSGN = 1.0
      RETURN
      END

      FUNCTION PSGNC (X)
      IMPLICIT REAL*8 (A-H,O-Z)
      IF (X .GT. 0.0) GO TO 10
      PSGNC = 1.0
      RETURN
   10 PSGNC = 0.0
      RETURN
      END
```

MOS TECHNOLOGY REVIEW

INTRODUCTION

This chapter is a brief review of the MOS technology. Since MOS circuits are the focus of this book, we shall emphasize those process variables that have significant impact on circuit performance.

Section 1 is a brief review of MOS process flow which provides a basis for later discussions. Section 2 is a more detailed discussion of the constraints of process and device design. We shall see that each process step is the result of compromise between conflicting requirements of device parameters. The results are then summarized in a brief discussion of the scaling principle in Section 3. Section 4 considers some recent advances in device design, namely, lightly doped drain (LDD) structure and trench isolation. The chapter concludes with a brief review of the evolution of MOS dynamic RAM cells.

Throughout this work we assume the reader is familiar with pn junction theory and basic process steps such as photolithography, ion implantation, annealing, diffusion drive-in, chemical vapor deposition, and reactive ion etching (RIE). Readers who are not familiar with these subjects are encouraged to consult the references listed at the end of the chapter.

1. BASIC PROCESS FLOW

In this section, basic process steps are described with an illustrative example. The process fabricates n-channel enhancement and depletion mode as well as p-channel enhancement mode devices. It is illustrative in the sense that both NMOS and CMOS devices are fabricated in the same process flow. In reality, due to cost and different optimization requirements, not all types of devices exist in any one process. Thus, a practical process may use a subset of the steps in this example. On the other hand, only the basic steps are covered in our discussion. In reality almost every step is followed by several minor or routine procedures such as annealing and wafer cleaning so as to maximize yield.

1. *Wafer Preparation and Gettering.* The process starts with a p-type wafer (substrate) with <100> orientation. The <100> orientation generates a minimum number of surface states. After scribing and initial cleaning, the wafer is subjected to gettering.

   Gettering is the process of introducing impurity traps in the substrate to absorb mobile ions that cause leakage currents. It can be accomplished by applying Ar^+ (argon) ion implant or focused laser beam to the back side of a wafer. Damages caused by this impact process will attract impurities from the surface in subsequent hot processes.

2. *N-well Definition.* In CMOS processes the n-well mask defines the n-well regions. The subsequent P (phosphorus) ion implant and drive-in creates the n-wells in which the p-channel devices will be formed (see Figure 1). The photoresist (PR) is an effective mask for ion implantation.

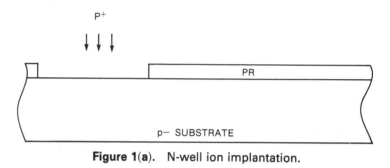

Figure 1(a). N-well ion implantation.

Figure 1(b). N-well formation.

3. *Gate Oxide Growth.* To grow gate oxide on devices, a layer of pad SiO_2 is first grown on the surface of the whole wafer, followed by a layer of Si_3N_4. A mask, called ROX or FOX, is used to remove the Si_3N_4 selectively as shown in Figure 2(a). The n-well mask is then used once more to apply photoresist covering all n-well regions. With the photoresist still on top of the Si_3N_4, a blanket B^+ (boron) implant is carried out. This is called *field tailoring* or *channel stopping*. It raises the threshold voltage of the parasitic n-channel transistors to be formed underneath the thick oxide.

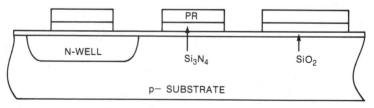

Figure 2(a). ROX definition.

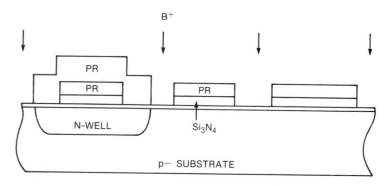

Figure 2(b). Channel stop implant.

After the PR has been stripped the wafer is subjected to oxidation. Areas covered by Si_3N_4 will remain undisturbed. Areas not protected by Si_3N_4 will be oxidized to form a thick layer of SiO_2. Si_3N_4 has very high tensile stress, so the pad SiO_2 helps relieve tension during the oxide growth. Since oxidation consumes silicon, the profile of the thick oxide is half-way "underground" as shown in Figure 3. This is called *semirecessed oxide* (SROX or simply ROX) or *field oxide* (FOX). The thick ROX forms parasitic MOS transistors with very high threshold voltages that can isolate active devices. This isolation technique is called the LOCOS (*local oxidation of silicon*) process. The semirecessed oxide also makes the wafer surface more planar.

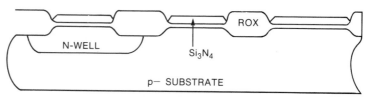

Figure 3. ROX formation.

The Si_3N_4 layer is subsequently removed by hot H_3PO_4 and the pad oxide removed by BHF (buffered HF) until bare silicon is exposed as shown in Figure 4. Right after the ROX etchback, a high-quality oxide layer is grown thermally over the whole wafer. This will be the gate oxide for all devices. The thickness and uniformity of the film must be carefully controlled.

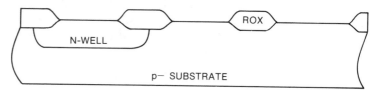

Figure 4. ROX etchback.

The wafer profile is now shown in Figure 5. Notice the bird's beak shapes extending from the ROX to the thin oxide. These formations and the boron's out-diffusion underneath the ROX are largely responsible for the large parasitic capacitances around the source/drain of the device and the so-called narrow-channel effect on threshold voltages.

Now that the gate oxide is formed, channel doping profiles of different devices such as the low-threshold enhancement mode and

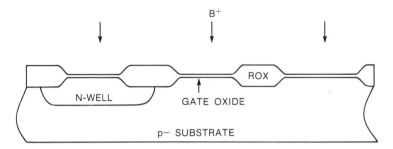

Figure 5.   Gate oxide growth and channel tailoring.

depletion mode devices can be achieved by ion implantation with appropriate kinds of dopant over areas selected by device masks. For example, the B^+ implant in Figure 5 adjusts the threshold voltages of all devices. The formation of depletion mode devices is described below.

4. *Depletion Mode Device Channel Tailoring and Buried Contact Openings.* As discussed in Chapter 1, the channel of a depletion mode device is formed by a shallow n-type layer underneath the gate. As shown in Figure 6 the depletion mask generates a PR pattern with openings over the areas where the devices are to be formed. A subsequent As^+ (arsenic) ion implant drives the threshold voltage of these regions to the required below zero level.

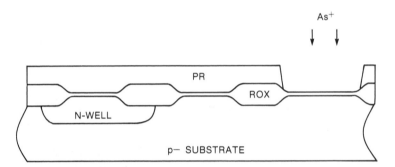

Figure 6.   Depletion mode device channel tailoring.

A buried contact is a contact between polysilicon and diffusion. The buried contact mask defines the areas where the thin oxide will be removed as shown in Figure 7. After the buried contact etch, a layer of polysilicon is grown across the whole wafer with a low-pressure chemical vapor deposition (LPCVD) process.

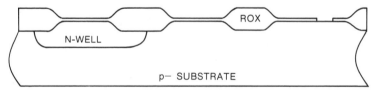

Figure 7.   Buried contact opening.

5. *Polysilicon Gate Definition.* A layer of polysilicon oxide is then grown over the first polysilicon layer by LPCVD to define device gates. Both layers are then doped by P^+ ion implant as shown in Figure 8. The gate

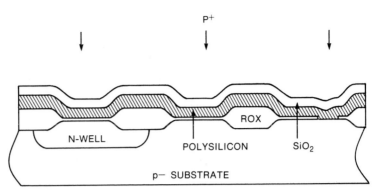

Figure 8. Polysilicon deposition and doping.

mask allows selective etching of the polysilicon oxide and polysilicon to define the gates. After that, the wafer is oxidized and a thin layer of SiO_2 covers the bare silicon areas as shown in Figure 9. Since the oxidation rate of polysilicon is three times faster than that of silicon, sidewalls of SiO_2 are formed around the gates. In a double polysilicon process, another polysilicon layer can now be deposited on top of the sidewalls. In practice up to three layers of polysilicon—P1, P2, and P3—have been used in some dynamic RAM cells.

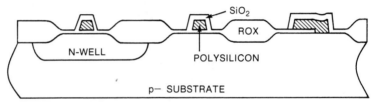

Figure 9. Polysilicon gate definition.

6. *Source/Drain (S/D) Formation.* After the device geometry has been defined, blanket implants of appropriate dopant form the S/D of devices. Since implants are stopped only by gates, the S/D of a device is actually defined by the gate itself. The process is therefore called *self-aligned silicon gate process.*

In CMOS technology two implants are required. A blocking mask blocks the p-channel device areas when the S/D of n-channel devices are formed by As^+ implant. Similarly, a complementary mask blocks the n-channel devices when the p-channel S/D are formed by B^+ implant. See Figures 10(a) and (b).

The wafer is then heated to 1000°C for S/D drive-in. The dopants out-diffuse toward the center of the channel, covering the gap overshadowed by sidewalls in previous ion implant. The high temperature also drives dopants in the polysilicon into the silicon through buried

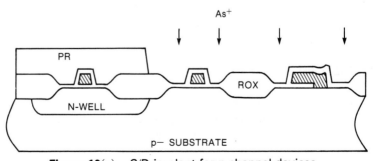

Figure 10(a). S/D implant for n-channel devices.

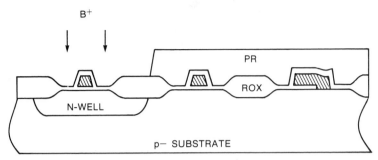

Figure 10(b). S/D implant for p-channel devices.

contact openings. In the end, the polysilicon is connected to the source of the depletion mode device. After drive-in, a layer of PSG (phosphosilicate glass) is deposited over the whole wafer. The PSG softens and flows at high temperatures. It covers the sharp steps of the polysilicon gates, creating a much smoother surface for subsequent metal films. The PSG layer also traps impurity ions such as Na^+, preventing them from contaminating the devices from above.

7. *Metalization.* To connect devices with metal, contacts defined by the contact hole mask are etched from the surface to the polysilicon and diffusion regions below. A subsequent Al (aluminum) vapor deposition covers the surface and fills up the contact holes. A mask for metal interconnect patterns is then used to remove all unused metal. In case there is more than one metal layer required, additional layers can be formed in the same fashion with composite dielectric layers of Si_3N_4 and polyimide as the insulating material. Up to three layers of metal—M1, M2, and M3—have been used in practical applications.

A contact hole that connects metal in different layers is called a *via* (hole). A via hole placed on top of another contact hole is a *stacked via*. Depending on the metal process a stacked via may or may not be allowed in a particular technology. Figure 11 shows profiles of devices completed with M1 contacts.

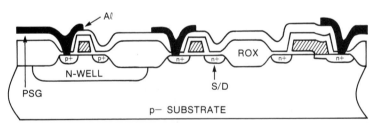

Figure 11. PSG and metalization.

8. *Passivation.* Finally, the wafer is coated with a layer of passivation material such as SiN or polyimide for protection.

2. PROCESS AND DEVICE DESIGN CONSTRAINTS

To optimize circuit performance, various requirements are demanded of device parameters. This, in turn, imposes constraints on the process. Since, in practice, circuit optimization does not lead to a consistent set of device parameters, the final selection of a process is always a compromise among conflicting factors.

For example, one device parameter that affects circuit performance significantly is S/D junction depth. The junction depth can be controlled tightly by ion implantation. As we shall see shortly, however, the S/D junction depth is not determined by its process controllability but by conflicting devices constraints such as the short-channel effect on the threshold voltage and the S/D series resistances. Thus, while the process does have limits of its own, most process constraints are set by device parameters.

2.1 Threshold Voltage

2.1.1 Substrate Sensitivity.   The threshold voltage is one of the most important device parameters controlled by the process. The expression of the threshold voltage V_T of an n-channel enhancement mode device is given by

$$V_T = V_{FB} + 2|\phi_F| + \frac{\sqrt{2\varepsilon_s q N_A (V_{SB} + 2|\phi_F|)}}{C_o'}$$

where N_A is the (uniform) doping concentration of the substrate and V_{SB} is the voltage bias from source to substrate. From this equation it is clear that V_T increases as V_{SB} increases. This is called *substrate sensitivity* or the *body effect*. To reduce substrate sensitivity, the doping concentration of the substrate, N_A, should be made as low as possible. Thus, lower substrate sensitivity favors a lightly doped substrate.

2.1.2 Punchthrough Effect.   Lowering the doping concentration of the substrate, however, affects the minimum channel length. A lightly doped substrate forms wide depletion regions around the device source and drain. If the channel length is too short, the S/D depletion regions can "meet" and *punchthrough* occurs. A small leakage current then has a way to flow into the channel and the device cannot be turned off. The gate loses control of the device current.

The punchthrough effect then favors a more heavily doped substrate. Substrate doping is limited by another factor as well. The fields across the S/D junctions and the substrate increase as the substrate doping increases. Heavily doped substrate, in addition to increasing substrate sensitivity, also lowers the S/D junction breakdown voltage.

2.1.3 Short-Channel Effect.   Something else that favors higher substrate doping is the *short-channel effect*. In the preceding equation, it is assumed that the field extending from the gate to the substrate is uniformly distributed from the source to the drain. In reality, however, the negative space charge around the source/drain is affected by the S/D n+ diffusion and hence not controlled by the gate voltage. For long-channel devices in which the S/D depletion width is a small portion of the total channel length, this effect is negligible. For short-channel devices, however, since the gate voltage does not have to impact as much charge in the channel, the threshold voltage decreases as the channel length is reduced. This is called short-channel effect.

The short-channel effect can also be explained by a lowered energy barrier caused by high voltage at the drain. Given the substrate and source voltage bias, threshold voltage decreases when the drain voltage goes up in short-channel devices.

The short-channel effect favors higher substrate doping and shallow S/D junction depth.

2.1.4 Narrow-Channel Effect.   The threshold voltage also depends on the device's channel width. The fringe fields of the boron implant under the bird's beaks increase the threshold voltage of the thin oxide devices as well. As the channel width decreases the threshold voltage increase becomes more pronounced. This is called the *narrow-channel effect*. Since the minimum channel width is seldomly used, the narrow-channel effect is not as severe as the short-channel effect.

2.1.5 Hot-Carrier Effect. The *hot-carrier effect* refers to the threshold voltage shift due to the trapping of high-energy (hot) carriers in gate oxide. In n-channel devices the hot carriers are electrons. There are two major sources of hot electrons. In short-channel devices electrons in the device current get much energy from the strong field across the channel while traveling from source to drain. Some electrons gain enough energy in the high field region around the drain to overcome the Si-SiO$_2$ barrier and enter into the gate oxide. The population of such electrons is proportional to the drain-source voltage drop V_{DS}. Another source is the leakage current thermally generated in the silicon bulk. Some of these electrons travel to the channel area and are accelerated by the strong vertical field of the depletion region. In the neighborhood of the S/D junction these electrons—and the multiplication current initiated by them in the local strong field—can generate more electrons "hot" enough to enter into the gate oxide. The population of such electrons is proportional to the source-substrate and drain-substrate voltage V_{SB} and V_{DB} and is a strong function of temperature.

Once inside the gate oxide, the electrons may be trapped there permanently, causing a positive threshold voltage shift. Since the trapping of electrons is a slow accumulative process, the hot electron effect is a long-term reliability concern.

The hot-electron effect limits the maximum power supply and substrate bias voltage. The leakage induced threshold shift can be alleviated by using lightly doped substrate.

Since the impact ionization energy of holes is much smaller than that of electrons, the hot-carrier effect is less severe for p-channel devices.

2.1.6 Threshold Voltage Adjustment: Channel Tailoring. In Section 1 we indicated that different doping profiles can be achieved by ion implantation after the gate oxide has been formed. In NMOS technology two implants in different channel regions create four types of devices: the low threshold enhancement mode, the normal enhancement mode, the weak depletion mode, and the normal depletion mode. The energy of the ion implantations is adjusted to concentrate the dopants in the depletion region below the surface when the source is grounded. In normal operation with a positive source voltage, the depletion region extends into the lightly doped substrate only slightly and substrate sensitivity is minimized.

In our process the silicon gate is n+ doped and the flat band voltage V_{FB} is large and negative (about -1 V in practice). Since the bulk charge term (the radical) is positive for p-type substrate, however, both positive and negative threshold voltages for n-channel devices can be achieved. For p-channel devices, the threshold voltage is given by

$$V_T = V_{FB} - 2|\phi_F| - \frac{\sqrt{2\varepsilon_s q N_D(-V_{SB} + 2|\phi_F|)}}{C_o'}$$

Notice that the bulk charge term is negative for an n-type substrate (the n-well). With $V_{FB} \ll 0$, V_T will be large and negative and the device will be very slow. To reduce the magnitude of V_T, the channel of p-channel devices received the same boron implant used in the n-channel devices in Figure 5. As a result, the channel surface in a p-channel device is slightly p-type. The surface is slightly depleted and the minimum potential occurs below the surface in normal bias conditions. As shown in Figure 12, p-channel devices transform into buried-channel devices.

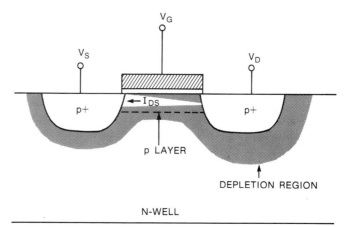

Figure 12. A buried-channel p-channel device.

A channel not on the surface is always more difficult to control. Buried channel devices are more susceptible to leakage and subthreshold current (Section 2.2) and short-channel effects in subthreshold regions. On the other hand, since the holes are in the bulk, hole mobility is higher and the device is faster. Unlike n-channel depletion mode devices, p-channel devices *are* required to turn off during normal operation. Large leakage and subthreshold currents can cause problems in components such as the transfer gate of a dynamic RAM cell.

2.2 Subthreshold Current

In Chapter 1 the device current, I_{DS}, was derived under the condition of strong inversion. Strong inversion occurs when band bending at the silicon surface reaches $2|\phi_F|$ and surface concentration of the carrier is equal to the bulk doping concentration. In strong inversion the device current in linear region is predominantly due to drift. I_{DS} is computed based on drift current alone. Under the same assumption, the carrier surface concentration is also assumed to be zero so that $I_{DS} = 0$ when $V_{GS} < V_T$.

In reality, however, the surface concentration does not decrease to zero abruptly at $V_{GS} = V_T$, and there indeed can be a small channel current even for $V_{GS} < V_T$. This current, designated as I_{st}, is called *subthreshold current*.

Since a detailed analysis of the subthreshold current will take us too far afield, some useful results will be stated shortly without derivation. Readers interested in more detail may consult the references listed at the end of this chapter.

Subthreshold current is due to diffusion of minority carriers from the source to the drain. It is given by

$$I_{st} = I_0 \frac{W}{L} e^{(V_{GS} - V_T)/V_C} \left(1 - e^{-qV_{DS}/kT}\right)$$

with

$$V_C = \frac{kT}{q}(1 + \alpha\sqrt{N_A}\,t_{ox})$$

where I_0 is linearly proportional to the temperature T and α is a constant. Since the contribution of V_{DS} becomes negligibly small for $V_{DS} > 4(kT/q) \cong 0.1$ V in room temperature, the subthreshold current is essentially independent of V_D for most practical purposes. Under such conditions we then have

$$I_{st} = I_0 \frac{W}{L} e^{(V_{GS}-V_T)/V_C}$$

To glue I_{DS} and I_{st} together it is customary to define a "strong turn-on" condition for the device by

$$V_{GS} \geq V_T + \Delta V$$

where ΔV is a small voltage in mV's. The subthreshold current is then given by

$$I_{st} = I_0 \frac{W}{L} e^{(V_{GS}-V_T-\Delta V)/V_C} \qquad (1)$$

where I_0 is calculated by

$$I_0 = \frac{I_{DS}|_{V_{GS}=V_T+\Delta V}}{W/L}$$

In practice I_{st} is also specified by mV/decade and Equation (1) can be written as

$$I_{st} = I_0 \frac{W}{L} 10^{(V_{GS}-V_T-\Delta V)/(\ln 10)V_C}$$

The value of $(\ln 10)V_C$ ranges from 60 to 120 mV/decade in practice. The subthreshold current depends on the temperature through I_0 and V_C. Since the threshold voltage is affected by channel length, channel width, and V_D, so is the subthreshold current.

2.3 Velocity Saturation Effect

One of the major effects limiting the performance of devices is the velocity saturation of carriers. In a high field the carrier velocity average approaches its thermal velocity limit (around 10^7 cm/S or 100 μ/nS). As shown in Figure 13 the average velocity of electrons starts to saturate beyond a field intensity around 3 V/μ. The average velocity of holes saturates a little later, but also approaches the same limit.

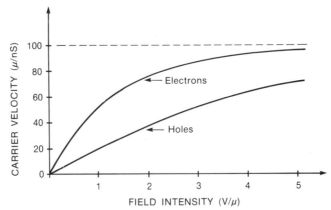

Figure 13. Carrier velocities as functions of field intensity at room temperature.

Figure 14 is a comparison of two devices with (solid curves) and without (dashed curves) the velocity saturation effect. Velocity saturation also causes currents to saturate at a much lower pinchoff voltage. Given the power supply, the benefit of better performance through reduced channel length diminishes quickly as the channel becomes too short.

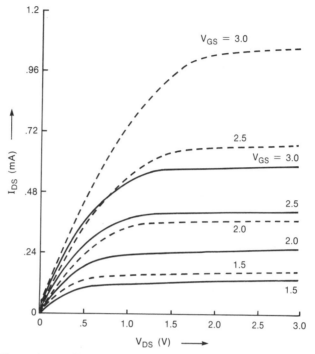

Figure 14.   Effect of velocity saturation to device currents.

The velocity of carriers is also adversely affected by the vertical field in the channel, but to a lesser degree for short channel devices.

2.4 Device Parasitic Capacitances and Resistances

Because of a characteristic low current, MOS circuits are very sensitive to parasitic capacitances. Beside gate capacitances, large parasitic capacitances are associated with S/D diffusions. Figure 15 is the cross section of a diffusion region bounded by ROX. The region is reversely biased by $V_R > 0$. The total capacitance associated with the diffusion region can be partitioned into two components: an area component C_{xbm} in the bottom and a perimeter component C_{xp} around the sidewall. For a heavily doped diffusion region inside a lightly doped substrate, the junction in the

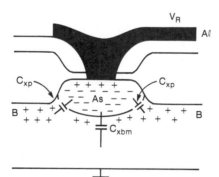

Figure 15.   Device S/D depletion capacitances.

bottom can be approximated by a one-sided abrupt pn junction. The capacitance *per unit area* is therefore given by

$$C'_{xbm} = \frac{C_{bm0}}{(\psi_O + V_R)^{1/2}}$$

where $\psi_O = kT/q \ln (N_A N_D / n_i^2)$ is the built-in potential and C_{bm0} is a constant. The modeling of C_{xp}, however, is much more complicated. The boron atoms implanted underneath the ROX redistribute during subsequent hot processes and move into the n+ diffusion region. The junction between the diffusion and the substrate along the sidewall is not well defined. Assuming a linearly graded junction for the sidewalls, for example, the perimeter capacitance *per unit length* is then given by

$$C'_{xp} = \frac{C_{p0}}{(\psi_O + V_R)^{1/3}}$$

where C_{p0} is a constant. In practice both perimeter and area capacitance are significant.

Minimizing the junction capacitance favors a shallow junction and a lightly doped substrate. The junction depth, however, is limited by the sheet resistance of the diffusion lines. In applications where long narrow diffusion lines such as the bit-lines in a dynamic RAM are used, reducing the capacitance at the expense of increasing the resistance does not improve performance. The high resistivity of the shallow S/D regions, however, does result in degrading resistances along the channel.

Since the depletion width around the drain changes with V_{DB}, the effective channel length decreases as V_{DB} increases. The device current therefore increases slightly as V_{DB} goes up. This effect, called *channel length modulation*, is largely responsible for the finite output resistances of short-channel devices.

2.5 Parasitic Bipolar Transistors: Snapback Breakdown and CMOS Latchup

With the source and drain acting as emitter and collector, an MOS device is also a bipolar structure with its channel as the base. As shown in Figure 16(a), in normal operation both S/D pn junctions are reverse biased so the bipolar transistor is cut off. Although the substrate is always biased so no pn junctions can become forward biased under normal conditions, the potential of the substrate near the channel region does vary with transient currents. If the voltage drop from the substrate to the source is high enough to forward bias the pn junction, the bipolar device will become active and a large emitter injection will occur. The device current can go up abnormally and the drain voltage sets back, clamping at some fixed level. A good example of a bipolar effect is the *snapback breakdown*.

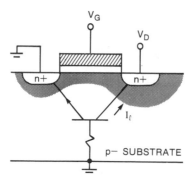

Figure 16(a). Parasitic bipolor transistor with leakage current I_l.

Snapback breakdown occurs when the device is biased in saturation with a high drain voltage. Figure 16(b) shows the device currents as a function of V_{DS} under various V_G. At first I_{DS} increases slightly after saturation due to channel length modulation. As V_{DS} increases beyond a certain limit, however, the leakage current in the depletion region around the drain initiates an avalanche breakdown. The large leakage current flowing from the substrate to the drain raises the substrate potential near the channel to such an extent that the *substrate-source junction* becomes forward biased. The bipolar transistor is then activated and electrons are emitted into the substrate, generating large current. This "collector current" then clamps the drain (collector) voltage to a constant level. The sustaining voltage, V_{SUS}, is therefore the maximum V_{DS} allowed in normal circuit operations.

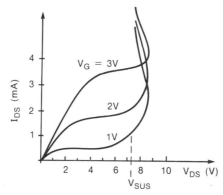

Figure 16(b). Snapback breakdown characteristics.

Another bipolar effect occurring in some CMOS circuits is the latchup. Figure 17 shows a potential latchup configuration in a CMOS layout. The p+ and n+ diffusions in the n-well and p− substrate are the sources of the p-channel and n-channel devices of an inverter. The p+ and n+ diffusions in the p− substrate and n-well are the contacts for ground and power. With the parasitic bipolar transistors connected as shown in Figure 17, the structure forms an SCR (silicon controlled rectifier) as shown in Figure 18. Also shown are the substrate and n-well resistances R_S and R_W, which play an important role in inducing the latchup.

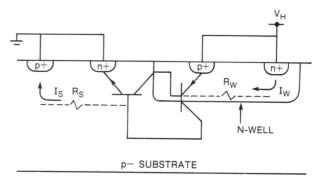

Figure 17. CMOS latchup configuration.

If properly "designed," an SCR has two stable DC solutions: the off state, in which both bipolar devices are cut off with $I_A = 0$ and $V_A = V_H$, and the on state, in which both bipolar devices are active with a large I_A and low V_A. The exact value of I_A and V_A depends on V_H and its output resistance R_O. Since both emitter-base junctions must be forward biased, a minimum value of I_A is needed to sustain the on state. Referring back to Figure 18, the current through R_W is given by

$$I_W = \frac{V_{BE(on)}}{R_W} = -I_{B1} + \beta_2 I_{B2} \qquad (2)$$

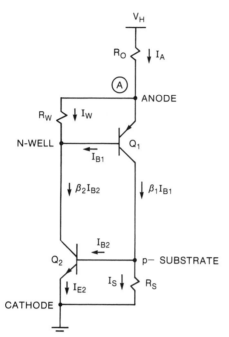

Figure 18. CMOS latchup SCR structure.

where $V_{BE(on)}$ is the turn-on voltage of the pn junction diode and β_2 is the forward current gain of Q_2. The current through R_S is given by

$$I_S = \frac{V_{BE(on)}}{R_S} = \beta_1 I_{B1} - I_{B2} \tag{3}$$

where β_1 is the forward current gain of Q_1. Multiplying Equation (2) by β_1 and substituting the result into Equation (3), we obtain

$$I_{B2} = \frac{V_{BE(on)}}{\beta_1 \beta_2 - 1} \left(\frac{\beta_1}{R_W} + \frac{1}{R_S} \right)$$

Since $I_{B2} > 0$,

$$\beta_1 \beta_2 > 1 \tag{4}$$

The total current I_A is the sum of I_S and I_{E2}:

$$\begin{aligned}
I_A &= \frac{V_{BE(on)}}{R_S} + (1 + \beta_2)I_{B2} \\
&= \frac{V_{BE(on)}}{\beta_1 \beta_2 - 1} \left[\frac{1}{R_S}(\beta_1 + 1)\beta_2 + \frac{1}{R_W}\beta_1(\beta_2 + 1) \right] \\
&= I_H
\end{aligned} \tag{5}$$

Thus I_H, called the *holding current,* is the minimum current generated from the power supply to sustain the on state.

In practice it is possible that both Equations (4) and (5) are satisfied by the circuit in Figure 18. In normal operations the SCR is in the off state with the bipolar transistors cut off. The SCR, however, can be triggered by large transient currents through R_S and R_W, which may forward bias the emitter-base junctions. Such transient currents can be caused by the power-on current, the junction leakage by radiation, or simply the large substrate current during switching. Once in the on state, the SCR will latch up and can only be turned off by removing the power supply. In many cases permanent damage results.

The latchup problem can be avoided by lowering the gain of the bipolar transistors or increasing the holding current. The former can be accomplished by more conservative layout rules or better device isolation. The latter can be achieved by special processes such as the retrograde doping of substrate, which will be discussed in Section 4.

3. SCALING THEORY

One general principle that has made a great impact on MOS technology is the scaling theory. The original scaling theory is based on the principle of constant field, that is, as the device dimensions decrease, the electric fields within the device remain constant.

The geometry of a device can be specified by four parameters: W, L, t_{ox}, and x_j. The channel width and length, W and L, are horizontal dimensions whereas the gate oxide thickness t_{ox} and the S/D junction depth x_j are vertical dimensions. According to scaling theory, all geometric parameters should be scaled down by the same factor $K > 1$. The device density therefore increases by a factor of K^2. To maintain the same fields, the power supply V_H, the substrate bias V_B, and the threshold voltage V_T must also be scaled down by the same factor K. The gate capacitance per unit area, C_o' becomes KC_o' and the device current $I_{DS} = \gamma C_o'(W/L)(V_{GS} - V_T)^2/2$ becomes I_{DS}/K. Since the total gate capacitance $C_o = C_o'WL$ decreases by a factor of K, the performance, which is proportional to $I_{DS}/C_o V_H$, improves by a factor of K. The total DC power per unit area, being proportional to $I_{DS} V_H/WL$, remains constant.

The scaling principle is only a first-order approximation. As the geometry shrinks in size many second-order effects discussed in the previous section become significant. For example, since the built-in potential does not scale, lowering the threshold voltage requires more than K times the increase of the substrate doping, which, in turn, affects the S/D junction depth and the minimum channel length. Also, because the temperature coefficient does not scale by K, a lower threshold voltage is more sensitive to temperature, which can severely affect the worst case noise margin.

Along with velocity saturation because of strong fields, another effect, the inversion charge modulation, becomes significant when t_{ox} becomes comparable to the thickness of the inversion layer. The finite thickness of the inversion layer reduces the charge stored in the channel and scales I_{DS} down by more than a factor of K. Circuit performance is further degraded because vertical scaling increases the resistance of interconnecting wires. Even though the capacitances decrease, the wire time constant, RC, stays the same.

Another serious detractor of scaling principle is the subthreshold current. Since I_{st} is proportional to the exponential of $(V_{GS} - V_T)/(kT/q)$, the subthreshold current grows rapidly as $(V_{GS} - V_T)$ approaches kT/q and, consequently, won't turn off.

From a system viewpoint, device design is not the only concern here. As the number of circuits increases more areas are occupied by interconnecting wires, and circuit loadings are predominantly wire capacitances and resistances. Smaller devices do not necessarily imply higher density, and performance improvement is by no means automatic. As a matter of fact, lowering the power supply as required by this theory renders the chip more vulnerable to outside noises.

Despite all these problems, however, the technology advances toward miniaturization. New device structures and processes are constantly being invented to overcome the "nonidealities." Some of these new developments are discussed in the next section.

4. ADVANCES IN PROCESS AND DEVICE DESIGN

In this section we shall discuss some recent developments in process and device design.

4.1 LDD Structure and Salicide Process

One of the major difficulties in applying the scaling principle is the required lowering of the power supply. In a complex system, MOS circuits must interface with circuits of various technologies such as bipolar and analog circuits. Since the built-in potentials of semiconductors do not scale, the power supply of bipolar circuits, for example, cannot scale indefinitely. The power supply of MOS circuits, therefore, does not always scale with geometric dimensions.

Maintaining the same power supply while shrinking device dimensions increases the internal fields. In a "conventional" device, the S/D regions are degenerately doped. As the channel length and S/D junction depth decrease with higher substrate doping, the strong field across the S/D junction lowers the breakdown voltage and intensifies the short-channel effects. To reduce the maximum field around the S/D junction, a lightly doped n-type region is created around the edges of the gate. This is called *the lightly doped drain* (LDD) structure. A typical n-channel LDD device is shown in Figure 19. As can be seen, the maximum field intensity around the drain is lowered considerably by spreading the field across the lightly doped region.

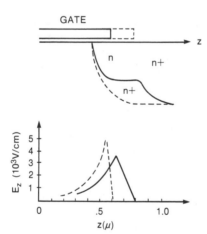

Figure 19. Comparison between LDD (solid curves) and conventional (dashed curves) devices © 1980 IEEE. [11]

Figure 20 shows a possible sequence of process steps for the LDD structure. Right after the polysilicon and polysilicon oxide depositions as described in Figure 8, a directional RIE etch defines the gate. A subsequent P^+ ion implant forms the lightly doped regions. A layer of CVD SiO_2 is then conformally deposited over the wafer. The oxide is then etched by directional RIE, leaving only sidewalls around the gate. These sidewalls, called *spacers,* block the subsequent As^+ ion implant forming the heavily doped S/D of the device. Notice that the width of the n− regions is very small ($\cong 0.3 \ \mu$). The thickness of the sidewalls, which depends on the thickness of the CVD oxide, must be carefully controlled.

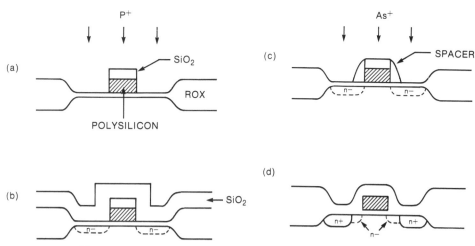

Figure 20. Sidewall-spacer technology © 1982 IEEE. [15]

Strictly speaking, the spacers are not necessary for the LDD process. A double implant of a light dose of P^+ followed by a heavy dose of As^+ as shown in Figure 10(a) can achieve the same result. Since phosphorus diffuses much faster than arsenic, an n− layer is formed around the heavily doped region after drive-in. As a matter of fact, because phosphorus forms a more gradual junction, the maximum field is even lower. The spacers, however, are needed for a salicide process described below.

A serious drawback of the LDD structure is the high resistivity of the n− regions—a resistivity that reduces V_{DS} for the "inner device." To lower the S/D resistivity a layer of metal, such as platinum (Pt) or titanium (Ti) is deposited across the whole wafer after the spacers formation as shown in Figure 21. With proper control of temperature and reaction time, the metal reacts with bare silicon only, forming a layer of silicide such as PtSi and $TiSi_2$. The unreacted metal over the SiO_2 is then removed by selective etching, leaving a layer of silicide on top of the gate and the S/D regions. After proper annealing the sheet resistance of the diffusion regions reduces from 20–50 $\Omega/\square$ down to 2–5 $\Omega/\square$. This process, which does not require an additional mask for alignment, is called *the self-aligned silicide*, or *salicide*, process. Since the spacers are used only to avoid creating a silicide bridge over the gate, its thickness is not as critical as that in Figure 20.

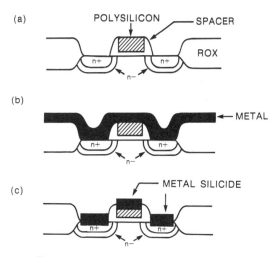

Figure 21. Salicide technology.

4.2 CMOS Processes

Since the n-well CMOS process of Section 1 is essentially an extension of an NMOS process, n-channel devices are optimized. P-channel devices are buried channel, making them more susceptible to leakage current and short-channel effects in sub-threshold regions. Furthermore, since the well concentration is an order of magnitude higher than the substrate for process control purposes, the junction capacitances of p-channel devices are also higher. On the other hand, however, placing the n-channel devices in p-wells leads to the same susceptibilities, with the roles of the n-channel and p-channel devices reversed of course.

A third alternative is to form both p- and n-wells on a lightly doped substrate and adjust the device characteristics independently. This is called the *twin-well* (*twin-tub*) process. The polysilicon gates can also be doped differently so that both types of devices operate with reasonably low threshold voltages.

Since each approach has its own merits, the choice of a particular process depends on application and cost.

4.2.1 Trench Isolations.   Spacing from the well edge is a critical parameter impacting the density of CMOS technology. Dopant out-diffusion around the well edges lowers the surface concentration and hence the threshold voltage of the field oxide transistors. As a result, devices must be well removed from the well border. While this may not impact the density of logic circuits where the border areas can be reserved for wiring, it is a severe restriction to memory cell layout.

To improve density, a deep trench $(5-6\mu)$ of dielectric isolation can be used to separate the p- and n-channel regions. A typical trench is shown in Figure 22. A deep groove is first etched into silicon by RIE, and then the sidewalls are oxidized. The oxide on the walls blocks diffusion of impurities in subsequent process steps. Next the trench is filled with SiO_2 or polysilicon and is capped with SiO_2.

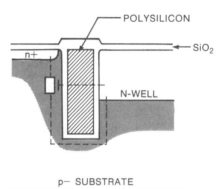

Figure 22.   CMOS trench isolation.

A particular problem associated with trench structure is trench surface inversion. With the parasitic FET shown in the figure, the n-well is usually biased at the most positive voltage or power supply, and the field extending from the n-well may penetrate through to the trench. This effect, aided by the fixed positive charges Q_{ss} along the sidewalls, may become strong enough to invert the p− substrate on the other side and turn on the parasitic FET. The n+ diffusion butted to the sidewall will then be shorted to the n-well along the trench surface, resulting in significant leakage current. Since the parasitic lateral npn transistor collects along the sidewall region, inversion of the sidewall enhances the bipolar collector efficiency and increases current gain, which further degrades the CMOS latchup immunity.

To eliminate trench inversion, a boron implant can be fixed to the bottom of the trench right after the trench groove is etched. This implant raises the threshold of the trench bottom which then blocks any current path from the n+ region to n-well. As circuit approaches, lower n-well bias (power supply) and more negative substrate bias also alleviate the problem.

4.2.2 Latchup Control. The latchup is a problem inherent to CMOS technology. As we saw in Section 2 the critical parameters that affect latchup sensitivity are R_W, R_S, and the current gain β_1 and β_2 of the parasitic bipolars.

(1) Control of R_W—the retrograde-wells: To reduce well resistivity a retrograde doping profile shown in Figure 23 can be formed through a high-energy implant. The high doping concentration in the bulk provides a low resistivity path for the lateral current, while the relatively low doping in the surface maintains high breakdown voltage at S/D junctions.

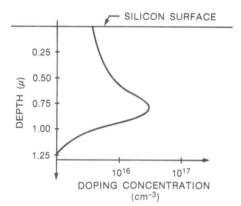

Figure 23. CMOS retrograde doping profile.

(2) Control of R_S—the epi-CMOS process: To reduce R_S, devices can be formed in a lightly doped epitaxial layer on a heavily doped substrate of the same type. Figure 24 shows an n-well process with a p− epi on a p+ substrate. The heavily doped substrate provides a low resistivity path for lateral substrate currents.

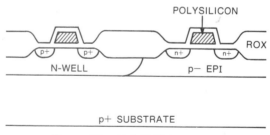

Figure 24. Epi-CMOS process.

The effectiveness of this approach depends on the thickness and bias voltage of the epitaxial layer. To ensure all lateral currents flow in the heavily doped substrate, deep trenches into the substrate are sometimes made.

(3) Control of the parasitic bipolars: One very effective method to eliminate the latchup is a combination technique of deep trench and epi with trench bottom implant. The boron implant at the trench bottom not only forces lateral currents into the substrate but also reduces the current gain of the bipolar as well.

The CMOS process is one of the most active areas of research and development. Interested readers are strongly encouraged to consult the references for more detailed information.

5. *MOS DYNAMIC RAM CELLS*

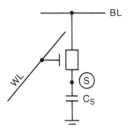

Figure 25. A DRAM cell circuit.

This section describes the evolution of MOS dynamic RAM cells: a good example of the advancement of MOS technology.

The one-device cell of a dynamic RAM (random access memory) consists of a capacitor and a device shown in Figure 25. The capacitor holds the information, storing different amounts of charge for logic 1's and 0's. The device, called a transfer or pass gate, controls the input/output (I/O) of the cell. When the wordline (WL) is high the device turns on and the capacitor can be accessed via the bitline (BL). When the WL is low, the device turns off and the capacitor is isolated from the outside world. Since the charge in the capacitor must be replenished to compensate for leakage when the device is off, the cell must be "refreshed" periodically by external circuitry; hence the name dynamic.

Figure 26 describes a straightforward implementation of the cell. A thin oxide capacitor C_O is formed between the storage node Ⓢ in the silicon surface and the storage plate. In NMOS technology, the storage plate is usually tied to the power supply V_H so the capacitor is always on.

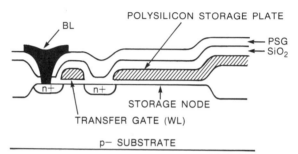

Figure 26. A single-poly DRAM cell.

The structure in Figure 26 was very popular in early dynamic memories. In silicon gate processes, as we saw in Section 1, a double-polysilicon process can be easily implemented by growing another layer of polysilicon after the formation of the polyoxide of the first polysilicon layer. With two layers of polysilicon available, a cell can now be constructed as shown in Figure 27. By eliminating the gap between the transfer gate and the storage plate, cell density is greatly improved.

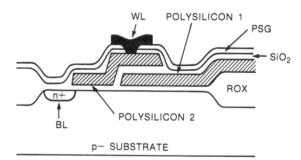

Figure 27. A double-poly DRAM cell.

One of the major considerations of the cell design is the amount of charge C_S can store. While the transfer gate is usually constructed to the minimum dimension allowed by the process, the capacitor must be large enough to hold charges adequate for signal sensing. Thus, a large thin oxide capacitor limits cell density.

The depletion capacitance C_B from the storage node to the bulk can also be used to hold charges in parallel with C_O. In a normal situation with lightly doped substrate, $C_B \ll C_O$ and $C_S \cong C_O$. To enhance C_B the substrate doping concentration of C_S can be selectively increased by ion implant. The p+ doped substrate narrows the depletion width and increases C_B. Increasing the substrate doping, however, increases the threshold voltage of C_S, which reduces the storage capability in C_O. To compensate for the bulk charge, another thin layer of n+ doping can be implanted in the surface. The result, shown in Figure 28, is called a *Hi-C* (high capacity) cell. It increases charge storage by 30–50% for the same area.

From a circuit designer's point of view a Hi-C cell is essentially a depletion mode capacitor with poor (high) substrate sensitivity. While a large C_B is to be avoided for normal devices, it is a desirable feature for storage capacitors.

Since the threshold voltage of C_O can be adjusted to a negative value, the storage plate of C_O can now be tied to the ground instead of V_H. Since in most systems the ground potential is much more stable than the power supply, a grounded plate enhances noise immunity. As we have discussed in Chapter 1, however, a grounded plate may deplete the surface and force the charge into the bulk. A capacitance C_C is then formed from the storage node to the surface, reducing the total capacitance to the storage plate as shown in Figure 28(b). This problem can be alleviated by adjusting the doping concentration and implant depth.

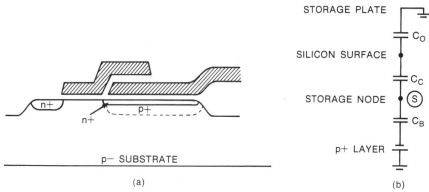

(a)(b)

Figure 28.A Hi-C DRAM cell.

While ion implantation is used to implement different devices in a silicon surface, reactive ion etching is used to build devices and capacitors in the vertical dimension. Parallel to the development of trench isolation in CMOS technology, trench capacitors also increase density. One such capacitor design, called the corrugated capacitor cell (CCC) is shown in Figure 29. The sidewalls of the capacitor are made of multilayers of dielectric materials creating a large equivalent dielectric constant. The trench is filled with a polysilicon "storage plate." The storage capacitance C_S consists mainly of the capacitances of the sidewalls and the trench bottom. Cell density is limited by the leakage current between adjacent trench capacitors.

Another trench capacitor structure, called the folded capacitor cell (FCC), is shown in Figure 30. A thick layer of "field oxide" at the bottom of the trench isolates the capacitors. The box with an exposed corner in the figure is the storage node. It is wrapped by thin oxide on all sides except the bottom. The storage plate is the polysilicon filling the trenches around the storage node.

Two trench capacitors that have storage nodes inside the trench are shown in Figures 31 and 32. The cell in Figure 31, the SPT (substrate-plate trench capacitor) cell, uses a conducting strap to connect the p+ doped polysilicon storage node to the source of the transfer gate. Notice that there is a parasitic p-channel device along the

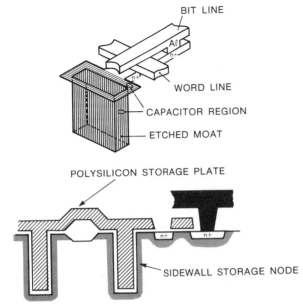

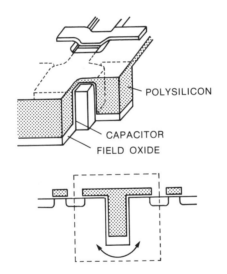

Figure 29.   A corrugated capacitor cell © 1982 IEEE. [16]

Figure 30.   A folded capacitor cell
© 1984 IEEE. [19]

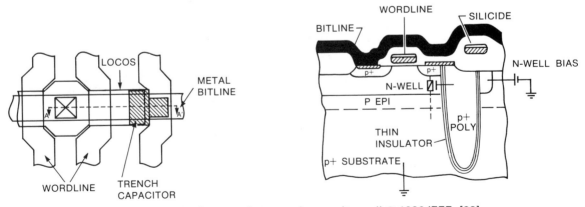

Figure 31.   A substrate-plate trench capacitor cell © 1986 IEEE. [23]

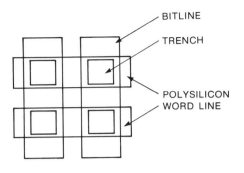

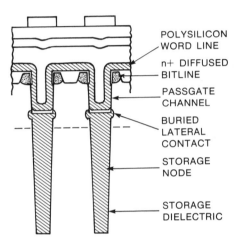

Figure 32. A trench transistor cell
© 1986 IEEE. [20]

sidewall of the trench that must be controlled. The cell in Figure 32, called a TTC (trench transistor cell), uses a vertical transfer gate on top of the storage node, thus achieving the ideal configuration of one cell per crosspoint of the bitline and word-line.

Both cells use the p+ substrate as the storage plate which can be either tied to a negative bias or grounded. Since the charge is stored inside the trench, the cells are less susceptible to α-particle disturbances (see Chapter 6).

6. CONCLUSION

MOS technology is an area of intensive research and development. New ideas and products are invented constantly. We have reviewed the basic processes and some recent developments in device design in this chapter. More advances in circuits and products will be discussed in later chapters.

REFERENCES

1. Pao, H. C., and C. T. Sah. "Effects of Diffusion Current on Characteristics of Metal-Oxide-Insulator Semiconductor Transistors." *Solid-State Electron,* Vol. 9, 1966, pp. 927–937.

2. Cobbold, R. S. C. *Theory and Applications of Field Effect Transistors.* New York; Wiley-Interscience, 1970.

3. Barron, M. B. "Low Level Currents in Insulated Gate Field Effect Transistors." *Solid-State Electron,* Vol. 15, March 1972, pp. 293–302.

4. Gregory, B. L., and B. D. Shafer. "Latch-up in CMOS Integrated Circuits." *IEEE Trans. on Nucl. Sci.,* Vol. NS-20, December 1973, pp. 293–299.

5. Lee, H. S. "An Analysis of the Threshold Voltage for Short-Channel IGFET's." *Solid-State Electron,* Vol. 16, 1973, pp. 1407–1417.

6. Troutman, R. R., and S. N. Chakravarti. "Subthreshold Characteristics of Insulated-Gate Field Effect Transistors." *IEEE Trans. on Circuit Theory,* Vol. CT-20, no. 6, November 1973, pp. 659–665.

7. Denard, R. H., F. H. Gaensslen, H. Yu, V. L. Rideout, E. Bassous, and A.R. LeBlanc. "Design of Ion-Implanted MOSFET's with Very Small Physical Dimensions." *IEEE J. of Solid State Circuits,* Vol. SC-9, no. 5, October 1974, pp. 256–268.

8. Rideout, V. L., F. H. Gaensslen, and A. LeBlanc. "Device Design Considerations for Ion-Implanted n-channel MOSFET's." *IBM J. of Research and Development,* Vol. 19, no. 1, January 1975, pp. 50–59.

9. Tasch, L. F., P. K. Chatterjee, H-S Fu, and T. C. Holloway. "The Hi-C RAM Concept." *IEEE Trans. on Electron Devices,* Vol. ED-25, no. 1, January 1978, pp. 33–41.

10. Murphy, B. T. "Unified Field-Effect Transistor Theory Including Velocity Saturation." *IEEE J. of Solid State Circuits,* Vol. SC-15, no. 3, June 1980.

11. Ogura, S., P. J. Tsang, W. W. Walker, D. L. Critchlow, and J. F. Shepard. "Design and Characteristics of the Lightly Doped Drain-Source (LDD) Insulated Gate Field Effect Transistor." *IEEE J. of Solid State Circuits,* Vol. SC-15, no. 4, Aug. 1980.

12. Parrillo, L. C., R. S. Payne, R. E. Davis, G. W. Reutlinger, and R. L. Field. "Twin-Tub CMOS—A Technology for VLSI Circuits." *IEDM Tech. Dig.,* 1980, pp. 752–755.

13. Rung, R. D., C. J. Dell'Oca, and L. G. Walker. "A Retrograde p-Well for High Density CMOS." *IEEE Trans. on Electron Devices,* Vol. ED-28, no. 10, October 1981.

14. El-Mansy, Y. A. and R. A. Burghard. "Design Parameters of the Hi-C DRAM Cell." *IEEE J. of Solid State Circuits,* Vol. SC-17, No. 5, October 1982.

15. Tsang, P. J., S. Ogura, W. W. Walker, J. F. Shepard, and D. L. Critchlow. "Fabrication of High-Performance LDDFET's with Oxide Sidewall-Spacer Technology." *IEEE J. of Solid State Circuits,* Vol. SC-17, no. 2, April 1982.

16. Sunami, H., T. Kure, N. Hashimoto, K. Itoh, T. Toyabe, and S. Asai. "A Corrugated Capacitor Cell (CCC) for Magbit Dynamic MOS Memories." *IEDM Tech. Dig.* 1982, pp. 806–808.

17. Cham, K. M., S. Y. Chiang, D. Wenocur, and R. D. Rung. "Characterization and Modeling of the Trench Surface Inversion Problem for the Trench Isolated CMOS Technology." *IEDM Tech. Dig.,* 1983, pp. 23–26.

18. Sunami, H., T. Kure, N. Hashimoto, K. Itoh, and T. Toyabe. "A Corrugated Capacitor Cell (CCC)." *IEEE Trans. on Electron Devices,* Vol. ED-31, no. 6, June 1984.

19. Wada, M., K. Hieda, and S. Watanabe. "A Folded Capacitor Cell (F.C.C.) for Future Megabit DRAMs." *IEDM Tech. Dig.,* 1984, pp. 244–247.

20. Richardson W. F., D. M. Bordelon, G. P. Pollack, A. H. Shah, S. D. S. Malhi, H. Shichijo, S. K. Banerjee, M. Elahy, R. H. Womack, C-P. Wang, J. Gallia, H. E. Davis, and P. K. Chatterjee. "Trench Transistor Cross-Point DRAM Cell." *IEDM Tech. Dig.,* 1985, pp. 714–717.

21. Yamaguchi, T., S. Morimoto, H. K. Park and G. C. Eiden. "Process and Device Performance of Submicrometer-Channel CMOS Devices Using Deep-Trench Isolation and Self-Aligned TiSi Technologies." *IEEE Trans. on Electron Devices,* Vol. ED-32, no. 2, Feb. 1985.

22. Chen, J. Y., "CMOS—the Emerging VLSI Technology." *IEEE Circuits Devices Magazine,* March 1986, pp. 16–31.

23. Lu, N. C-C, P. E. Cottrel, W. J. Craig, S. Dash, D. L. Critchlow, R. L. Mohler, B. J. Machesney, T. H. Ling, W. P. Noble, R. M. Parent, and R. E. Scheuerlein, E. J. Sprogis, and L. M. Terman, "A Substrate-Plate Trench-Capacitor (SPT) Memory Cell for Dynamic RAM's." *IEEE J. of Solid State Circuits,* Vol. SC-21, no. 5, Oct. 1986.

24. Troutman, R. R., *Latchup in CMOS Technology*. Hingham, MA: Kluwer Academic, 1986.

DIGITAL INVERTERS —DC ANALYSIS

INTRODUCTION

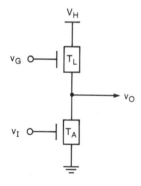

Figure 1. A simple inverter.

The inverter is one of the basic building blocks of MOS circuits. As shown in Figure 1, a simple inverter consists of two devices. The upper device, T_L, is called the load (in linear circuitry) or *the pullup* (in digital circuitry), and the lower device, T_A, is called the active device or driver (in linear circuitry) or *the pulldown* (in digital circuitry). While the pulldown is always an enhancement mode device, the pullup can be either a depletion or an enhancement device. The former is called an E/D (enhancement/depletion) inverter whereas the latter is called an E/E (enhancement/enhancement) inverter.

In Chapter 1 we saw that from a circuit point of view, a device can be described by a quadruple (γ, k_1, V_{T0}, λ), where $\gamma = \mu_0 C_o'$ as defined in Equation (9) in Chapter 1. The substrate sensitivity is k_1, V_{T0} is the threshold voltage, and λ measures the channel length dilation. The device current is determined by the gate-drives:

$$x_S = (v_{GS} - V_{TS})\ \text{sgn}\ (v_{GS} - V_{TS}) \quad \text{and} \quad x_D = (v_{GD} - V_{TD})\ \text{sgn}\ (v_{GD} - V_{TD})$$

For example, for an n-channel device,

$$i_{DS} = \begin{cases} \dfrac{1}{2}\beta\ \dfrac{x_S^2 - x_D^2}{1 + \lambda\,|x_S - x_D|}, & x_D > x_P \qquad (1) \\[4mm] \dfrac{1}{2}\beta\ \dfrac{x_S^2 - x_P^2}{1 + \lambda\,|x_S - x_P|}, & x_D \le x_P \qquad (2) \end{cases}$$

where $x_P = x_S + (1 - \sqrt{1 + 2\lambda x_S})/\lambda$ is the pinchoff point and

$$\beta = \gamma\,\frac{W}{L}$$

is a design parameter controlling the magnitude of i_{DS}. By convention, a device is said to be operating in the linear region whenever $x_D > x_P$ and Equation (1) applies. It is said to be operating in saturation region whenever $x_D \leq x_P$ and Equation (2) applies. Once the device enters the saturation region, the current i_{DS} becomes independent of v_D, hence the name saturation.

EXERCISE

Write the corresponding current equations for a p-channel device.

The expressions of i_{DS} for T_L and T_A depend on the operating condition defined by the terminal voltages applied to the device. Referring to Figure 1 again, for the pullup T_L to operate in the saturation region, substitute $x_D \leq x_P$ for T_L,

$$\Rightarrow \quad v_G - V_H - V_{TD} \leq v_G - v_O - V_{TS} + \frac{1 - \sqrt{1 + 2\lambda x_S}}{\lambda}$$

Approximating $V_{TX} = V_{T0} + k_1 v_X$, where $x = s$ or D, we obtain

$$-(1 + k_1)V_H \leq -(1 + k_1)v_O + \frac{1 - \sqrt{1 + 2\lambda[v_G - V_{T0} - (1 + k_1)v_O]}}{\lambda}$$

The foregoing inequality is satisfied if either

$$v_G - V_{T0} \leq (1 + k_1)V_H$$

or

$$v_G - V_{T0} > (1 + k_1)V_H$$

but

$$v_O \leq V_H - \frac{2[v_G - V_{T0} - (1 + k_1)V_H]}{\lambda(1 + k_1)}$$

Thus, the device is saturated when either the drain is completely depleted or the voltage drop from drain to source is large enough to cause channel pinchoff because of velocity saturation. If these conditions aren't met, the device is in the linear region. Now, on defining

$$X_D \triangleq v_G - V_{T0} - (1 + k_1)V_H$$

we conclude that

$$T_L \text{ is in} \begin{cases} \text{saturation region,} & \text{if } \begin{cases} X_D \leq 0, & \text{or} \\ X_D > 0, & \text{but } v_O \leq V_H - \dfrac{2X_D}{\lambda(1 + k_1)} \end{cases} \\ \text{linear region,} & \text{if } X_D > 0, \text{ but } v_O > V_H - \dfrac{2X_D}{\lambda(1 + k_1)} \end{cases} \quad (3)$$

Similarly, it can be shown that for the pulldown T_A

$$T_A \text{ is in} \begin{cases} \text{saturation region,} & \text{if } v_O > \dfrac{-1 + \sqrt{1 + 2\lambda(v_I - V_{T0})}}{\lambda(1 + k_1)} \\ \text{linear region,} & \text{otherwise} \end{cases} \quad (4)$$

In this chapter inverters will be analyzed from the DC point of view. The transfer characteristics of different inverters will be derived. We then discuss two commonly used procedures to determine the β ratio of inverters based on noise considerations. We will see that the concept of noise margin and minimum logic levels are closely related to the stability margin of an unbalanced latch. Finally, the transfer characteristics and noise margins of NMOS and CMOS circuits will be compared.

1. THE TRANSFER CHARACTERISTIC

Once the device sizes of an inverter have been determined, the relation between the input, v_I and the output, v_O can be derived by making the device currents equal. The equation

$$i_{DS}(T_L) = i_{DS}(T_A)$$

defines a functional relationship $v_O = \hat{v}_O(v_I)$ describing a continuous curve in the v_I–v_O plane called the *transfer characteristic* or simply the TC plot of the inverter. Many circuit design concepts, such as the unity gain points and logic thresholds, can be readily interpreted with the help of transfer characteristic curves.

A typical TC plot of an E/D inverter is shown in Figure 2. When the input is low, that is, $v_I \leq V_{T0} (T_A)$, device T_A is off and the output $v_O = V_H$. As soon as v_I exceeds $V_{T0} (T_A)$, device T_A starts to draw current from T_L, pulling v_O toward the ground. Initially v_O falls only gradually with v_I. After v_I reaches a critical level, V_{ts}, at which the slope of the TC plot equals -1, however, v_O begins to drop at a much steeper rate. It reaches a specified downlevel V_{OL} (to be explained shortly) quickly and levels off for $v_I \gg V_{ts}$. The transition voltage, V_{ts}, also called the *unity gain point*, therefore, signifies the logic transition of the output from a high level to a low level.

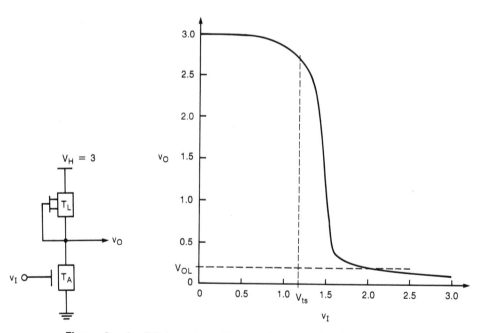

Figure 2. An E/D inverter with a typical transfer characteristic.

For the E/E inverter shown in Figure 3, however, the output v_O starts to fall at $v_I = V_{T0} (T_A)$ with a discontinuous change of slope. A meaningful V_{ts} can only be defined for the logic transition from the low level to the high.

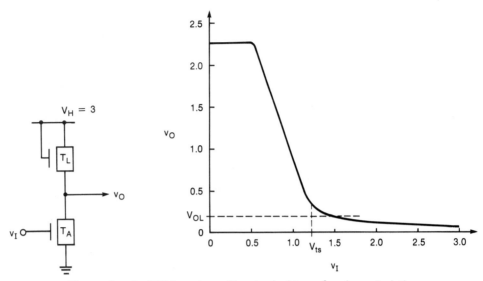

Figure 3. An E/E inverter with a typical transfer characteristic.

For the CMOS inverter in Figure 4, there are two well-defined unity gain points that can be used to signify logic transitions: $\overline{V}_{ts}$ for the LH (output rises from low to high) transition, and $\underline{V}_{ts}$ for the HL (output falls from high to low) transition.

Thus, the concept of unity gain points, being widely used as logic thresholds in discrete TTL logic circuits, makes sense in our case only when the particular inverter type is also specified. As a matter of fact, as we shall see in the following discussions, the notion of unity gain points is not as useful with integrated MOS circuits as it is with discrete TTL logics.

Now, since i_{DS} is directly proportional to the β of a device, both V_{ts} and V_{OL} are affected by the β ratio between T_A and T_L defined by

$$\beta_r \triangleq \frac{\beta(T_A)}{\beta(T_L)} = \frac{\beta_A}{\beta_L}$$

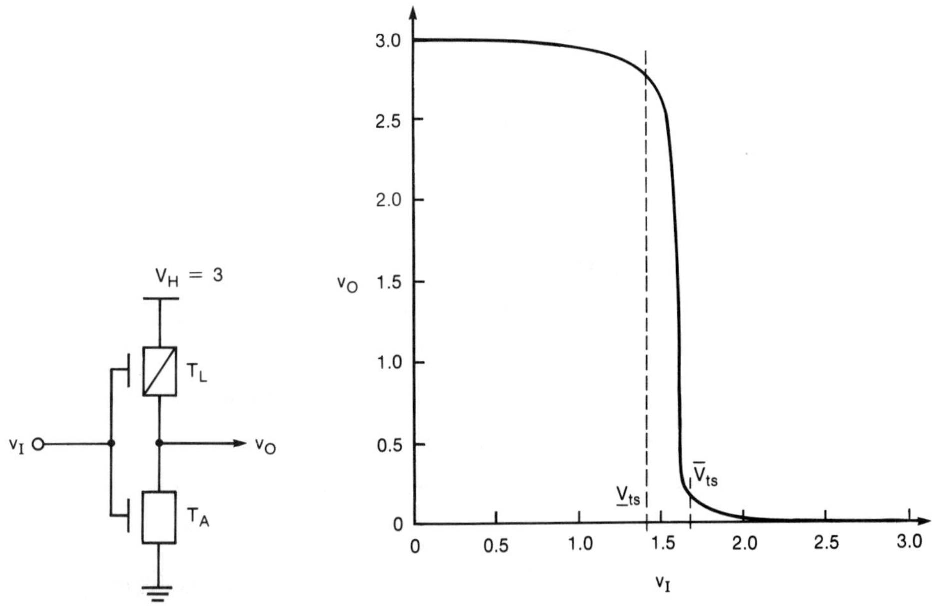

Figure 4. A CMOS inverter with a typical transfer characteristic.

Figure 5 shows the TC plots of a family of E/D inverters with different β ratios. As is evident, the bigger the β ratio, the lower the value of V_{ts} and the sharper the TC plot. Since lower V_{ts} and a sharper TC plot allow lower V_{OL} and better DC stability, conservative designers tend to use large β ratios. Large β ratios, however, require bigger pulldowns. Since the gate of T_A is a capacitive load to its previous driver, an overly conservative design with a large β ratio not only has poorer density, but also results in lower speeds. On the other hand, a small β ratio implies higher V_{OL} and less distinct transition of the output. An extremely aggressive design with a very small β ratio is, therefore, more noisy and less stable.

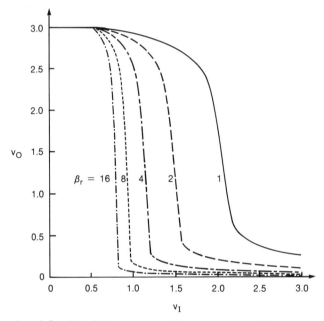

Figure 5. A family of TC plots of E/D inverters of different β ratios.

In practice, there are two commonly used procedures for determining the β ratios, depending on the environment of the circuit: (1) the maximum downlevel design and (2) the DC stability design.

2. THE MAXIMUM DOWNLEVEL DESIGN

In this approach, the β ratio is calculated to ensure a specified downlevel V_{OL}, which, in turn, is determined by the circuit environment. The procedure is best illustrated by an example.

EXAMPLE 1

Consider the network of E/D inverters shown in Figure 6. The circuits are to be operated under the following conditions:

$$\Delta V_H = \text{tolerance of power supply}$$

$$= 10\% \text{ of } V_H$$

$$\delta V_H = \text{maximum } V_H \text{ drop from one inverter to another}$$

$$= 0.2$$

$$\delta V_G = \text{maximum ground shift from one inverter to another}$$

$$= 0.2$$

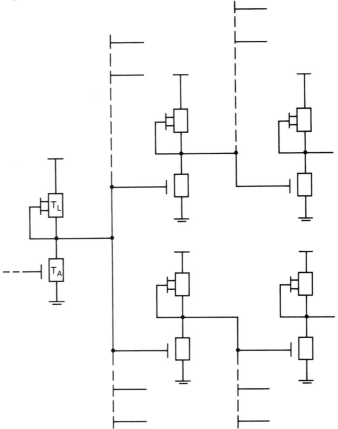

Figure 6(a).   A network of E/D inverters.

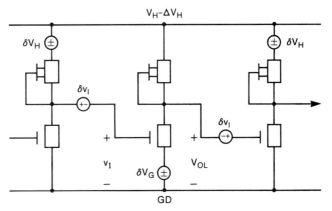

Figure 6(b).   A chain of inverters under the worst case noise condition.

δv_I = random noise from one inverter to another

$\quad = 0.05$

Given the device parameters

$$V_{T0}(T_A) \in (0.45, 0.55)$$

$$V_{T0}(T_L) \in (-1.4, -1.2)$$

$$k_1 = 0.1, \ \theta = 0.5, \text{ and } \gamma = 0.082 \text{ for both devices}$$

and assuming the channel lengths $L(T_A) = 0.5$ and $L(T_L) = 1.0$, determine the β ratio of the inverters so that no inverter can be turned on by noise.

Solution

Since the minimum value of V_{T0} (T_A) is 0.45 and the maximum random noise on the line is 0.05, the maximum downlevel V_{OL} should be specified as

$$V_{OL} = 0.45 - 0.05 = 0.4$$

so no inverter will be turned on. This downlevel must be maintained at the worst case (lowest) input:

$$v_I = V_H - \Delta V_H - \delta V_H - \delta v_I$$
$$= 3.0 - 0.3 - 0.2 - 0.05$$
$$= 2.45$$

Since V_{OL} is low, T_A is in the linear region. Now from Equation (1)

$$i_{DS}(T_A) = \frac{1}{2}\beta_A \frac{x_S^2 - x_D^2}{1 + \lambda_A |x_S - x_D|}$$

where

$$\lambda_A = \frac{\theta}{L(T_A)} = \frac{0.5}{0.5} = 1.0$$

$$x_S = v_{GS} - V_{T0} = v_I - \delta V_G - V_{T0}$$
$$= 2.45 - 0.2 - 0.55 = 1.70$$

and

$$x_D = v_{GD} - V_{TD} = (v_I - V_{OL}) - (V_{T0} + k_1 V_{OL})$$
$$= (2.45 - 0.4) - (0.55 + 0.1 \times 0.4)$$
$$= 1.46$$

Note that $V_{T0} = 0.55$ is the worst case for $i_{DS}(T_A)$. Substitution of the above values into $i_{DS}(T_A)$ yields

$$i_{DS}(T_A) = 0.31 \, \beta_A$$

On the other hand, it is easy to see that T_L is in the saturation region so that

$$i_{DS}(T_L) = \frac{1}{2}\beta_L \frac{x_S^2 - x_P^2}{1 + \lambda_L |x_S - x_P|}$$

where

$$\lambda_L = \frac{0.5}{1.0} = 0.5$$

$$x_S = -V_{TS} = -(V_{T0} + k_1 V_{OL}) = 1.4 - 0.1 \times 0.4 = 1.36$$

and

$$x_P = 1.36 + \frac{1 - \sqrt{1 + 2 \times 0.5 \times 1.36}}{0.5} = 0.29$$

Here $V_{T0} = -1.4$ for the worst case. Substituting these data into $i_{DS}(T_L)$, we obtain

$$i_{DS}(T_L) = 0.58 \, \beta_L$$

Now, since $i_{DS}(T_L) = i_{DS}(T_A)$, we have

$$\beta_r = \frac{\beta_A}{\beta_L} = \frac{0.58}{0.31} = 1.87$$

Since $\beta = \gamma(W/L)$, with $\gamma(T_L) = \gamma(T_A) = 0.082$ we obtain, for the "ratio of ratios"

$$\rho \triangleq \frac{(W/L)(T_A)}{(W/L)(T_L)} = 1.87 \times \frac{0.082}{0.082} = 1.87$$

In this case, the ratio of ratios $\rho = \beta_r$ because $\gamma(T_A) = \gamma(T_L)$. In case $\gamma(T_A) \neq \gamma(T_L)$ as in a CMOS inverter, $\rho \neq \beta_r$ in general.

3. DC STABILITY DESIGN

In a network where the β ratio is determined by the maximum downlevel design, any inverter would be able to drive any other inverter without degrading the logic levels. Remember, no inverter can be turned on unintentionally even under the worst case V_H tolerance and ground shift. Maximum downlevel design, however, is overly conservative at times. The β ratio obtained from maximum downlevel considerations is usually larger than necessary. Since the circuit performance is adversely affected by large β ratios, it is important to use a β ratio as small as possible. Sometimes the pulldowns of an inverter that should be off can be allowed to be slightly on as long as the inverter output is high enough to drive the output of the next stage low. This criterion is called DC stability design.

The concept of DC stability design is best understood by considering the DC stability problem of a network. Consider a network of inverters in a noisy environment. A device that should be off may be "slightly on" if the noise picked up by its input is large and persistent. An inverter whose pulldown should be off but is turned on by noise cannot drive its output to V_H and so the logic level degrades. This effect may be amplified in subsequent stages causing logic errors. Similarly, noise picked up by the input of an inverter whose pulldown is on may degrade the output downlevel to such an extent that its next stages are turned on slightly. Again, if the noise is large and persistent, logic errors may occur.

To evaluate the logic level degradation and the corresponding stability problem across the whole chip, consider the infinite chain of inverters shown in Figure 7. This circuit consists of two kinds of inverters: the odd-numbered ones have unity fanin, and the even-numbered ones have nine inputs, which is the maximum possible fanin of any inverter in the chip. A *degraded* uplevel, v_{I1}, coming from the previous stage, is applied to inverter I1, pulling down its output v_{o1}. Since I1 drives I2, the input of I2 is given by

$$v_{I2} = v_{DS}(T_{A1}) + \delta v_l + \delta V_G$$

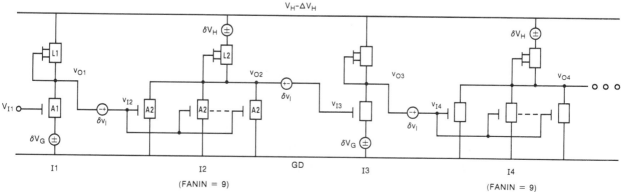

Figure 7. An infinite chain of E/D inverters under the worst case noise and fanin conditions. Physical location of each inverter can be anywhere in the chip.

where δv_l represents the maximum noise on the wire connecting I1 to I2 and δV_G is the maximum ground shift between the inverters. With δv_l and δV_G as shown in the figure, v_{I2} is in the worst case noise condition. Now v_{I2}, even at downlevel, can be high enough to turn on device T_{A2}. With all T_{A2}'s slightly on, the drain-source voltage drop $v_{DS}(T_{L2}) > 0$ because of the finite current flowing though T_{L2}. Thus, the output of I2 will not reach $V_H - \Delta V_H$ and the uplevel degrades. Hence, at the input of I3

$$v_{I3} = (V_H - \Delta V_H) - \delta V_H - v_{DS}(T_{L2}) - \delta v_l$$

Again, in this worst case noise condition, the uplevel of v_{I3} can be considerably lower than $V_H - \Delta V_H$.

As these voltage levels propagate along the chain, one of two possibilities will occur. If $v_{I3} > v_{I1}$ so that $v_{O3} < v_{O1}$ and $v_{I4} < v_{I2}$, the up- and downlevels eventually converge to some values V_{UL} and V_{DL}, respectively, with $V_{UL} > V_{DL}$. The logic levels then are distinct, and the network is said to be DC stable. On the other hand, if $v_{I3} < v_{I1}$, the logic levels will degrade progressively along the chain and finally become completely indistinguishable. In this case, the network is DC unstable and a logic error will occur.

Given a noise environment, the minimum uplevel that the network can "sustain" without causing an error is called the *least positive uplevel* (LPUL) of the circuits. Similarly, the maximum downlevel a network can "sustain" without causing an error is called the *most positive downlevel* (MPDL) of the circuits. Both LPUL and MPDL are determined by the β ratio of the inverters. In a maximum downlevel design, for example, the MPDL and LPUL are automatically determined once the β ratio is found through V_{OL}.

In DC stability design, either the LPUL or the MPDL is first specified by noise and chip interface considerations. A proper β ratio is then determined to ensure that all uplevels in the network are higher than the LPUL specified, or all downlevels in the network are lower than the MPDL specified. Thus the DC stability design guarantees the worst case logic levels in a given noise environment.

In practice, the LPUL is preferred most often because of the ease of its specification. Let LPUL $= \bar{V}$ be given. It should be apparent from the above discussion that at the required β ratio, the voltage levels must satisfy

$$v_{I3} = v_{I1} = \bar{V}$$

under the worst case conditions so that the inverter chain operates on the border line between the stability and instability regions. Since $v_{I3} = v_{I1}$, we can truncate the infinite chain at inverter I3 and feed v_{I3} back to the input of inverter I1, as shown in Figure 8. The problem then becomes

$$\text{Find } \beta_r \text{ and } v_{O1}$$

subject to
$$i_{DS}(T_{A1}) = i_{DS}(T_{L1})$$
$$9 i_{DS}(T_{A2}) = i_{DS}(T_{L2}) \tag{5}$$
$$v_{I3} = v_{I1} = \bar{V}$$

where the first two conditions describe the inverters and the last condition comes from the feedback. Upon defining the device currents "per β,"

$$j_{DS} = \frac{i_{DS}(x_S, x_D, \lambda)}{\beta/2}$$

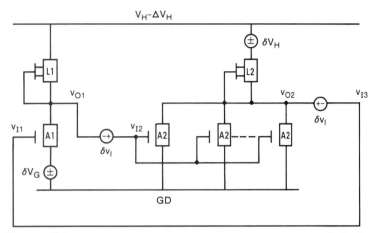

Figure 8(a). A latch circuit equivalent to the infinite chain for v_{I1} = LPUL.

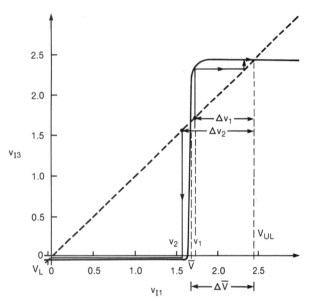

Figure 8(b). DC solutions of above circuit.

we can write

$$i_{DS}(T_{A1}) = (1/2)\beta_A \, j_{DS}[\bar{V} - V_{T0} - (1 + k_1)\delta V_G, \bar{V} - V_{T0} - (1 + k_1)v_{O1}, \lambda_A]$$

$$i_{DS}(T_{L1}) = (1/2)\beta_L \, j_{DS}[-V'_{T0} - k_1 V_{O1}, v_{O1} - V'_{T0} - (1 + k_1)(V_H - \Delta V_H), \lambda_L]$$

$$i_{DS}(T_{A2}) = (1/2)\beta_A \, j_{DS}[v_{O1} - V_{T0} + \delta v_l, v_{O1} - V_{T0} - (1 + k_1)\bar{V} - k_1 \delta v_l, \lambda_A]$$

$$i_{DS}(T_{L2}) = (1/2)\beta_L \, j_{DS}[-V'_{T0} - k_1(\bar{V} + \delta v_l),$$
$$\bar{V} + \delta v_l - V'_{T0} - (1 + k_1)(V_H - \Delta V_H - \delta V_H), \lambda_L]$$

Substituting the previous equations into Equation (5) and noting that $\beta_r = \beta_A/\beta_L$, two new equations are obtained in two unknowns: β_r and v_{O1}. Numerical methods such as the Newton-Raphson method can then be used to solve these equations. A FORTRAN program, OPTBETA, has been written for calculating the required β_r and the corresponding v_{O1} or v_{I1} for a specified LPUL or MPDL, respectively. The following examples illustrate the use of this program. Listings of the program can be found in the appendix to this chapter.

EXAMPLE 2

Find the β ratio for the chain network with an LPUL = 2. Assume the same device parameters as in Example 1.

Solution

With LPUL = $\bar{V}$ = 2, the program OPTBETA yields

$$\beta_r = 1.79 \quad \text{and} \quad v_{o1} = 0.61$$

The ratio of ratios is therefore given by

$$\rho = \frac{(W/L)(\text{T}_\text{A})}{(W/L)(\text{T}_\text{L})} = 1.79 \times \frac{0.082}{0.082} = 1.79$$

In this case the β ratio is almost the same as that obtained by the maximum downlevel design.

EXAMPLE 3

Repeat Example 2 with a new LPUL = 2.25.

Solution

In this case OPTBETA yields β_r =1.53 and v_{o1} = 0.57. The ratio of ratios is

$$\rho = \frac{(W/L)(\text{T}_\text{A})}{(W/L)(\text{T}_\text{L})} = 1.53 \times \frac{0.082}{0.082} = 1.53$$

which is about 20% smaller than in Example 1. It is "saved," however, by a considerable sacrifice of noise margins, as we will see in the next section.

EXAMPLE 4

Calculate the β ratio for the same network with MPDL = 0.66.

Solution

For MPDL = 0.66, v_{o1} = MPDL $- \delta v_l$ = 0.66 $-$ 0.05 = 0.61. Program OPTBETA then yields

$$\beta_r = 1.79 \quad \text{and} \quad v_{l1} = 2$$

which checks with Example 2.

Notice that as β ratio decreases, the LPUL increases and the MPDL decreases. Variation of MPDL is, however, much smaller than LPUL. Notice also that the values of LPUL and MPDL are not independent of each other. Once the LPUL is specified, the MPDL is automatically determined by the β ratio, and vice versa.

4. THE NOISE MARGINS

Both the LPUL and the MPDL can be explained graphically from transfer characteristics. Refer to Figure 8. Without feedback, the voltage v_{l3} follows v_{l1} as shown by the solid TC curve in Figure 8(b). Since the feedback forces v_{l3} to equal to v_{l1}, the circuit solutions are the intersections of the transfer characteristic and the straight line $v_{l3} = v_{l1}$. As shown in the figure, there are three intersection points: V_L, $\bar{V}$, and V_{UL}. Since the feedback network starts with $v_{l3} = v_{l1}$ as an uplevel, only $\bar{V}$ and V_{UL}

are relevant to our analysis. Why? Assume $v_{I1} = v_{I3} = V_{UL}$. Let Δv_1 be a noise level displacing v_{I1} to a new point v_1 as shown. Since v_1 is to the right of $\overline{V}$, the network will return to its original state by "zigzagging" between the transfer characteristic and the feedback line. Any noise level, therefore, with a magnitude not large enough to displace v_{I1} beyond $\overline{V}$ cannot force the latch to switch states. Equivalently, the effect of a noise level not large enough to displace v_{I1} to the left of $\overline{V}$ will eventually be absorbed by the infinite inverter chain without causing any logic errors.

On the other hand, let Δv_2 be a noise level displacing v_{I1} to a new point v_2, which is to the left of $\overline{V}$. In this case, the latch will zigzag in the opposite direction as shown in the figure and eventually change state. In the infinite inverter chain, then, any noise level that moves v_{I1} to the left of $\overline{V}$ will cause logic errors.

By defining the LPUL as the minimum uplevel the input of a logic network can take without causing an error, it is obvious that LPUL $= \overline{V}$ and the worst case noise margin of the uplevel is $\Delta\overline{V} = V_{UL} - \overline{V}$.

Similarly, by considering I2 and I3 with feedback as shown in Figure 9, we obtain the MPDL $= \underline{V}$ and the worst case noise margin $\Delta\underline{V} = \underline{V} - V_{DL}$.

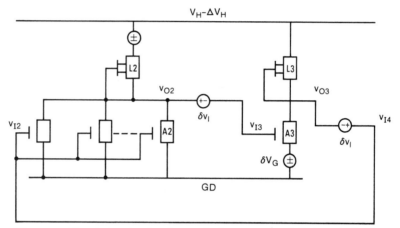

Figure 9(a). A latch circuit equivalent to the infinite chain for $v_{I2} = $ MPDL.

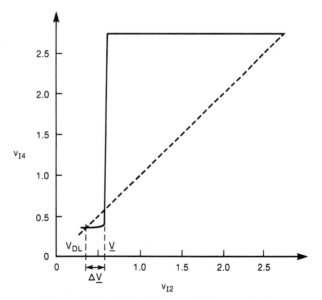

Figure 9(b). DC solutions of above circuit.

EXAMPLE 5

Calculate the worst case noise margins of the circuits in Examples 2 and 3.

Solution

To obtain the transfer characteristics, analyze the circuits in Figures 8 and 9. The TC curves for the LPUL and the MPDL are obtained by plotting the node voltages v_{I3} versus v_{I1} and v_{I4} versus v_{I2}, respectively. As shown in Figure 10, the worst case noise margins are given by

$$\Delta \overline{V} = 450 \text{ mV} \quad \text{and} \quad \Delta \underline{V} = 180 \text{ mV}$$

$$\text{for} \quad \text{LPUL} = 2.00 \quad \text{and} \quad \text{MPDL} = 0.66$$

$$\Delta \overline{V} = 180 \text{ mV} \quad \text{and} \quad \Delta \underline{V} = 90 \text{ mV}$$

$$\text{for} \quad \text{LPUL} = 2.25 \quad \text{and} \quad \text{MPDL} = 0.62$$

Generally speaking, the β ratio obtained from DC stability design is smaller

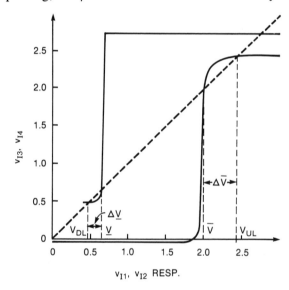

Figure 10(a). Calculation of $\Delta \overline{V}$ and $\Delta \underline{V}$ for the case $\beta_r = 1.79$.

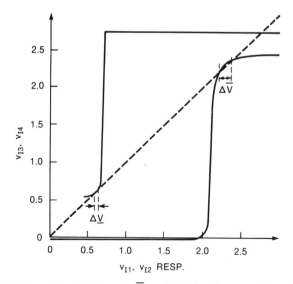

Figure 10(b). Calculation of $\Delta \overline{V}$ and $\Delta \underline{V}$ for the case $\beta_r = 1.53$.

than that obtained from the maximum downlevel design because it allows some of the pulldowns to be slightly on. The maximum downlevel design, however, gives better stability. In practice the maximum downlevel design is mostly used for inverters that can drive any circuits in the chip. Global drivers, such as the powerful clock drivers in memory chips, are always designed with ample downlevel noise margins. On the other hand, inverters operating in controlled environments, such as a custom designed ALU circuit, allow a smaller β ratio as long as the internal voltage levels are within reasonable limits. In these cases the DC stability design is preferred most often. It is advisable to "restore" logic levels to full voltage swings at the last stage with circuits of large β ratios before they leave the controlled environment.

The previous stability analysis can also be used to analyze the cross-coupled latches. Figure 11 shows a perfectly balanced latch, together with its TC plot from input to output. There are three DC solutions called *equilibrium states*. The two outer states, V_{DL} and V_{UL}, are stable in the sense that small perturbations around them will eventually die out with the latch returning to its original state. The center state, V_M, however, is unstable because any trajectory in its neighborhood will move away toward V_{DL} or V_{UL} and the latch can never settle on V_M. Obviously the noise margins of this circuit are $\Delta \overline{V} = V_{UL} - V_M$ and $\Delta \underline{V} = V_M - V_{DL}$.

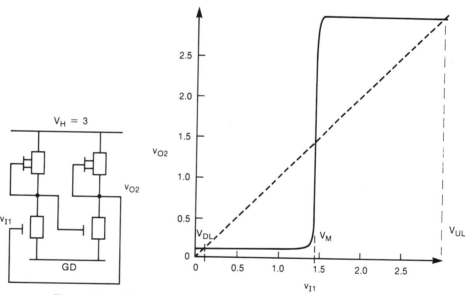

Figure 11. A balanced latch and its equilibrium states for $\beta_r = 3.0$.

If the latch is unbalanced, the center state V_M will move either to the right or to the left, depending on the bias. By biasing the latch with the worst case circuit parameters and noises against V_{DL} or V_{UL}, V_M can be moved to $\underline{V}$ to the left or to $\overline{V}$ to the right, respectively. The noise margins are then given by $\Delta \underline{V} = \underline{V} - V_{DL}$ and $\Delta \overline{V} = V_{UL} - \overline{V}$. Thus, the concepts of minimum logic levels and noise margins as discussed in previous sections can all be interpreted in terms of the stability of an unbalanced latch.

5. COMPARISON: NMOS VERSUS CMOS

In previous sections we have mainly considered NMOS inverters. It is clear that the most important DC design parameter for NMOS circuits is the β ratio. When the output v_O is at its downlevel, the gatedrive of T_L is large, causing a large DC current.

The pulldown T_A, however, is in a linear region with $v_{DS} = v_O$ which is very small. Since T_L and T_A must pass the same current, β_A must be at least β_r times larger than β_L. Circuits that must be designed with proper β ratios are, therefore, called *ratio-type designs*.

Since the DC current in a CMOS inverter should be zero, the DC logic levels of CMOS inverters are not subject to β ratios. Figure 12 shows a family of TC plots with different ratio of ratios. While the shapes of the TC curves are different, the up/downlevel of the inverter is practically independent of the β ratio. For this reason, CMOS circuits are commonly regarded as *"ratioless" designs*.

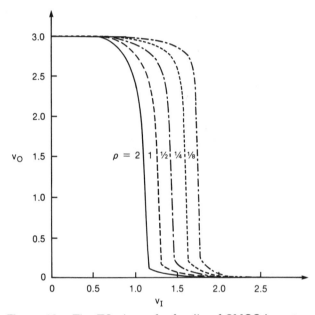

Figure 12. The TC plots of a family of CMOS inverters.

The worst case noise margins of CMOS circuits are defined in the same way as NMOS circuits. Whereas CMOS inverters are generally more forgiving in β ratios, CMOS *networks are* subject to noise problems as shown in the following example.

EXAMPLE 6

Consider the CMOS logic circuit shown in Figure 13. Calculate the noise margins under the same noise conditions outlined in Example 5.

Solution

The TC plots of v_{I3} and v_{I4} versus v_{I1} and v_{I2}, respectively, are shown in Figure 13(b). The worst case noise margins $\Delta\overline{V} = 550$ mV and $\Delta\underline{V} = 330$ mV are of the same order of magnitude as those in Example 5.

6. DEVICE TRACKING AND TEMPERATURE EFFECTS

In the previous analysis, the threshold voltages V_{T0} and V'_{T0} were varied across the full range of process specification: $(0.45, 0.55)$ and $(-1.4, -1.2)$, respectively. The threshold voltages of T_{A1} and T_{A2} in Figure 8, for example, were set to 0.55 and 0.45, respectively, to account for the worst case noise condition.

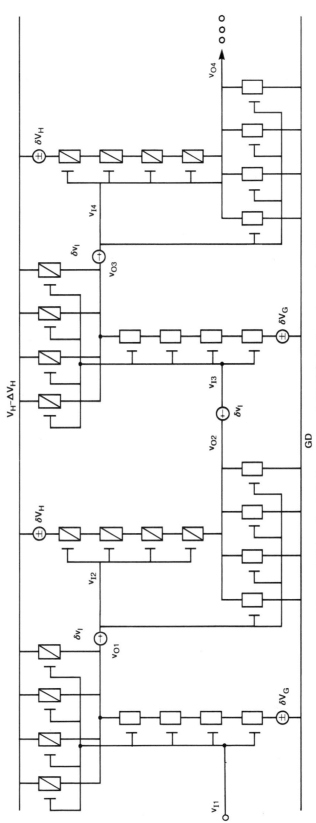

Figure 13(a). An infinite chain of CMOS logic gates under the worst case noise condition. Device sizes are all equal.

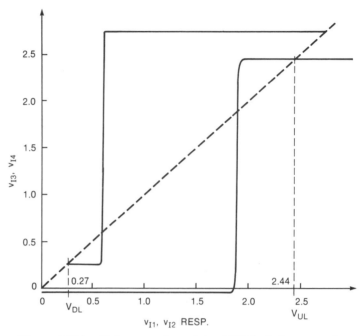

Figure 13(b).   Noise margins of the CMOS chain network.

While these extreme values do occur in different chips, the actual values of the threshold voltages of the same type of device track with each other in the same chip. When $V_{T0}(T_{A1})$ is low, so is $V_{T0}(T_{A2})$, and vice versa. The tracking between $V_{T0}(T_{A1})$ and $V_{T0}(T_{A2})$ is defined by: $|V_{T0}(T_{A1}) - V_{T0}(T_{A2})|/\{[V_{T0}(T_{A1}) + V_{T0}(T_{A2})]/2\}$, measured in percentage. Thus, letting V_{T0} vary from 0.45 to 0.55 in the same chip is equivalent to $(0.55 - 0.45)/[(0.45 + 0.55)/2] = 20\%$ tracking. Since 5–10% on-chip threshold voltage tracking is readily achievable in practical processes, our calculation of the noise margins was too pessimistic with respect to tracking.

On the other hand, our analysis did not include temperature effect either. Since V_{T0} decreases by $1.5 - 2$ mV per degree Centigrade as temperature goes up, $V_{T0}(T_{A1})$ and $V_{T0}(T_{A2})$ should be calculated at the highest temperature. Since the "spec" (0.45, 0.55) is defined at room temperature (22°C), our calculation was therefore too optimistic with respect to temperature. Thus, detailed calculation of noise margins depends heavily on the process tolerance as well as on the chip operating environment. The β ratio is a design parameter which enables the chip designer to optimize circuit performance for a particular application.

7. CONCLUSION

We have discussed the DC analysis of digital inverters in detail in this chapter. In particular, we have defined the concepts of minimum logic levels and noise margins that are important parameters in logic chip design. A FORTRAN program OPTBETA has been written for calculating the minimum β ratio for a specified LPUL or MPDL under given noise conditions. Basically, the program finds the unstable equilibrium state of an unbalanced latch by solving a pair of nonlinear algebraic equations. While OPTBETA gives correct answers for most problems, it may adopt the corresponding results of a nearby stable equilibrium state if the specifications are too close to the limit. For example, for an LPUL = 2.38, OPT-BETA gives $\beta_r = 1.57$ with an MPDL = 0.55 for the circuit of Example 2. We simulate the circuit with $\beta_r = 1.57$ and plot its TC curves in Figure 14. As is evident in the figure, the result obtained actually corresponds to the stable equilibrium state

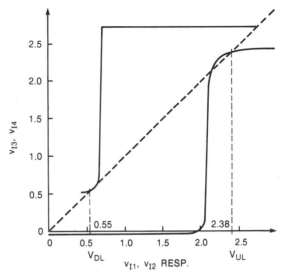

Figure 14. TC plots of the circuit in Example 2 with $\beta_r = 1.57$ that illustrate the misconvergence problem of OPTBETA.

$V_{DL} = 0.55$ and $V_{UL} = 2.38$. This problem can be detected by either plotting the TC curves as in Figure 14 or by comparing the results in Example 2 or 3 to see that β_r is no longer a monotonic function of LPUL or MPDL anymore.

REFERENCES

1. Stern, T. E., *Theory of Nonlinear Networks and Systems,* Reading, MA: Addison-Wesley, 1965.
2. Chua, L. O., *Introduction to Nonlinear Network Theory,* New York: McGraw-Hill, 1969.

PROBLEMS

1. Consider the single device shown in Figure 15. For simplicity, assume the channel length of the device is long enough so $\lambda \cong 0$. Prove the device currents are

$$i_{DS} = \frac{\beta}{2}[v_G - V_{T0} - (1 + k_1)v_S]^2$$

when T is in saturation and

$$i_{DS} = \beta\left[v_G - V_{T0} - \frac{1}{2}(1 + k_1)v_S - \frac{1}{2}(1 + k_1)v_D\right](1 + k_1)(v_D - v_S)$$

when T is in linear region.

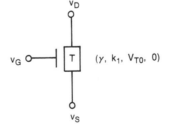

Figure 15.

2. Derive the current equations for a long-channel PMOS device similar to Problem 1.

3. Consider the E/E inverter in Figure 16. Use the equations in Problem 1 for the following problems:

   (a) If both T_L and T_D are in saturation, prove that the slope of the TC plot is given by $A_v = -\sqrt{\beta_r}/(1 + k_1')$. Thus, the inverter is a linear amplifier with a small-signal voltage gain approximately equal to the square root of the β ratio.

   (b) Define the parameters m and k_r as

$$m = \frac{(1 + k_1')V_H}{v_G - v_{T0}'}, \qquad k_r = \frac{1 + k_1}{1 + k_1'}$$

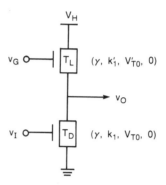

Figure 16.

Prove the expressions of β_r in the following table:

m	Normalized Input V_I	Normalized Output V_O
≥ 1	$\dfrac{v_I - V_{T0}}{v_G - V_{T0}'}$	$\dfrac{(1 + k_1)v_O}{v_G - V_{T0}'}$
	$\beta_r = \dfrac{(1 - V_O/k_r)^2}{2V_O(V_I - V_O/2)}$	
<1	$\dfrac{v_I - V_{T0}}{(1 + k_1)V_H}$	$\dfrac{v_O}{V_H}$
	$\beta_r = \dfrac{(1/m - \frac{1}{2} - \frac{1}{2}V_O)(1 - V_O)}{k_r^2 V_O(V_I - \frac{1}{2}V_O)}$	

4. Consider the E/D inverter shown in Figure 16. Define the normalized variables as follows:

$$V_I = \frac{v_I - V_{T0}}{-V_{T0}'}, \qquad V_O = \frac{(1 + k_1)v_O}{-V_{T0}'}$$

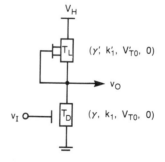

Figure 17.

(a) Let

$$m = \frac{(1 + k_1')V_H}{-V_{T0}'} \quad \text{and} \quad k_r = \frac{1 + k_1}{1 + k_1'}$$

Use the equations in Problem 1 to prove that

$$\beta_r = \frac{\{1 - [k_1'/(1 + k_1)]V_O\}^2}{2V_O(V_I - V_O/2)}$$

(b) Assuming both T_L and T_D are in saturation region, prove that

$$1 - \frac{k_1'}{1 + k_1}V_O = \sqrt{\beta_r}\, V_I$$

and the small signal voltage gain is given by

$$A_v = -\frac{\sqrt{\beta_r}}{k_1'}$$

(c) Plot V_O versus V_I as described by above equation and prove that

$$V_{ts} \cong -\frac{V_{T0}'}{\sqrt{\beta_r}}$$

provided that β_r is reasonably large and $k_1 \ll 1$.

5. Find the β ratio and noise margins for the network in Example 2 under the following conditions: $\delta V_H = 0.1$, $\delta V_G = 0.1$, $\delta v_i = 0.05$.

6. Under the noise conditions of Problem 5, find the β ratio and the corresponding v_{o1} for different LPULs ranging from 1.5 to 2.25. Plot LPUL and its corresponding v_{o1} versus the β ratio. Comment on the slow variation of v_{o1}.

7. Consider the "V_T generating circuit" shown in Figure 18. Assume that both λ and k_1 are negligibly small so that a perfect square law can be applied for the device currents. Let $k_{r1} = \sqrt{\beta_2/\beta_1}$ and $k_{r2} = \sqrt{\beta_4/\beta_3}$. Prove that if $k_{r2} = 1 + k_{r1}$, then as long as all devices are in saturation, the output v_o is equal to V_{T0}, independent of the power supply.

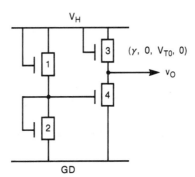

Figure 18.

8. In this chapter we studied the DC stability of inverters via an infinite chain. By trying a few examples and studying the TC plots, convince yourself that the "infinite" chain can be as short as five stages in practice.

9. Calculate the worst case noise margins of the CMOS gate chain in Example 6 assuming that $(W/L)(\text{PMOS})/(W/L)(\text{NMOS}) = 2.5$.

10. Modify OPTBETA (see the appendix) to handle CMOS inverters. Assume that the current is inversely proportional to the number of devices in series and directly proportional to the number of devices in parallel. Modify subroutine PJDS to include PMOS devices.

APPENDIX

PROGRAM OPTBETA

OPTBETA is a program that solves the simultaneous Equations (5) by the Newton-Raphson method. Because there are two equations in two unknowns, OPTBETA gives the β ratio and the corresponding v_{O1} for a specified v_{I1} = LPUL or the β ratio and the corresponding v_{I1} for a specified v_{O1} = MPDL $- \delta v_l$. These two cases are distinguished by parameter LEVEL. When v_{I1} is specified, set LEVEL = 1, and when v_{O1} is specified, set LEVEL = -1. Depending on the value of LEVEL, the program will assume the logic level specified and ignore the value of the other. Thus, if LEVEL = 1, the value of v_{I1} will be held constant, whereas the input value of v_{O1} will be destroyed; therefore, it can be arbitrary.

The device parameters, noise conditions, minimum logic level, and the fanin are inputted to the program as a NAMELIST /PARM/ at program's end. Since the solution is very sensitive to the input conditions, particularly MPDL, OPTBETA diverges whenever an improper logic level is specified. For example, when the message "DIVERGENCE, IMPROPER MPDL" appears in the output, change v_{O1} or noise conditions and rerun. Notice also that the program may give the solution corresponding to a stable equilibrium state when the value specified is too close to the limit, as explained in the conclusion of this chapter.

The following is a list of definitions of the parameters:

Electrical Parameters

$$\text{PLMBDL} = \lambda_L$$

$$\text{PLMBDA} = \lambda_A$$

PK1 = Substrate sensitivity k_1

(VTOEL, VTOEH) = Range of threshold voltage of enhancement mode devices

(VTODL, VTODH) = Range of threshold voltage of depletion mode devices

VH = Power supply

CDVH = ΔV_H, tolerance of the power supply

DVH = δV_H, maximum V_H drop from one inverter to another

DVG = δV_G, maximum ground shift from one inverter to another

DVL = δv_1, random noise from one inverter to another

Programming Parameters

PNORM = ABSOLUTE VALUE OF THE DIFFERENCE BETWEEN TWO CONSECUTIVE ITERATES, DEFAULT AT 1.0D-15 FOR CONVERGENCE TEST

MAX = MAXIMUM NUMBER OF ITERATIONS, DEFAULT 50

SS = STEP SIZE FOR THE CALCULATION OF DERIVATIVES, DEFAULT 1.0D-15

```
C******************************************************************
C*                                                              *
C*                    PROGRAM OPTBETA                           *
C*                                                              *
C*     CALCULATION OF THE BETA RATIO FOR A SPECIFIED LPUL OR MPDL *
C*                                                              *
C******************************************************************
      IMPLICIT REAL*8 (A-H, O-Z)
```

```
          COMMON PLMBDL, PLMBDA, PK1, VH, CDVH, DVH, DVG, DVL, VTOEL, VTOEH,
         *        VTODL, VTODH
          NAMELIST /PARAM/ PLMBDL, PLMBDA, PK1, VTODL, VTODH, VTOEL, VTOEH,
         *                 VH, CDVH, DVH, DVG, DVL, LEVEL, VI1, VO1, FANIN

C --- PLMBDL = LAMBDA L
C --- PLMBDA = LAMBDA A
C --- PK1 = SUBSTRATE SENSITIVITY K1
C --- THRESHOLD VOLTAGE VTOE = (VTOEL, VTOEH)
C --- THRESHOLD VOLTAGE VTOD = (VTODL, VTODH)
C --- VH = POWER SUPPLY
C --- CDVH = CAPITAL DELTA VH, TOLERANCE OF THE POWER SUPPLY
C --- DVH = DELTA VH, MAXIMUM VH DROP FROM ONE INVERTER TO ANOTHER
C --- DVG = DELTA VG, MAXIMUM GROUND SHIFT FROM ONE INVERTER TO ANOTHER
C --- DVL = DELTA VL, RANDOM NOISE FROM ONE INVERTER TO ANOTHER

C --- F = THE FUNCTION TO BE SOLVED
C --- FJ = JACOBIAN OF FUNCTION F
C --- MAX = MAXIMUM NUMBER OF ITERATIONS SPECIFIED BY USER

          READ (5,PARAM)
          WRITE(6,1)
        1 FORMAT (1H1,/ ' THIS PROGRAM CALCULATES THE REQUIRED BETA RATIO'/)
          IF (LEVEL .LT. 0) GO TO 3
          WRITE(6,2)
        2 FORMAT (' FOR A SPECIFIED LPUL')
          GO TO 9
        3 WRITE(6,4)
        4 FORMAT (' FOR A SPECIFIED MPDL')

        9 MAX = 50

          WRITE(6,903)  PLMBDL
      903 FORMAT (///' LUMBDA L = ',E10.3,)
          WRITE(6,904)  PLMBDA
      904 FORMAT (/' LUMBDA A = ',E10.3,)
          WRITE(6,905)  PK1
      905 FORMAT (/' K1       = ',E10.3,)
          WRITE(6,906)  VTODL, VTODH
      906 FORMAT (/' VTODL  = ',E10.3,'      VTODH   = ',E10.3,)
          WRITE(6,907)  VTOEL, VTOEH
      907 FORMAT (/' VTOEL    = ',E10.3,'      VTOEH   = ',E10.3,)
          WRITE(6,908)  VH, DVH
      908 FORMAT (/' VH       = ',E10.3,'      DVH     = ',E10.3,)
          WRITE(6,909)  DVG
      909 FORMAT (/' DVG      = ',E10.3,)
          WRITE(6,902)  DVL
      902 FORMAT (/' DVL      = ',E10.3,)
          WRITE(6,901)  FANIN
      901 FORMAT (/' FANIN    = ',E10.3,)

          IF (LEVEL .GT.  0)  VO1 = VTOEH
          IF (LEVEL .LT.  0)  VI1 = VH/1.5

C --- ITERATION COUNT

          ITRE = 0
          JTER = 0

        5 ITER = ITER + 1
```

```
      WRITE (6,1000) ITER
1000 FORMAT (///' NEW ITERATION = ', I3)

      IF (ITER .GT. MAX)  STOP

      IF (LEVEL .LT. 0) GO TO 401

      IF (VO1-VT0EL)  300, 300, 315
 300 VO1 = VT0EL + .5*(VT0EH-VT0EL)
      JTER = JTER + 1
      IF (JTER - 5)  315, 310, 310
 310 WRITE (6,311)
 311 FORMAT (///' DIVERGENCE, IMPROPER LPUL '/)
      STOP
 315 Y = VI1-.05
      IF (VO1 .LT. Y)  GO TO 405
      WRITE (6,311)
      STOP

 401 IF (VH-CDVH-VI1)  400, 400, 405
 400 VI1 = VH - (CDVH+DVH)
      JTER = JTER + 1
      IF (JTER - 5) 415, 410, 410
 410 WRITE (6,411)
 411 FORMAT (///' DIVERGENCE, IMPROPER MPDL '/)
      STOP
 415 Z = VO1+.05
      IF (VI1 .GT. Z)  GO TO 405
      WRITE (6,411)
      STOP

 405 WRITE (6,100)  VO1
 100 FORMAT (/' VO1 =  ',E10.3)
      WRITE (6,200)  VI1
 200 FORMAT (/' VI1 =  ',E10.3,//)

C --- DEFINITION OF FUNCTION F AND ITS DERIVATIVE FJ

C --- 1ST HALF OF F AND FJ, PART G

      CALL FG (LEVEL, 1.0D-15, VI1, VO1,  G, DG)

      WRITE (6,105)  VI1, VO1,  G, DG
 105 FORMAT (' PART G ',/, 4(3XE10.3)//)

      F = G

      FJ = DG

C --- 2ND HALF OF F AND FJ, PART H

      CALL FH (LEVEL, 1.0D-15, VI1, VO1, H, DH)

      WRITE (6,120)  VI1, VO1, H, DH
 120 FORMAT (' PART H ',/, 4(3XE10.3)//)

      F = F - (1.0/FANIN)*H
```

```
            WRITE(6,125)   F
      125 FORMAT (' F FUNCTION = 'E10.3,//)

C --- FILL IN FJ

            FJ = FJ - (1.0/FANIN)*DH

            WRITE(6,130) FJ
      130 FORMAT (' FJ = 'E10.3,//)

C --- ITERATION

            IF (DABS(FJ) .LT. 1.0D-10)  FJ = 1.0D-10

            IF (LEVEL .GT.  0)  VO1P = VO1 - F/FJ
            IF (LEVEL .LT.  0)  VI1P = VI1 - F/FJ

C --- TEST FOR CONVERGENCE

            IF (LEVEL .GT.  0)  PNORM = DSQRT((VO1P-VO1)**2)
            IF (LEVEL .LT.  0)  PNORM = DSQRT((VI1P-VI1)**2)

            WRITE(6,140) PNORM
      140 FORMAT (' PNORM = 'E10.3)

            IF (LEVEL .GT.  0)  VO1 = VO1P
            IF (LEVEL .LT.  0)  VI1 = VI1P
            IF (DABS(PNORM) .GT. 1.0D-15)  GO TO 5

C --- OUTPUT

            IF (LEVEL .LT. 0)  GO TO 160
            WRITE(6,145)  VI1
      145 FORMAT (/// ' LPUL SPECIFIED = ',E15.8,/)
            WRITE(6,150)  G
      150 FORMAT ('   REQUIRED BETA RATIO = ',E15.8,/)
            WRITE(6,155)  VO1
      155 FORMAT ( '   VO1 = ',E15.8,/)
            GO TO 185

      160 PMPDL = VO1 - DVL
            WRITE(6,165)  PMPDL
      165 FORMAT (/// ' MPDL SPECIFIED = ',E15.8,/)
            WRITE(6,170)  G
      170 FORMAT ('   REQUIRED BETA RATIO = ',E15.8,/)
            WRITE(6,175)  VI1
      175 FORMAT ( '   VI1 = ',E15.8,/)

      185 WRITE(6,190) ITER
      190 FORMAT ( '   NUMBER OF ITERATIONS = ', I3)

            STOP
            END




            FUNCTION PJDS (XS, XD, PLMBD)
```

```
C --- FUNCTION PJDS AS DEFINED IN THE BOOK
C --- GIVEN THE GATEDRIVES XS AND XD, CALCULATES THE DEVICE CURRENT

      IMPLICIT REAL*8 (A-H,O-Z)

      PD=(XD*PSGN(XD))
      PS=(XS*PSGN(XS))

      XPD=(PS+(1-DSQRT(1+2*PLMBD*PS))/PLMBD)
      PINCHD=(PSGN(XPD-PD))
      XPS=(PD+(1-DSQRT(1+2*PLMBD*PD))/PLMBD)
      PINCHS=(PSGN(XPS-PS))

      XDP=(PD*(1-PINCHD) + XPD*PINCHD)
      XSP=(PS*(1-PINCHS) + XPS*PINCHS)
      PJDS=(XSP**2-XDP**2)/(1.0+PLMBD*DABS(XSP-XDP))

      RETURN
      END

      FUNCTION PSGN (X)
      IMPLICIT REAL*8 (A-H,O-Z)
      IF (X .GT. 0.0)  GO TO 10
      PSGN = 0.0
      RETURN
   10 PSGN = 1.0
      RETURN
      END



      SUBROUTINE GDI (VI1, VO1,   XSA1, XDA1, XSL1, XDL1)

C --- CALCULATION OF THE GATEDRIVES OF INVERTER I (HIGH INPUT)

      IMPLICIT REAL*8 (A-H,O-Z)

      COMMON PLMBDL, PLMBDA, PK1, VH, CDVH, DVH, DVG, DVL, VTOEL, VTOEH,
     *       VTODL, VTODH

      XSA1 = VI1-VTOEH-(1.+PK1)*DVG
      XDA1 = VI1-VTOEH-(1.+PK1)*VO1
      XSL1 = -VTODL-PK1*VO1
      XDL1 = VO1-VTODL-(1.+PK1)*(VH-CDVH)

      RETURN
      END



      SUBROUTINE GDII  (VI1, VO1,   XSA2, XDA2, XSL2, XDL2)

C --- CALCULATION OF THE GATEDRIVES OF INVERTER II (LOW INPUT)

      IMPLICIT REAL*8 (A-H,O-Z)

      COMMON PLMBDL, PLMBDA, PK1, VH, CDVH, DVH, DVG, DVL, VTOEL, VTOEH,
     *       VTODL, VTODH

      XSA2 = VO1-VTOEL+DVL
      XDA2 = VO1-VTOEL-(1.+PK1)*VI1-PK1*DVL
```

```
               XSL2 = -VT0DH-PK1*(VI1+DVL)
               XDL2 = VI1+DVL-VT0DH-(1.+PK1)*(VH-CDVH-DVH)


               RETURN
               END



               SUBROUTINE FG (LEVEL, SS,  VI1, VO1,  G,  DG)

C --- CALCULATION OF FUNCTION G AND ITS 1ST ORDER DERIVATIVE DG
C --- W.R.T. VO1 OR VI1 DEPENDING ON LEVEL= 1 OR -1, RESPECTIVELY
C --- SS = THE 'INFINITESIMAL' INCREMENT SPECIFIED BY THE USER
C --- DG = PDG/PDVO1 OR PDG/PDVI1

               IMPLICIT REAL*8(A-H,O-Z)
               COMMON PLMBDL, PLMBDA, PK1, VH, CDVH, DVH, DVG, DVL, VT0EL, VT0EH,
             *          VT0DL, VT0DH

C --- CALCULATION OF THE DIFFERENTIALS

               IF (LEVEL .LT. 0) GO TO 10
               X = VO1
               CALL GDI (VI1, X,    XSA1, XDA1, XSL1, XDL1)
               DO  = PJDS(XSL1, XDL1, PLMBDL) / DMAX1(PJDS(XSA1, XDA1, PLMBDA),
             *                                       1.0D-10)
               GO TO 20
           10 X = VI1
               CALL GDI (X, VO1,    XSA1, XDA1, XSL1, XDL1)
               DO  = PJDS(XSL1, XDL1, PLMBDL) / DMAX1(PJDS(XSA1, XDA1, PLMBDA),
             *                                       1.0D-10)


           20 IF (LEVEL .LT. 0) GO TO 30
               X = VO1+SS
               CALL GDI (VI1, X,    XSA1, XDA1, XSL1, XDL1)
               DGP = PJDS(XSL1, XDL1, PLMBDL) / DMAX1(PJDS(XSA1, XDA1, PLMBDA),
             *                                       1.0D-10)
               GO TO 40
           30 X = VI1+SS
               CALL GDI (X, VO1,    XSA1, XDA1, XSL1, XDL1)
               DGP = PJDS(XSL1, XDL1, PLMBDL) / DMAX1(PJDS(XSA1, XDA1, PLMBDA),
             *                                       1.0D-10)

           40 IF (LEVEL .LT. 0) GO TO 50
               X = VO1-SS
               CALL GDI (VI1, X,    XSA1, XDA1, XSL1, XDL1)
               DGM = PJDS(XSL1, XDL1, PLMBDL) / DMAX1(PJDS(XSA1, XDA1, PLMBDA),
             *                                       1.0D-10)
               GO TO 60
           50 X = VI1-SS
               CALL GDI (X, VO1,    XSA1, XDA1, XSL1, XDL1)
               DGM = PJDS(XSL1, XDL1, PLMBDL) / DMAX1(PJDS(XSA1, XDA1, PLMBDA),
             *                                       1.0D-10)

C --- FUNCTION G

           60 G = DO

C --- THE 1ST ORDER DERIVATIVE OF G

               DG = (DGP - DGM) / (2.0*SS)
```

```
          RETURN
          END



          SUBROUTINE FH (LEVEL, SS,  VI1, VO1,  H,  DH)

C --- CALCULATION OF FUNCTION H AND ITS 1ST ORDER DERIVATIVE DH
C --- W.R.T. VO1 OR VI1 DEPENDING ON LEVEL= 1 OR -1, RESPECTIVELY
C --- SS = THE 'INFINITESMAL' INCREMENT SPECIFIED BY THE USER
C --- DF = PDF/PDVO1 OR PDF/PDVI1

          IMPLICIT REAL*8 (A-H,O-Z)

          COMMON PLMBDL, PLMBDA, PK1, VH, CDVH, DVH, DVG, DVL, VTOEL, VTOEH,
         *        VTODL, VTODH

C --- CALCULATION OF THE DIFFERENTIALS

          IF (LEVEL .LT. 0) GO TO 10
          X = VO1
          CALL GDII (VI1, X,   XSA2, XDA2, XSL2, XDL2)
          D0 = PJDS(XSL2, XDL2, PLMBDL) / DMAX1(PJDS(XSA2, XDA2, PLMBDA),
         *                                        1.0D-10)
          GO TO 20
   10 X = VI1
          CALL GDII (X, VO1,   XSA2, XDA2, XSL2, XDL2)
          D0 = PJDS(XSL2, XDL2, PLMBDL) / DMAX1(PJDS(XSA2, XDA2, PLMBDA),
         *                                        1.0D-10)
   20 IF (LEVEL .LT. 0) GO TO 30
          X = VO1+SS
          CALL GDII (VI1, X,   XSA2, XDA2, XSL2, XDL2)
          DHP = PJDS(XSL2, XDL2, PLMBDL) / DMAX1(PJDS(XSA2, XDA2, PLMBDA),
         *                                        1.0D-10)
          GO TO 40
   30 X = VI1+SS
          CALL GDII (X, VO1,   XSA2, XDA2, XSL2, XDL2)
          DHP = PJDS(XSL2, XDL2, PLMBDL) / DMAX1(PJDS(XSA2, XDA2, PLMBDA),
         *                                        1.0D-10)
   40 IF (LEVEL .LT. 0) GO TO 50
          X = VO1-SS
          CALL GDII (VI1, X,   XSA2, XDA2, XSL2, XDL2)
          DHM = PJDS(XSL2, XDL2, PLMBDL) / DMAX1(PJDS(XSA2, XDA2, PLMBDA),
         *                                        1.0D-10)
          GO TO 60
   50 X = VI1-SS
          CALL GDII (X, VO1,   XSA2, XDA2, XSL2, XDL2)
          DHM = PJDS(XSL2, XDL2, PLMBDL) / DMAX1(PJDS(XSA2, XDA2, PLMBDA),
         *                                        1.0D-10)
C --- FUNCTION H

   60 H = D0
C --- THE 1ST ORDER DERIVATIVE OF H

          DH = (DHP - DHM) / (2.0*SS)

          RETURN
          END
//GO.SYSIN DD *
```

```
$PARAM PLMBDL = 0.5,
       PLMBDA = 1.0,
       PK1 = .1,

       VT0DL = -1.4,
       VT0DH = -1.2,

       VT0EL = .45,
       VT0EH = .55,

       VH = 3.0,
       CDVH = .1,
       DVH = .2,
       DVG = .2,
       DVL = .05,

       LEVEL = 1,
       VI1 = 1.8,
       VO1 = 0.2,

       FANIN = 9.0   $END
```

4

DIGITAL INVERTERS— TRANSIENT ANALYSIS

INTRODUCTION

In Chapter 3 we analyzed inverters from a DC point of view. Particularly, procedures to determine the β ratios were discussed. The β ratio, however, specifies only the relative sizes of the pullup and pulldown. The actual sizes of the devices have to be determined by the performance requirement.

In this chapter we will consider the transient responses of inverters. We shall calculate the risetime and falltime of an inverter in both charging and discharging a capacitor. Since the device equations are very complicated, transient responses with step inputs are analyzed first. Then, the rise/falltimes of the step responses are used to calculate the actual circuit delays.

In practice, the designing stage proceeds in the opposite direction to the previous outline. The delay of a circuit is first determined by the overall performance requirement of the chip. Once the delay is specified we can determine the corresponding rise/falltime of the circuit for the capacitive load it has to drive and then determine the physical sizes of the devices in terms of Ws and Ls. It is always a good practice to check the DC power after determining the size of an inverter because sometimes a tradeoff between power and performance has to be made before the design is finished.

Throughout this chapter we shall use $L = 1.0$ (1.4 mask) and $L = 0.5$ (0.9 mask) for pullups and pulldowns, respectively. The longer channel length of the pullups allows better control of the DC power when the process varies from lot to lot (practically, from day to day). As a result the coefficient of the channel length dilation $\lambda = 0.5$ for n-channel pullups and $\lambda = 1.0$ for n-channel pulldowns. $L = 0.5$ will be used for p-channel devices. The coefficient of the channel length dilation $\lambda = 0.5$ for p-channel devices because the hole mobility is only about 40% of the electron mobility.

Since the "capacitive loading" comes mostly from the gate capacitances of the pulldowns, a longer pullup channel length does not significantly degrade the performance.

1. CHARGING UP A CAPACITOR

1.1 Constant Pullups

Let us consider the circuit shown in Figure 1, where a constant voltage v_G is applied to the gate of a pullup T_X, which is an n-channel enhancement or depletion mode device. At time $t = 0$, the switch is opened instantaneously allowing T_X to charge up its load capacitor C. Defining

$$X_D \triangleq v_G - V_{T0} - (1 + k_1)V_H$$

then from Equation (3) of Chapter 3, three cases can occur:

1. $X_D \leq 0$,
   T_X operates in the saturation region for all $t \geq 0$.
2. $X_D > 0$ and $V_H \leq \sqrt{2X_D/\lambda}/(1 + k_1)$,
   T_X operates in linear region for all $t \geq 0$.
3. $X_D > 0$ and $V_H > \sqrt{2X_D/\lambda}/(1 + k_1)$,
   T_X starts in saturation region and enters the linear region at some later time $T_s > 0$.

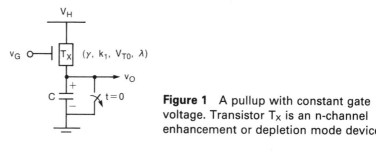

Figure 1 A pullup with constant gate voltage. Transistor T_X is an n-channel enhancement or depletion mode device.

In cases 2 and 3 we have substituted 0 for v_O in Equation (3) of Chapter 3. Let us now discuss these cases in detail.

(1) Operation in the saturation region: Since $X_D \leq 0$ the drain of T_X is completely depleted, and the device operates entirely in the saturation region. Now, since $i_C = i_{DS}(T_X)$, we have

$$C \frac{d}{dt} v_O = \frac{1}{2} \beta \frac{x_S^2 - x_P^2}{1 + \lambda(x_S - x_P)} \tag{1}$$

where $x_S = v_G - V_{T0} - (1 + k_1)v_O$ is the "gatedrive" of T_X and $x_P = x_S + (1 - \sqrt{1 + 2\lambda x_S})/\lambda$ is the pinchoff point. To simplify (1), let

$$\xi \triangleq \sqrt{1 + 2\lambda x_S} - 1 \tag{2}$$

$$\Rightarrow \qquad (\xi + 1)^2 = 1 + 2\lambda x_S = 1 + 2\lambda[v_G - V_{T0} - (1 + k_1)v_O]$$

hence

$$dv_O = -\frac{1}{\lambda(1 + k_1)}(\xi + 1)\, d\xi$$

and (1) becomes

$$-C \frac{(\xi + 1)}{(1 + k_1)} \frac{d}{dt} \xi = \frac{1}{2} \beta \frac{\xi^2}{\lambda}$$

Integrating the above equation from $(0, \xi_0)$ to $(t, \xi(t))$, we have

$$\int_{\xi_0}^{\xi} \frac{(\xi' + 1)}{\xi'^2} d\xi' = -\frac{1 + k_1}{2\lambda} \frac{\beta}{C} \int_0^t dt'$$

where $\xi_0 = \xi(0) = \sqrt{1 + 2\lambda x_s(0)} - 1$. The risetime, then, is

$$T_R = 2 \frac{\lambda}{1 + k_1} \left[\ln \left(\frac{\xi_0}{\xi} \right) + \frac{\xi_0 - \xi}{\xi_0 \xi} \right] \frac{C}{\beta} \tag{3}$$

Thus, the risetime for charging up a capacitor is directly proportional to the load C and inversely proportional to the device β, which is characteristic of all types of drivers. Notice that the dimensionless variable ξ decreases monotonically from ξ_0 to 0 as v_o increases from 0 to $(v_G - V_{T0})/(1 + k_1)$. Since $T_R \to \infty$ as $\xi \to 0$ in Equation (3), $v_o(\infty) = (v_G - V_{T0})/(1 + k_1)$; that is, the ultimate level the output v_o can reach is *one threshold lower* than the gate voltage. Clearly, to maintain the uplevel a pullup with a nonpositive threshold voltage is required, which, among other reasons, demands for the introduction of depletion mode devices and CMOS circuits.

EXAMPLE 1

Consider the circuit shown in Figure 2(a) where the enhancement mode device is described by $(\gamma, k_1, V_{T0}, \lambda) = (0.082, 0.08, 0.5, 0.5)$. Determine the device size raising the output v_o from 0 to 90% of its final value in 1 nS.

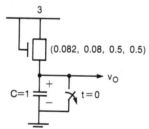

Figure 2(a) An enhancement mode pullup in diode connection.

Solution

By definition,

$$x_s(0) = v_G - V_{T0} = 3 - 0.5 = 2.5$$

Let T_R be the 90% risetime. Since $v_o(\infty) = (v_G - V_{T0})/(1 + k_1)$ and $v_o(T_R) = 0.9 v_o(\infty)$,

$$x_s(T_R) = 0.1 \times (v_G - V_{T0}) = 0.25$$

Substituting the above into Equation (2), we have $\xi_0 = 0.87$ and $\xi(T_R) = 0.12$. From Equation (3), we obtain

$$T_R = 8.49 \frac{C}{\beta} \tag{4}$$

Now with $C = 1$ and $T_R = 1$, $\beta = 8.49$ and the size of T_X is

$$\frac{W}{L} = \frac{\beta}{\gamma} = \frac{8.49}{0.082} = 103$$

Therefore, with $L = 1$, the channel width $W = 103 \times 1 = 103$.

The acute reader may have noticed that the device size was rounded up to a convenient number. This practice is customary because the finite accuracy of models does not warrant a superaccurate calculation of the device sizes. The following discussion, however, indicates that the significance of accuracy also depends on the application.

EXAMPLE 2

Consider the same circuit as above, determine the device size for a more realistic case with $C = 0.5$ pF and $T_R = 2$ nS.

Solution

Since

$$T_R = 8.49 \frac{C}{\beta}, \qquad \beta = \frac{8.49 \times 0.5}{2} = 2.12$$

The channel width is therefore given by $W = 1 \times (2.12/0.082) = 26$. A plot of the output waveform is shown in Figure 2(b). Notice that the final level of v_O is only 2.3 V.

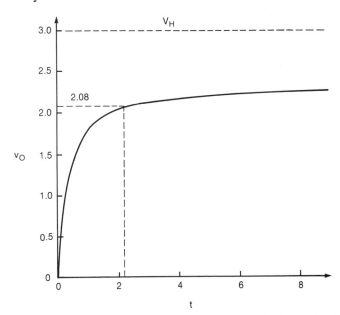

Figure 2(b) Waveform of $v_O(t)$ with $W/L = 26/1$. Since the final level is 2.3 V, the 90% point is 2.08 V.

A careful inspection of Figure 2(b) shows, however, that the 90% risetime of v_O is 2.18, instead of 2. This is because when v_O rises a small current flows through the gate capacitance C_{gs} of T_X. Let us estimate C_{gs} by

$$C_{gs} = \alpha C_o' WL = \alpha C_o' \frac{\beta}{\gamma} L^2$$

where α is a constant depending on the operating condition of T_X. Since T_X is in the saturation region, $\alpha \cong 2/3$. Now adding C_{gs} to C in Equation (4),

$$T_R = 8.49 \frac{C + C_{gs}}{\beta} = 8.49 \frac{C + \alpha C_o'(\beta/\gamma)L^2}{\beta}$$

Solving for β, we obtain

$$\beta = 8.49 \frac{C}{T_R - 8.49\alpha C_o' L^2/\gamma} \triangleq 8.49 \frac{C}{T_R - T_{R0}}$$

Let $C_o' = 1.38 \times 10^{-3}$ and $L = 1.4$ (mask dimension). $T_{R0} = 0.19$, and the new device size becomes

$$\frac{W}{L} = 8.49 \times \frac{0.5}{2 - 0.19} \times \frac{1}{0.082} = \frac{29}{1}$$

Waveforms of both cases are shown in Figure 2(c).

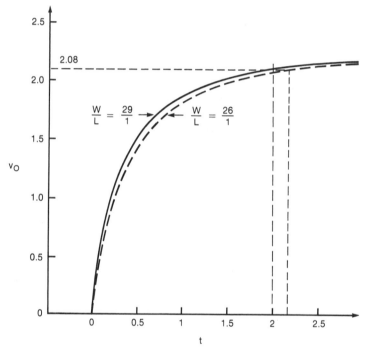

Figure 2(c) Comparison between waveforms with $W/L = 26/1$ versus $29/1$.

This example illustrates the "inherent" or "intrinsic" time delay T_{R0} associated with the device:

$$T_{R0} = \frac{8.49\alpha C_o' L^2}{\gamma} \tag{5}$$

Since T_{R0} is independent of β, the inherent delay remains the same for any risetime T_R. Whereas in this case T_{R0} is less than 10% of T_R, it can become a significant part of the total circuit delay in high-speed logic applications where T_R is in the subnanosecond range.

(2) Operation in the linear region: This case occurs when the drain voltage V_H is low enough so that the device stays in the linear region for all $t \geq 0$. Again, since the capacitor and the device pass the same current

$$C \frac{d}{dt} v_O = \frac{1}{2} \beta \frac{x_S^2 - X_D^2}{1 + \lambda(x_S - X_D)}$$

where $X_D = v_G - V_{T0} - (1 + k_1)V_H$ is a constant and $x_S = v_G - V_{T0} - (1 + k_1)v_O$. Replacing $dv_O = -dx_S/(1 + k_1)$ in the above equation, we have

$$-C \frac{dx_S}{(1 + k_1)\,dt} = \frac{1}{2} \beta \frac{x_S^2 - X_D^2}{1 + \lambda(x_S - X_D)}$$

Integrating the equation from $(0, x_S(0))$ to $(t, x_S(t))$ yields

$$\frac{1}{2X_D} \ln \frac{[x_S(t) - X_D][x_S(0) + X_D]}{[x_S(t) + X_D][x_S(0) - X_D]} + \lambda \ln \left[\frac{x_S(t) + X_D}{x_S(0) + X_D} \right] = -\frac{1 + k_1}{2C} \beta t$$

The above equation can be simplified considerably by substituting $x_S(t) - X_D = (1 + k_1)(V_H - v_O)$ and $x_S(t) + X_D = (1 + k_1)(V_H - v_O) - 2X_D$ and solving for the risetime T_R:

$$T_R = \frac{1}{1 + k_1} \left[\frac{1}{X_D} \ln \frac{K - \eta}{K(1 - \eta)} + 2\lambda \ln \frac{K}{K - \eta} \right] \frac{C}{\beta} \qquad (6)$$

where

$$K \triangleq 1 + \frac{2X_D}{(1 + k_1)V_H} \quad \text{and} \quad \eta \triangleq \frac{v_O}{V_H}$$

This case often occurs in the transfer (pass) gate when the device is turned on hard to let the drain (source) charge (discharge) the source (drain).

(3) T_X starts in the saturation region and enters the linear region at some later time $T_s > 0$: At the transition time T_s, the output $v_O = V_H - \sqrt{2\lambda X_D}/(1 + k_1)$. It can be shown that $\xi_s \triangleq \xi(T_s) = \sqrt{2\lambda X_D}$ where ξ was defined in case (1). The time T_X stays in the saturation region can be obtained from Equation (3):

$$T_s = 2 \frac{\lambda}{1 + k_1} \left[\ln \left(\frac{\xi_0}{\xi_s} \right) + \frac{\xi_0 - \xi_s}{\xi_0 \xi_s} \right] \frac{C}{\beta} \qquad (7)$$

At $t = T_s$ the device enters the linear region, and it is easy to show that the voltage ratio $\eta = v_O/V_H$ is given by

$$\eta_s \triangleq \eta(T_s) = 1 - \frac{\xi_s}{(1 + k_1)\lambda V_H}$$

The time T_X spends in the linear region can then be obtained from Equation (6):

$$T_l = \frac{1}{1 + k_1} \left[\frac{1}{X_D} \ln \frac{(K - \eta)(1 - \eta_s)}{(1 - \eta)(K - \eta_s)} + 2\lambda \ln \left(\frac{K - \eta_s}{K - \eta} \right) \right] \frac{C}{\beta} \qquad (8)$$

The total risetime T_R is therefore given by

$$T_R = T_s + T_l \qquad (9)$$

EXAMPLE 3

Examine the pullup devices in Figure 3 where device T_X is a "weak" depletion mode device and T_Y a "regular" or "normal" depletion mode device. Let us make a comparison between the two devices in terms of power and density for the same performance.

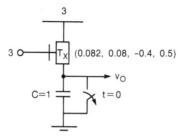

Figure 3(a)   A weak depletion mode pullup.

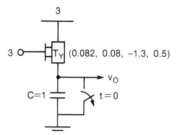

Figure 3(b)   A regular depletion mode pullup.

Solution

For T_X the constants are

$$X_D = X = 3 - (-0.4) - (1 + 0.08) \times 3 = 0.16$$

and

$$K = 1 + 2 \times \frac{0.16}{(1 + 0.08) \times 3} = 1.10$$

At $t = 0$,

$$x_S(0) = 3 - (-0.4) = 3.4$$

$$\Rightarrow \qquad \xi_0 = \sqrt{1 + 2 \times 0.5 \times 3.4} - 1 = 1.1$$

and at $t = T_s$,

$$\xi_s = \sqrt{2\lambda X_D} = \sqrt{2 \times 0.5 \times 0.16} = 0.4$$

$$\Rightarrow \qquad \eta_s = 1 - \frac{0.4}{1.08 \times 0.5 \times 3} = 0.75$$

From Equation (7) we have

$$T_s = 2 \times \frac{0.5}{1.08} \left[\ln\left(\frac{1.1}{0.4}\right) + \left(\frac{1.1 - 0.4}{1.1 \times 0.4}\right) \right] \frac{C}{\beta} = 2.41 \frac{C}{\beta}$$

From Equation (8) we have

$$T_l = \frac{1}{1.08}\left[\frac{1}{0.16}\ln\frac{(1.1-0.9)(1-0.75)}{(1-0.9)(1.1-0.75)} + 2\times 0.5\times \ln\left(\frac{1.1-0.75}{1.1-0.9}\right)\right]\frac{C}{\beta}$$
$$= 2.58\frac{C}{\beta}$$

Here, we set $\eta = 0.9$.

Therefore, the 90% risetime T_R is given by

$$T_R(\mathrm{T_X}) = T_s + T_l = 4.99\frac{C}{\beta} \tag{10}$$

For comparison purposes, let $C = 1$ and $T_R = 1$. The device size will be

$$\frac{W}{L} = \frac{4.99}{0.082} = \frac{61}{1}$$

with an inherent delay

$$T_{R0} = \frac{(2.41\times\frac{2}{3} + 2.58\times\frac{1}{2})\times 1.38\times 10^{-3}\times 1.4^2}{0.082} = 0.096$$

which is small.

Now for $\mathrm{T_Y}$, the constants are

$$X_D = X = 3 - (-1.3) - (1+0.08)\times 3 = 1.06$$

and

$$K = 1 + \frac{2\times 1.06}{(1+0.08)\times 3} = 1.65$$

At $t = 0$,

$$x_S(0) = 3 - (-1.3) = 4.3$$
$$\Rightarrow \qquad \xi_0 = \sqrt{1 + 2\times 0.5\times 4.3} - 1 = 1.3$$

and at $t = T_s$,

$$\xi_s = \sqrt{2\lambda X_D} = \sqrt{2\times 0.5\times 1.06} = 1.03$$
$$\Rightarrow \qquad \eta_s = 1 - \frac{1.03}{1.08\times 0.5\times 3} = 0.36$$

From Equation (7) we have

$$T_s = 2\times\frac{0.5}{1.08}\left[\ln\left(\frac{1.3}{1.03}\right) + \left(\frac{1.3-1.03}{1.3\times 1.03}\right)\right]\frac{C}{\beta} = 0.40\frac{C}{\beta}$$

From Equation (8) we have

$$T_l = \frac{1}{1.08}\left[\frac{1}{1.06}\ln\frac{(1.65-0.9)(1-0.36)}{(1-0.9)(1.65-0.36)} + 2\times0.5\times\ln\left(\frac{1.65-0.36}{1.65-0.9}\right)\right]\frac{C}{\beta}$$

$$= 1.65\frac{C}{\beta}$$

Therefore, the 90% risetime T_R is given by

$$T_R(T_Y) = T_s + T_l = 2.05\frac{C}{\beta} \tag{11}$$

Again, let $C = 1$ and $T_R = 1$. The device size will be

$$\frac{W}{L} = \frac{2.05}{0.082} = \frac{25}{1}$$

The inherent delay of T_Y is

$$T_{R0} = \frac{(0.4\times\frac{2}{3} + 1.65\times\frac{1}{2})\times1.38\times10^{-3}\times1.4^2}{0.082} = 0.036$$

Both devices are widely used as output stages of pushpull drivers as shown in Figure 4 where the pullup T_L can be either T_X or T_Y. To reduce the DC power (and hence the β ratio), an inverter S1 is used that not only drives the gate of T_L to V_H when the input is low, but also discharges the gate to a downlevel when the input is high. This inverter, commonly referred to as the "predriver," is a small circuit whose sole function is to charge and discharge the gate capacitance of T_L.

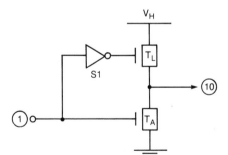

Figure 4 Configuration of a typical pushpull driver.

To calculate the DC power, assume that $v_G = v_o = 0$. The validity of this assumption will be discussed in the next chapter in detail.

First for T_X:

$$x_S = -(-0.4) = 0.4 \quad\text{and}\quad x_P = 0.4 + \frac{1-\sqrt{1+2\times0.5\times0.4}}{0.5} = 0.034$$

Since the DC power is the product of V_H and $i_{DS}(T_X)$,

$$P_{DC}(T_X) = V_H \times \frac{1}{2}\beta \frac{x_S^2 - x_P^2}{1 + \lambda(x_S - x_P)}$$

$$= 3 \times 0.5 \times 4.99 \times \frac{0.4^2 - 0.034^2}{1 + 0.5 \times (0.4 - 0.034)}$$

$$= 1.01$$

Then for T_Y:

$$x_S = -(-1.3) = 1.3 \quad \text{and} \quad x_P = 1.3 + \frac{1 - \sqrt{1 + 2 \times 0.5 \times 1.3}}{0.5} = 0.27$$

and the DC power is given by

$$P_{DC}(T_Y) = 3 \times 0.5 \times 2.05 \times \frac{1.3^2 - 0.27^2}{1 + 0.5 \times (1.3 - 0.27)}$$

$$= 3.28$$

Thus, even though a normal depletion mode device is much faster than a weak depletion mode device for the same device size, it burns much more DC power and requires a larger β ratio. On the other hand, since the size of a weak depletion mode device is much larger than the size of a regular one for the same performance, it requires a more "powerful" predriver that can offset the power advantage to some extent. The net gain of the weak depletion mode drivers is therefore a moderate 15–30% delay-power product saving, which circuit designers have to justify for the more complicated process.

We shall return to this subject in the next chapter in a study of high-power drivers.

1.2 Depletion Mode Pullups

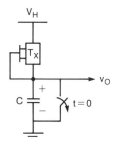

Figure 5(a) A depletion mode pullup.

In the previous example we discussed the advantage of depletion mode pullups with constant voltage applied to the gates. To use their driving power, a small predriver that consumes less power but pulls its output all the way up to V_H is needed. Because the threshold voltage of a depletion mode device is negative, the circuit shown in Figure 5 is widely used. It can be used either as a "standalone" inverter for driving a moderate load or as a predriver to drive large output devices in a pushpull driver (like circuit S1 in Figure 4).

To understand the operation of T_X, assume that before $t = 0$ the capacitor C is uncharged. Since V_{T0} is negative, T_X starts to charge up C at $t = 0$. At this time, T_X is in the saturation region because the gate voltage is still much lower than V_H, and the drain is completely depleted. The device will get out of "depletion saturation" at some later time, when $x_D = 0$. The drain of T_X is strongly inverted at this point, but the current changes very little because of velocity saturation. T_X finally enters the linear region only when $x_D = x_P$, which occurs toward the last 20–30% of the rise-time. Thus, since $x_S \cong -V_{T0}$ and the device is in the saturation region most of the time, T_X behaves like a constant current source whose output current is independent of v_O. This is a very desirable feature from a circuit point of view.

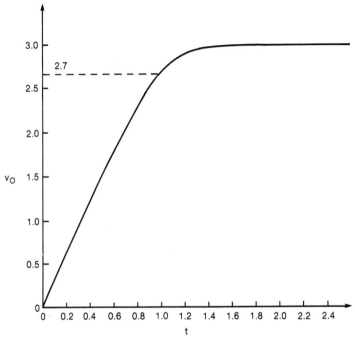

Figure 5(b)   Waveform of $v_O(t)$. Since the final level is 3 V, the 90% point is 2.7 V.

Refer to Figure 5(a) again. At $t = 0$ the circuit equation is

$$C \frac{d}{dt} v_O = \frac{1}{2} \beta \frac{x_S^2 - x_P^2}{1 + \lambda (x_S - x_P)}$$

where $x_S = v_G - V_{T0} - (1 + k_1)v_O = -V_{T0} - k_1 v_O$ and $x_D = v_O - V_{T0} - (1 + k_1)V_H$. Following the procedure outlined in the previous subsection, it can be shown that

$$T_s = 2 \frac{\lambda}{k_1} \left[\ln \left(\frac{\xi_0}{\xi_s} \right) + \frac{\xi_0 - \xi_s}{\xi_0 \xi_s} \right] \frac{C}{\beta} \qquad (12)$$

where $\xi_s = \xi(T_s)$ is obtained by solving the equation $x_D(T_s) = x_P(T_s)$,

$$\xi_s = -\frac{1}{1 + k_1} + \sqrt{\left(\frac{1}{1 + k_1} \right)^2 + 2\lambda X} \qquad (13)$$

with

$$X = x_S(\infty) = x_D(\infty) = -V_{T0} - k_1 V_H$$

For $t > T_s$ the device is in the linear region and the circuit is described by

$$-C \frac{1}{1 + k_1} \left(\frac{d}{dt} x_S \right) = \frac{1}{2} \beta \frac{x_S^2 - x_D^2}{1 + \lambda (x_S - x_D)}$$

with x_S and x_D both functions of v_O. Now it is more convenient to work with the voltage ratio $\eta = v_O / V_H$, and transform the above equation into

$$\frac{1 - \lambda (1 + k_1)V_H (\eta - 1)}{(\eta - K)(\eta - 1)} d\eta = -\frac{\beta}{2C} (1 - k_1^2)V_H \, dt \qquad (14)$$

where the constant

$$K \triangleq 1 - \frac{2X}{(1 - k_1)V_H} \tag{15}$$

As in the previous section, at $t = T_s$

$$\eta_s = 1 - \frac{\xi_s}{(1 + k_1)\lambda V_H} \tag{16}$$

Integrating Equation (14) from (T_s, η_s) to $(t, \eta(t))$, we obtain

$$t - T_s = \left[\frac{1}{(1 + k_1)X} \ln \frac{(\eta - K)(\eta_s - 1)}{(\eta - 1)(\eta_s - K)} + \frac{2\lambda}{1 - k_1} \ln \left(\frac{\eta - K}{\eta_s - K} \right) \right] \frac{C}{\beta} \tag{17}$$

which accounts for the rest of the risetime.

EXAMPLE 4

Let T_X be described by $(\gamma, k_1, V_{T0}, \lambda) = (0.082, 0.08, -1.3, 0.5)$. Calculate the device size required to charge up $C = 1$ pF to 90% of $V_H = 3$ V in $T_R = 1$ nS.

Solution

For $t < T_s$, we have

$$x_S(0) = -V_{T0} = 1.3 \qquad \Rightarrow \qquad \xi_0 = 0.52$$

and

$$X = -V_{T0} - k_1 V_H = 1.06 \qquad \Rightarrow \qquad \xi_s = 0.46$$

Substituting the above into Equation (12) we obtain

$$T_s = 2 \times \frac{0.5}{0.08} \left(\ln \frac{0.52}{0.46} + \frac{0.52 - 0.46}{0.52 \times 0.46} \right) \frac{C}{\beta}$$

$$= 4.68 \frac{C}{\beta}$$

For $t > T_s$, we have, from Equations (15) and (16), $K = 0.23$ and $\eta_s = 0.72$. Now, with $\eta = 0.9$ Equation (17) yields

$$T_l = \left[\frac{1}{1.08 \times 1.06} \ln \frac{(0.72 - 1) \times (0.9 - 0.23)}{(0.9 - 1) \times (0.72 - 0.23)} \right.$$

$$\left. + \frac{2 \times 0.5}{1 - 0.08} \ln \left(\frac{0.9 - 0.23}{0.72 - 0.23} \right) \right] \frac{C}{\beta}$$

$$= 1.51 \frac{C}{\beta}$$

Finally the total risetime T_R is given by

$$T_R = T_s + T_l = (4.68 + 1.51) \frac{C}{\beta} = 6.19 \frac{C}{\beta}$$

and the device size

$$\beta = 6.19 \quad \Rightarrow \quad \frac{W}{L} = \frac{\beta}{\gamma} = \frac{6.19}{0.082} = \frac{76}{1}$$

Figure 5(b) displays the waveform of v_O with $W/L = 76/1$. Again, if we estimate the gate capacitance C_{gd} by

$$C_{gd} \cong \alpha C_o' WL = \alpha C_o' \frac{\beta}{\gamma} L^2$$

where $\alpha \cong 0.5$ when T_X is in the linear region, the inherent time delay T_{R0} of T_X is given by

$$T_{R0} = \frac{1.51 \times 0.5 \times 1.38 \times 10^{-3} \times 1.4^2}{0.082} = 0.025$$

which is quite negligible.

1.3 Bootstrap Pullups

Depletion mode pullups have the advantage of high speed and high uplevels. They do require more processing steps, however, and hence are more expensive. In many cases the performance of simple enhancement mode drivers can be greatly improved by "bootstrapping" the gate of the pullup. Bootstrapping is a valuable technique every MOS circuit designer should master. Even when depletion mode or p-channel devices are available, bootstrapping is often used to enhance the performance of a circuit or to amplify the magnitude of a signal or even to implement simple logic functions, as we shall see in later chapters.

The basic idea of bootstrapping is rather straightforward. Consider the simple circuit shown in Figure 6 where a capacitor C_B is added between the gate and the source of T_X. Before the switch is opened at $t = 0$, C_B has been precharged to a certain level $v_B(0) = V_{B0} > V_{T0}$. At $t = 0$, v_O starts to rise. If node $Ⓖ$ is not connected to any conductive path and does not discharge, the gatedrive x_S of T_X will remain constant, independent of v_O. As v_O increases the gate voltage $v_G = v_O + v_B$ will be boosted higher and eventually drive T_X into the linear region. In other words v_O bootstraps itself up by pushing the gate of T_X via the coupling of C_B; thus, bootstrapping not only allows v_O to reach V_H but also greatly improves the pullup speed.

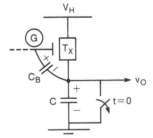

Figure 6 Basic configuration of a bootstrap inverter.

Ideally, the voltage across C_B should stay constant for $t \geq 0$. In reality, however, $v_B(t)$ decreases as v_O increases. First, the parasitic capacitances associated with $Ⓖ$ will always provide a "sneak path" to let C_B discharge to its neighborhood. As soon as T_X enters the linear region the gate capacitance C_{gd} turns on, providing a conductive path to V_H that allows C_B to discharge even more. As a result, the final level of v_G not only depends on the initial voltage V_{B0} but also on the initial amount of charge stored, hence the area of C_B.

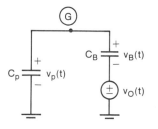

Figure 7 Discharge of the bootstrap capacitor.

To analyze the discharge of C_B in more detail, consider the circuit in Figure 7. Here capacitor C_p represents all "sneak paths" associated with node Ⓖ. Applying Kirchhoff's voltage law (KVL) around the loop, we have

$$v_p(t) = v_B(t) + v_O(t)$$

$$\Rightarrow \qquad \frac{d}{dt} v_p = \frac{d}{dt} v_B + \frac{d}{dt} v_O \qquad (18)$$

Applying Kirchhoff's current law (KCL) to Ⓖ, we have

$$C_p \frac{d}{dt} v_p = -C_B \frac{d}{dt} v_B \qquad (19)$$

Substituting Equation (19) into (18)

$$\frac{d}{dt} v_B = -\frac{C_p}{C_B + C_p} \frac{d}{dt} v_O$$

Integrating, we obtain

$$v_B(t) = V'_{B0} - \frac{C_p}{C_B + C_p} v_O(t)$$

where

$$V'_{B0} \triangleq V_{B0} + \frac{C_p}{C_B + C_p} v_O(0)$$

Thus as $v_O(t)$ increases, $v_B(t)$ decreases with v_O. The proportionality constant

$$k_2 \triangleq \frac{C_p}{C_B + C_p}$$

represents its voltage loss. The ratio

$$1 - k_2 = \frac{C_B}{C_B + C_p}$$

is called *the bootstrap efficiency*. In practice bootstrap efficiency is designed to be around 80–90%. Apparently, the higher the efficiency, the greater the initial charge required and hence the larger the area of C_B.

Analysis of bootstrap pullups is very similar to that of depletion mode pullups except that at $t = 0$ device T_X may be either in saturation or linear region depending on the precharge voltage of node $\textcircled{G}$. Now, refer back to Figure 6. Assume T_X starts in the saturation region, the circuit is described by

$$C \frac{d}{dt} v_o = \frac{1}{2} \beta \frac{x_S^2 - x_P^2}{1 + \lambda(x_S - x_P)}$$

where $x_S = v_G - V_{T0} - (1 + k_1)v_o = V'_{B0} - V_{T0} - kv_o$, $x_D = v_G - V_{T0} - (1 + k_1)V_H = V'_{B0} - V_{T0} - (1 + k_1)V_H + (1 - k_2)v_o$ and $k \triangleq k_1 + k_2$. Now, following the procedure used in the previous subsection, the time interval T_X spends in the saturation region is

$$T_s = 2\frac{\lambda}{k} \left[\ln\left(\frac{\xi_0}{\xi_s}\right) + \frac{\xi_0 - \xi_s}{\xi_0 \xi_s} \right] \frac{C}{\beta} \tag{20}$$

where $\xi_s = \xi(T_s)$ is obtained by solving the equation $x_D(T_s) = x_P(T_s)$:

$$\xi_s = -\frac{1 - k_2}{1 + k_1} + \sqrt{\left(\frac{1 - k_2}{1 + k_1}\right)^2 + 2\lambda X} \tag{21}$$

with

$$X = x_S(\infty) = x_D(\infty) = V'_{B0} - V_{T0} - kV_H$$

Once T_X enters the linear region at $t = T_s$, it is more convenient to work with the voltage ratio $\eta = v_o/V_H$. It can be shown that in the linear region the circuit is described by

$$C \frac{d}{dt} \eta = -\frac{1}{2} \beta \frac{(1 + k_1)(1 - k_1 - 2k_2)V_H(\eta - 1)(\eta - k)}{1 - \lambda(1 + k_1)V_H(\eta - 1)} \tag{22}$$

where

$$K \triangleq 1 - \frac{2X}{(1 - k_1 - 2k_2)V_H} .$$

Integrating Equation (22) from (T_s, η_s) to (t, η) yields

$$t - T_s = \left[\frac{1}{(1 + k_1)X} \ln \frac{(\eta_s - 1)(\eta - K)}{(\eta_s - K)(\eta - 1)} + \frac{2\lambda}{1 - k_1 - 2k_2} \ln\left(\frac{\eta - K}{\eta_s - K}\right) \right] \frac{C}{\beta} \tag{23}$$

where, as usual,

$$\eta_s = 1 - \frac{\xi_s}{(1 + k_1)\lambda V_H}$$

Once η is specified we can calculate the time T_X spends in the linear region from Equation (23).

If T_X starts in the linear region at $t = 0$, Equation (23) still applies, except that now η_s should be replaced by $\eta_0 = v_o(0)/V_H$. A few examples follow.

EXAMPLE 5

Consider the simple bootstrap driver shown in Figure 8. Prior to $t = 0$, the input is high. Device T_3 is on, allowing T_1 to precharge C_B. Then at $t = 0$ device T_3 is turned off, and the output device T_2 starts to charge up the load capacitor.

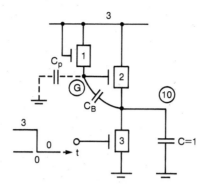

Figure 8 A simple bootstrap inverter.

As v_{10} goes up, v_G is coupled up through C_B. Since T_1 is a diode, Ⓖ is isolated, allowing v_G to be boosted to a very high level. Now assume the β ratio is such that $v_{10} = 0.2$ V when T_3 is on. Calculate the size of T_2 and C_B so that v_{10} rises to 90% of 3 V within 1 nS.

Solution

We shall design for 80% bootstrap efficiency so $k_2 = 0.2$. First, the constants are

$$V_{B0} = \frac{3 - 0.5}{1 + 0.08} - 0.2 = 2.11 \quad \Rightarrow \quad V_{B0}' = 2.11 + 0.2 \times 0.2 = 2.15$$

Hence

$$X = 2.15 - 0.5 - (0.08 + 0.2) \times 3 = 0.81$$

and

$$K = 1 - \frac{2 \times 0.81}{(1 - 0.08 - 2 \times 0.2) \times 3} = -0.04$$

Now, at $t = 0$,

$$x_S(0) = 2.15 - 0.5 - 0.28 \times 0.2 = 1.59$$

$$\Rightarrow \quad \xi_0 = \sqrt{1 + 2 \times 0.5 \times 1.59} - 1 = 0.61$$

and at $t = T_s$,

$$\xi_s = -\frac{1 - 0.2}{1 + 0.08} + \sqrt{\left(\frac{1 - 0.2}{1 + 0.08}\right)^2 + 2 \times 0.5 \times 0.81} = 0.43$$

$$\Rightarrow \quad \eta_s = 1 - \frac{0.43}{1.08 \times 0.5 \times 3} = 0.73$$

Therefore,

$$T_s = \frac{2 \times 0.5}{0.28}\left(\ln \frac{0.61}{0.43} + \frac{0.61 - 0.43}{0.61 \times 0.43}\right)\frac{C}{\beta} = 3.71 \frac{C}{\beta}$$

and

$$T_l = \left[\frac{1}{1.08 \times 0.81} \ln \frac{(0.73 - 1)(0.9 + 0.04)}{(0.73 + 0.04)(0.9 - 1)}\right.$$

$$\left. + \frac{2 \times 0.5}{1 - 0.08 - 0.4} \ln\left(\frac{0.9 + 0.04}{0.73 + 0.04}\right)\right]\frac{C}{\beta}$$

$$= 1.74 \frac{C}{\beta}$$

$$\Rightarrow \qquad T_R = (3.71 + 1.74)\frac{C}{\beta} = 5.45 \frac{C}{\beta}$$

Hence, the size of T_2 is

$$\frac{W}{L} = \frac{5.45}{0.082} = \frac{66}{1}$$

To determine the size of C_B, we must know the physical layout of T_2 to estimate C_P. Once C_p is known we can calculate the size of C_B by charge conservation as follows:

At $t = 0$, the total charge on the gate of T_2 is

$$Q(0) = \left(C_B + \frac{2}{3}C_o'WL\right)V_{B0} + C_p[V_{B0} + v_{10}(0)]$$

and at $t = T_R$,

$$Q(T_R) = (C_B + \tfrac{1}{2}C_o'WL)v_B(T_R) + \tfrac{1}{2}C_o'WL[v_B(T_R) + 2.7 - 3]$$
$$+ C_p[v_B(T_R) + 2.7]$$

with $V_{B0} = 2.11$, $v_O(0) = 0.2$, and $v_B(T_R) = 1.61$. Since $Q(0) = Q(T_R)$, we have

$$C_B = 0.1C_o'WL + 4C_P$$

If the first term is negligible, the bootstrap capacitor C_B is four times larger than the total parasitic capacitance associated with the gate. This is, of course, precisely what the 80% bootstrap efficiency implies.

The curious reader may wonder about the size of T_1. In most cases it is a very small device because its only mission is to charge the gate of T_2 prior to $t = 0$ and then replenish the charge slightly every time the bootstrap action takes place.

EXAMPLE 6

In the previous example the gate capacitance is precharged to V_{B0}, which is slightly less than $(V_H - V_T)$. As a result, the size of T_2 is only slightly smaller than the depletion mode device in Example 4.

Figure 9 displays a "double bootstrap" circuit that uses a complete pulse to increase the precharge of the gate of T_3 to a full V_H. This increase greatly enhances the effect of the bootstrap action. This circuit operates as follows. Prior to $t = -T$, the input is low, allowing T_1 to charge up C_{B1} to approximately $(V_H - V_T)$. The input goes up to V_H at $t = -T$, boosting the potential

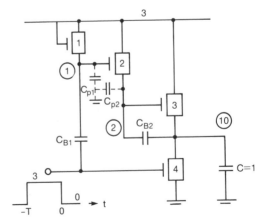

Figure 9 A double bootstrap inverter.

at ① to almost $(2V_H - V_T)$ (slightly lower actually because the bootstrap efficiency is always less than 1). The increase drives T_2 into the linear region quickly. Since T_4 is also turned on hard, T_2 will charge the bootstrap capacitor C_{B2} to almost full V_H (a little lower than V_H because the down level of ⑩ is slightly above the ground). Then, at $t = 0$ the input goes down, turning off T_4. Now device T_3, with its gate precharged to V_H, will be able to pull up the load much faster.

Let us now calculate the size of T_3 under the same condition as Example 5. First, the constants are

$$V_{B0} = 3 - 0.2 = 2.8 \quad \Rightarrow \quad V'_{B0} = 2.8 + 0.04 = 2.84$$

Therefore,

$$X = 2.84 - 0.5 - 0.28 \times 3 = 1.5$$

and

$$K = 1 - \frac{2 \times 1.5}{0.52 \times 3} = -0.92$$

At $t = 0$,

$$x_s(0) = 2.84 - 0.5 - 0.28 \times 0.2 = 2.28$$

$$\Rightarrow \qquad \xi_0 = \sqrt{1 + 2 \times 0.5 \times 2.28} - 1 = 0.81$$

then at $t = T_s$:

$$\xi_s = -0.74 + \sqrt{0.74^2 + 2 \times 0.5 \times 1.5} = 0.69$$

$$\Rightarrow \qquad \eta_s = 1 - \frac{0.69}{1.08 \times 0.5 \times 3} = 0.57$$

Therefore,

$$T_s = \frac{2 \times 0.5}{0.28}\left[\ln\frac{0.81}{0.69} + \left(\frac{0.81 - 0.69}{0.81 \times 0.69}\right)\right]\frac{C}{\beta} = 1.34\frac{C}{\beta}$$

and

$$T_l = \left[\frac{1}{1.08 \times 1.5}\ln\frac{(0.57 - 1)(0.9 + 0.92)}{(0.57 + 0.92)(0.9 - 1)}\right.$$

$$\left. + \frac{2 \times 0.5}{1 - 0.08 - 0.4}\ln\left(\frac{0.9 + 0.92}{0.57 + 0.92}\right)\right]\frac{C}{\beta}$$

$$= 1.40\frac{C}{\beta}$$

$$\Rightarrow \qquad T_R = (1.34 + 1.40)\frac{C}{\beta} = 2.74\frac{C}{\beta}$$

Hence, the size of T_3

$$\frac{W}{L} = \frac{2.74}{0.082} = \frac{33}{1}$$

which is much smaller than the previous case.

Both T_1 and T_2 should be minimum devices if possible. The sizes of C_{B1} and C_{B2} should be about four times the sizes of the parasitic capacitances C_{p1} and C_{p2}, respectively.

EXAMPLE 7

Before concluding this section, let us consider another interesting case displayed in Figure 10. Here the inverter (T_1, T_2) is a dedicated circuit whose sole function is to boost the gate of T_4. If the parasitic capacitance C_{p2} is small (as it usually is), node ② rises much faster than ⑩ as the input goes down at $t = 0$. As v_2 rises toward V_H, node ① is boosted very high, driving T_4 into the linear region quickly. With its gate potential much higher than V_H, device T_4 then pulls up its load with a large gatedrive. In this case, we can assume ① reaches a very high level at $t = 0+$, which then drives T_4 as a constant potential for all $t > 0$.

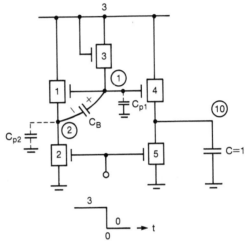

Figure 10   A bootstrap inverter with constant gate boost.

Analyze the circuit under the condition of a 0.2 V downlevel and 80% bootstrap efficiency. For ①, since $v_1 = v_B + v_2$, the effective level of boosted gate potential is given by

$$v_1(0+) = v_2(90\%) + V'_{B0} - k_2 v_2(90\%)$$

$$= 2.7 + \left(\frac{3 - 0.5}{1.08} - 0.2 + 0.2 \times 0.2\right) - 0.2 \times 2.7 = 4.31$$

Now with $v_1(0+) = 4.31$ device T_4 is in the linear region. Equation (6) applies here. First, the constants are

$$X = X_D = 4.31 - 0.5 - 1.08 \times 3 = 0.57$$

$$K = 1 + \frac{2 \times 0.57}{1.08 \times 3} = 1.35$$

The risetime T_R can then be obtained from Equation (6):

$$T_R = \frac{1}{1.08}\left[\frac{1}{0.57}\ln\frac{1.35 - 0.9}{1.35 \times (1 - 0.9)} + 2 \times 0.5 \times \ln\left(\frac{1.35}{1.35 - 0.9}\right)\right]\frac{C}{\beta}$$

$$= 2.97\frac{C}{\beta}$$

The size of T_4 is therefore

$$\frac{W}{L} = \frac{2.97}{0.082} = \frac{36}{1}$$

for $T_R = 1$ nS and $C = 1$ pF.

Again devices T_1 and T_3 should be minimal if possible. The gate capacitances of T_4 should be considered part of the "parasitic capacitances" associated with ① when calculating the size of C_B.

In this example we have neglected the "internal delay" of (T_1, T_2). Overlooking the internal delay may no longer be possible for high-power drivers with very large output devices, as we shall see in the next chapter.

2. DISCHARGING A CAPACITOR

In the previous section we have discussed different kinds of circuits for charging up capacitors. We were able to relate the risetime and the size of a pullup in terms of Ws and Ls for a given load. This section focuses on the discharge of capacitors with pulldowns. The falltime and the size of the pulldown will be related for a given load.

Since the equations and operations are similar for both n-channel and p-channel devices as we shall see shortly, a p-channel device used as a complementary n-channel device will also be discussed in this section, even though it is really charging up a capacitor. This classification should benefit the reader in the long run.

2.1 NMOS Pulldowns

Consider the circuit shown in Figure 11. Let us assume for a moment that T_X is an n-channel enhancement mode device. Assume further that the initial output voltage $v_O(0) = V_O$ is high enough that T_X is in saturation. Equating the currents of T_X and C, we have

$$C \frac{d}{dt} v_O = -\frac{1}{2}\beta \frac{x_S^2 - x_P^2}{1 + \lambda(x_S - x_P)}$$

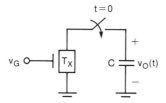

Figure 11(a) An n-channel pulldown.

where $x_S = v_G - V_{T0} = X$ and $x_P = X + (1 - \sqrt{1 + 2\lambda X})/\lambda$ are both constants. Integrating, we obtain

$$t = \frac{2\lambda^2}{\xi^2} V_O(1 - \zeta)\frac{C}{\beta}$$

where

$$\xi = \sqrt{1 + 2\lambda X} - 1 \quad \text{and} \quad \zeta = \frac{v_O}{V_O}$$

T_X will stay in saturation until $x_D = x_P$ at $t = T_s$, at which x_D becomes a function of v_O. The time T_X spends in saturation is therefore given by

$$T_s = \frac{2\lambda^2}{\xi^2} V_O(1 - \zeta_s)\frac{C}{\beta} \tag{24}$$

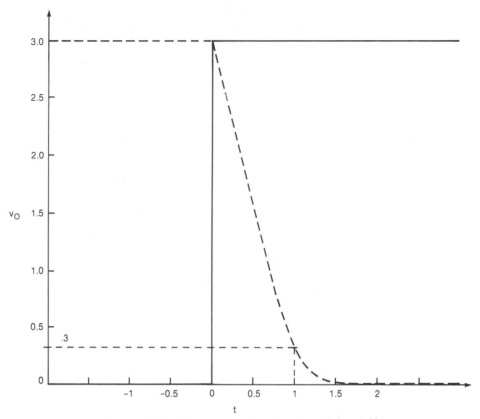

Figure 11(b)   Waveform of $v_O(t)$ with $v_O(0) = 3$ V.

with

$$\zeta_s = \frac{\xi}{(1 + k_1)\lambda V_O}$$

For $t > T_s$ it can be shown that the circuit is described by

$$\frac{(1 + k_1)^2 V_O \zeta(\zeta - K)}{1 + \lambda(1 + k_1)V_O \zeta} = -\frac{1}{2}\frac{C}{\beta}\frac{d}{dt}\zeta$$

Solving the equation

$$t - T_s = \frac{1}{1 + k_1}\left[\frac{1}{X}\ln\frac{(\zeta - K)\zeta_s}{(\zeta_s - K)\zeta} + 2\lambda \ln\left(\frac{\zeta - K}{\zeta_s - K}\right)\right]\frac{C}{\beta} \tag{25}$$

with

$$K \triangleq \frac{2X}{(1 + k_1)V_O}$$

The falltime T_F for a specified ζ is therefore given by

$$T_F = T_s + T_l$$

where $T_l = t(\zeta) - T_s$ is obtained from Equation (25).

EXAMPLE 8

Consider the pulldown in Figure 11 where the device characteristics are given by $(\gamma, k_1, V_{T0}, \lambda) = (0.082, 0.08, 0.5, 1.0)$. Calculate the device size of T_X for the output to fall from $V_O = 3$ V to 10% of V_O in $T_F = 1$ nS. Assume $v_G = 3$ V.

Solution

The constants are

$$X = 3 - 0.5 = 2.5 \qquad \Rightarrow \qquad \xi = \sqrt{1 + 2 \times 1 \times 2.5} - 1 = 1.45$$

$$\zeta_s = \frac{1.45}{1 \times 1.08 \times 3} = 0.45 \quad \text{and} \quad K = \frac{2 \times 2.5}{1.08 \times 3} = 1.54$$

The time T_X spends in the saturation region is given by Equation (24):

$$T_s = \frac{2 \times 1}{1.45^2} \times 3 \times (1 - 0.45)\frac{C}{\beta} = 1.57\frac{C}{\beta}$$

The time T_X spends in the linear region is given by Equation (25):

$$T_l = \frac{1}{1.08}\left[\frac{1}{2.5}\ln\frac{(0.1 - 1.54) \times 0.45}{(0.45 - 1.54) \times 0.1} + 2 \times 1 \times \ln\left(\frac{0.1 - 1.54}{0.45 - 1.54}\right)\right]\frac{C}{\beta}$$

$$= 1.18\frac{C}{\beta}$$

Therefore,

$$T_F = T_s + T_l = 2.75\frac{C}{\beta}$$

and the device size

$$\frac{W}{L} = \frac{2.75}{0.082} = \frac{17}{0.5}$$

The waveform of v_O is shown in Figure 11(b).

2.2 PMOS Pullups

As mentioned earlier the operation of a PMOS pullup is quite similar to the operation of NMOS devices; however, since up until now PMOS cases have been left as exercises, we will work out a more detailed analysis here so readers who have not tried the exercises will have a chance to appreciate the analogy between n-channel and p-channel devices.

Consider then the p-channel device in Figure 12 where v_G is a constant close to 0 and the source of T_X is tied to V_H. Assuming that the substrate of T_X is connected to V_H, the threshold voltage of T_X can be approximated by

$$V_{TX} = V_{T0} - k_1(V_H - v_X) \tag{26}$$

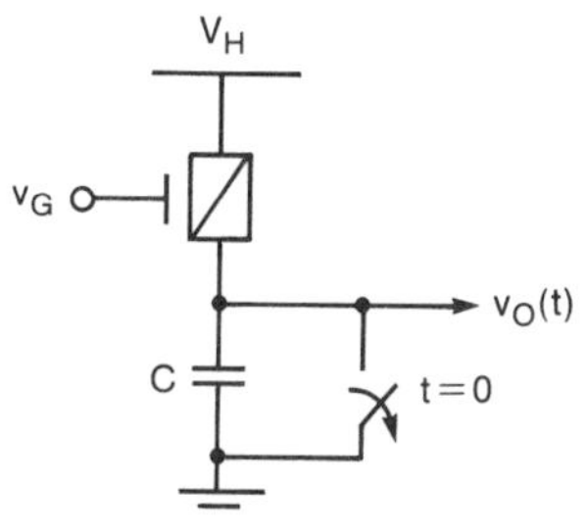

Figure 12 A p-channel pullup.

110

Chap. 4 / Digital Inverters—Transient Analysis

where $x = s$ or D. The coefficient of substrate sensitivity, k_1, is a positive constant ranging from 10% to 20%, which is usually much higher than the n-channel devices in strict NMOS technology. The device current i_{DS} flowing from ⑩ to V_H is given by

$$i_{DS} = \frac{1}{2} \beta \frac{x_D^2 - x_S^2}{1 + \lambda(x_D - x_S)}$$

where

$$x_S = (v_{GS} - V_{TS}) \, \text{sgn} \, (V_{TS} - v_{GS})$$

and

$$x_D = (v_{GD} - V_{TD}) \, \text{sgn} \, (V_{TD} - v_{GD})$$

In case T_X is in the saturation region

$$x_D = x_P = x_S - \frac{1 - \sqrt{1 - 2\lambda x_S}}{\lambda}$$

Using the linear approximation for V_T in Equation (26), we obtain, for the circuit in Figure 12

$$x_S = v_G - (V_{T0} + V_H)$$

and

$$x_D = v_G - V_{T0} + k_1 V_H - (1 + k_1)v_O$$

Notice that x_S is constant and

$$x_D(\infty) = x_S \triangleq X = v_G - (V_{T0} + V_H)$$

Now, equating the currents through C and T_X, we have

$$C \frac{d}{dt} v_O = -\frac{1}{2} \beta \frac{x_P^2 - x_S^2}{1 + \lambda(x_P - x_S)}$$

Since $v_O(0) = 0$, T_X starts in the saturation region so that $x_D = x_P$. By defining the voltage ratio $\eta = v_O/V_H$ and integrating from $(0, \eta_0)$ to (T_S, η_s) as in NMOS case, we obtain

$$T_s = \frac{2\lambda^2}{\xi^2} V_H \eta_s \frac{C}{\beta} \tag{27}$$

where

$$\eta_s = 1 + \frac{\xi}{(1 + k_1)\lambda V_H} \quad \text{and} \quad \xi = 1 - \sqrt{1 - 2\lambda X}$$

T_X will leave the saturation region at $t = T_s$. For $t > T_s$ the circuit equation becomes

$$-\frac{(1 + k_1)^2 V_H(\eta - 1)(\eta - K)}{1 + \lambda(1 + k_1)V_H(1 - \eta)} = 2\frac{C}{\beta}\frac{d}{dt}\eta$$

where

$$K = 1 + \frac{2X}{(1 + k_1)V_H}$$

Solving the above equation yields

$$t - T_s = \frac{1}{1 + k_1}\left[\frac{1}{X}\ln\frac{(\eta - 1)(\eta_s - K)}{(\eta - K)(\eta_s - 1)} + 2\lambda\ln\left(\frac{\eta - K}{\eta_s - K}\right)\right]\frac{C}{\beta} \qquad (28)$$

which accounts for the time T_X spends in the linear region.

Again, in case T_X starts in the linear region at $t = 0$, Equation (28) applies with η_s replaced by $\eta_0 \triangleq v_o(0)/V_H$.

EXAMPLE 9

Let us consider the PMOS pullup shown in Figure 12. The device characteristics are given by $(\gamma, k_1, V_{T0}, \lambda) = (0.032, 0.2, -0.5, 0.5)$. Calculate the size of T_X so that v_o rises to 2.7 V in 1 nS for $C = 1$ pF. Assume $v_G = 0$.

Solution

First, the constants are

$$X = -(-0.5 + 3) = -2.5 \quad \Rightarrow \quad \xi = 1 - \sqrt{1 - 2 \times 0.5 \times (-2.5)}$$

$$= -0.87$$

$$\eta_s = 1 - \frac{0.87}{1.2 \times 0.5 \times 3} = 0.52 \quad \text{and} \quad K = 1 - \frac{2 \times 2.5}{1.2 \times 3} = -0.39$$

The time T_X spends in the saturation region can be obtained from Equation (27):

$$T_s = \frac{2 \times 0.5^2}{0.87^2} \times 3 \times 0.52\frac{C}{\beta} = 1.03\frac{C}{\beta}$$

The time T_X spends in the linear region before v_o reaches 2.7 V is given by Equation (28) with $\eta = 0.9$:

$$T_l = \frac{1}{1.20}\left[-\frac{1}{2.5}\ln\frac{(0.9 - 1)(0.52 + 0.39)}{(0.9 + 0.39)(0.52 - 1)} + 2 \times 0.5 \times \ln\left(\frac{0.9 + 0.39}{0.52 + 0.39}\right)\right]\frac{C}{\beta}$$

$$= 0.93\frac{C}{\beta}$$

Therefore, $T_R = T_s + T_l = 1.96\ C/\beta$ and the device size

$$\frac{W}{L} = \frac{1.96}{0.032} = \frac{31}{0.5}$$

Notice that in this example the mobility and hence the value of γ of the p-channel device is about 40% that of the n-channel device. Also the substrate sensitivity k_1 is much higher. This is because in strict NMOS technology, channel tailoring (Chapter 2) and the negative substrate bias (Chapter 5) reduce substrate sensitivity substantially. In CMOS technology, however, the substrate biases for NMOS and PMOS are usually GD and V_H, respectively. This difference, together with the fact that it is more difficult to "tailor" the threshold voltages of both n-channel and p-channel devices simultaneously, results in a large k_1.

3. DESIGN PRACTICE: THE STATISTICAL ANALYSIS

In the previous sections we have derived the relationship between the size of a device and its performance through a single simple equation:

$$T_D = DK \frac{C}{\beta} \tag{29}$$

where T_D is the rise/falltime and DK is a constant. Since in practice the process varies from day to day, both the "constant" DK and the device β fluctuate. Since all device parameters are basically statistical in nature, both DK and β are random variables, and so is T_D.

The performance of a circuit is also affected by environmental variables like the power supply V_H and the device junction temperature T_j. Since the circuit has to operate under the most adverse conditions, the effects of these environmental constraints are evaluated separately from the process parameters. The performance of the circuit is simply evaluated at the extreme points of temperature and power supply with process parameters as random variables. At each corner of the rectangle bounded by the extreme values of T_j and V_H in the T_j–V_H plane, a distribution of T_D is then obtained. By comparing the end points of these distributions, we can determine the worst case (WC) and the best case (BC) performances of the circuit.

Thus, the performance T_D of a circuit is most properly described statistically by an interval (T_{min}, T_{max}) instead of a single number. Here the lower bound, T_{min}, is referred to as the best case value and the upper bound, T_{max}, is called the worst case value. As long as the environmental conditions remain within the specification, the actual value of T_D will satisfy the relationship $T_{min} < T_D < T_{max}$.

The term "typical value" is also used to indicate the nominal value of T_D in a normal distribution. If all parameters and constraints are normally distributed, the typical value will be in the center of its distribution. Today, however, due to the fierce competition in the industry, quoted typical values always seem to be closer to the best case value than the middle of the distribution.

The above methodology is best illustrated by an example.

EXAMPLE 10

Consider the depletion mode pullup of Example 4 in Figure 13. Assume the process parameters are distributed as follows:

(a) $V_{T0} \in (-1.4, -1.2)$

(b) $\Delta W \in (-0.25, 0.25)$

(c) $\Delta L \in (-0.15, 0.15)$

where $X \in (a, b)$ means the values of X are normally distributed with the 3σ points a and b.

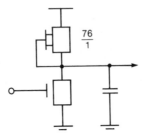

Figure 13(a) An E/D inverter under statistical analysis.

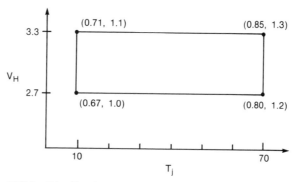

Figure 13(b) Distributions of T_R at T_j–V_H corners. Constant 90% point.

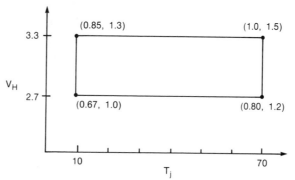

Figure 13(c) Distributions of T_R at T_j–V_H corners. Variable 90% point.

Calculate the 3σ BC and WC risetime T_R of the circuit in the following ranges of power supply and temperature:

(a) $V_H \in (2.7, 3.3)$

(b) $T_j \in (10, 70)$

Solution

As we have pointed out in Section 6 of Chapter 1, both the electron mobility and the threshold voltage of the device depend on the temperature. As temperature goes up, both μ and V_{T0} decrease, but the effect of mobility is stronger than the effect of V_{T0} and the device slows down. Figure 13(b) shows the distributions of T_D at each corner of the rectangle in the T_j–V_H plane. These distributions are obtained by Monte Carlo simulation on Equation (29). In Figure 13(b), T_R is defined as the risetime required for the output to reach $2.7 \times 90\% = 2.43$ V. We conclude that T_R (BC) is given by $T_{min} = 0.67$ nS and T_R (WC) is given by $T_{max} = 1.3$ nS.

The risetime can also be defined as the time required for the output to reach 90% of the final value. These results are shown in Figure 13(c). In this case $T_{min} = 0.67$ nS and $T_{max} = 1.5$ nS.

Notice that in both cases the worst case performance is obtained from the high end of the distribution of both temperature and power supply. This is characteristic of depletion mode devices. For enhancement mode devices, however, the worst cases usually occur at the highest temperature and the lowest power supply.

4. *CONCEPT OF CIRCUIT DELAY*

Before applying the results of the previous sections to the delay calculations that will concern us for the rest of this chapter, we will introduce the concept of circuit delay. Consider the inverter chain shown in Figure 14 in which all stages are identical with the same load. An input is applied to the first stage and its effect propagates along the chain. We shall examine the waveforms at nodes ②, ③, and ④ which are "buffered" by circuits. As shown in Figure 14(b), as v_2 goes down v_3 goes up and v_4 goes down. Since all circuits are the same, the waveforms of v_2 and v_4 are identical except that v_4 follows v_2 with a time lag. The exact duration of the time lag can be measured by comparing the time instants at which the waveforms cross a particular voltage level. While there are many "system considerations" such as chip interfacing that tend to make certain choices more preferable, the time lag stays relatively constant for measuring levels around half V_H. Since we have had no special requests from "system people" yet, we use half V_H at 1.5 V to measure the delay. As shown in the figure the time lag measured at 1.5 V from points ⓐ to ⓒ is about 3.5 nS.

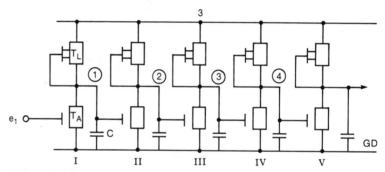

Figure 14(a)   An inverter chain for the definition of circuit delays.

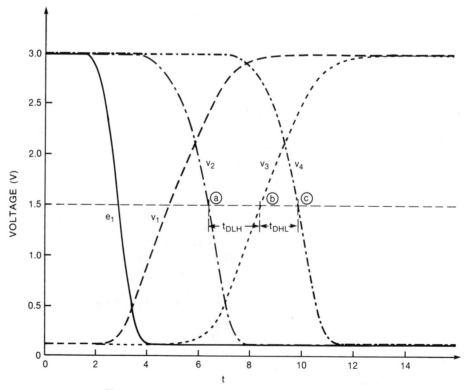

Figure 14(b)   Waveforms of the internal nodes.

Since there are two identical stages involved, the time delay "per stage," denoted by t_D, is therefore defined as $3.5/2 = 1.75$ nS.

The time interval ⓐ–ⓒ actually consists of two parts: ⓐ–ⓑ, when v_2 goes down and v_3 goes up at the output, and ⓑ–ⓒ, when v_3 goes up and v_4 goes down at output. *Following the outputs,* the delay of v_3 from v_2 is called the rise delay, denoted t_{DLH}, and the delay of v_4 from v_3 is called the fall delay, denoted t_{DHL} where *LH* and *HL* stand for transitions from low to high and high to low, respectively. Obviously the time delay t_D is equal to $(t_{DLH} + t_{DHL})/2$, which is sometimes also referred to as the "average delay" of the inverter. Notice that whereas the average t_D stays relatively constant, the individual components t_{DLH} and t_{DHL} are strong functions of the voltage level of delay measurement.

Computation of delays for complex logic circuits is by no means trivial. Since it is impractical to simulate large logic networks with detailed waveforms and device models, most timing analysis on the chip level is performed with simple "delay calculators" that characterize each circuit by a block delay. Once the signal path is identified (or, in logic design language, "sensitized"), the total delay along the path can be obtained by simply adding the block delays of each individual circuit. Our next task, therefore, is the construction of delay calculators for inverters.

5. CALCULATION OF DELAYS

The previous sections have concentrated on the transient responses of pullups and pulldowns as single devices. Since inverters are composed of pullups and pulldowns in pairs, we have tacitly neglected the current flowing from V_H to *GD* through both devices during switching. As will be explained in the following discussion, the effect of this "transient current" is more serious for pulldowns than for pullups. For E/D inverters, for example, current through the depletion device increases as the output falls. If the β ratio is kept small, this current can become a significant portion of the total discharging current, and our simple analysis must be modified. Since the total charge flowing through the depletion device is proportional to the falltime, which is, in turn, proportional to the load capacitor C, a factor α_c where $0 < \alpha_c < 1$ can be added to the total capacitive loading. Since the value of α_c also depends on the β ratio, it is best determined empirically by circuit simulation.

To make the problem more mathematically tractable, step functions were used for the inputs. Examples were then given for some "typical cases": the 90% risetime and the 10% falltime of pullups and pulldowns, respectively. We hoped that these results could be related to circuit delays even though in reality all input waveforms have finite transition times. We will now investigate this problem more closely.

The acute reader may have noticed that for step inputs, the risetime T_R is much longer than the falltime T_F for the same device size. This is easily explained by comparing the gatedrives of pullups and pulldowns. The pullup response is less sensitive to the waveform of the input than the pulldown response. Indeed, results derived in the previous sections for pullups can be applied directly to cases with finite input transition times. For pulldowns, however, those simple equations will have to be supplemented with more elaborate solutions.

For example, let us reexamine the pullup response of inverter III in Figure 14, redrawn in Figure 15 with its step response. As the figure makes obvious, the output v_3 has hardly moved before the input v_2 reaches its downlevel. The risetime T_R is actually independent of the falltime of the input, and so the step response does represent a good approximation. As a matter of fact the reader should be able to see that the rise delay t_{DLH} measured at $V_H/2$ is very close to the value $T_R(90\%)/2$ of the step response. Measuring the corresponding time intervals in the figure settles the case.

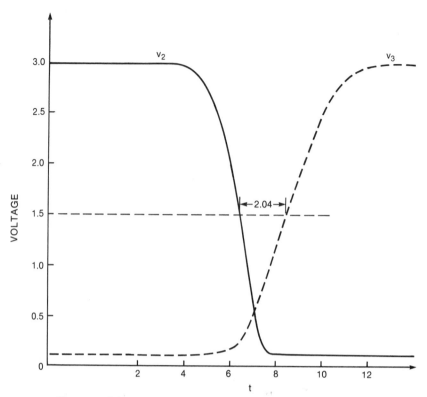

Figure 15(a) Pullup response of inverter III in Figure 14.

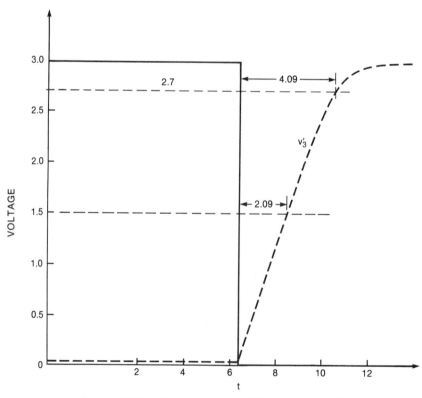

Figure 15(b) Step response of the same circuit.

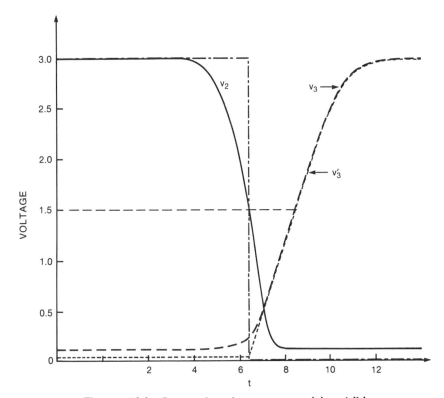

Figure 15(c) Comparison between case **(a)** and **(b)**.

We shall therefore estimate t_{DLH} by

$$t_{DLH} = \frac{T_R(90\%)}{2}$$

By aligning the step input with the actual input at $V_H/2$ and comparing the corresponding responses as shown in Figure 15(c), we see that this is a very good approximation. This happy coincidence, however, applies only to depletion mode pullups with reasonable input falltimes.

For other cases such as the diode pullup in Figure 2, it is more accurate to estimate the half V_H delay by

$$t_{DLH} = \frac{\tau_F}{2} + T_R(\eta)$$

where τ_F is the falltime of the input and η is the ratio of $V_H/2$ to the final level of the output. The calculation of $T_R(50\%)$ for different kinds of pullups, for example, is left as a simple exercise (see Problem 8).

It is also obvious from Figure 2(c) that the risetime of v_O can be more accurately estimated as $10[T_R(50\%) - T_R(40\%)]$ instead of $T_R(90\%)$ of the step response because v_O levels off and slows down considerably when it approaches the 90% point.

The relationship between the fall delay t_{DHL} and T_F, however, is much more complicated. Since the pulldown runs very fast, it "follows" the input closely. Delay t_{DHL} is therefore highly dependent on the risetime of the input. The step response does not serve as a good approximation any more. As shown in Figure 16, the fall delay t_{DHL} at $V_H/2$ is considerably longer than $T_F(10\%)$ of the step response. The rela-

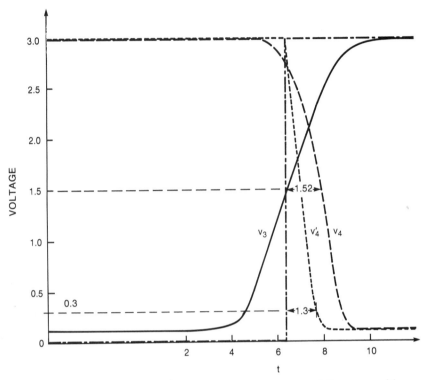

Figure 16 Comparison between step response and the case with finite input risetime of inverter IV in Figure 14 showing the inadequacy of the step input "approximation" for pulldown responses.

tionship between the two, if any, is not clearly defined. A more accurate representation of the input is needed.

Now refer to Figure 17 in which a simple enhancement mode pulldown is shown. Since the next simplest member in the family of "singular functions" is the ramp function, use a function of the form

$$v_G = \begin{cases} \dfrac{V_H}{\tau_R}t, & 0 < t < \tau_R \\ V_H, & \tau_R < t \end{cases} \tag{30}$$

as the input where V_H is the uplevel and τ_R is the risetime. Example 8 shows that the device stays in the saturation region for $\zeta = 1$ to 0.45, and since we are only interested in the delay for $\zeta = 0.5$ at the $V_H/2$ point, consider the saturation case only. Now from Equation (24) we obtain

$$\frac{1}{\tau_R}\frac{C}{\beta}\,d\zeta = \frac{1}{(\lambda V_H)^2}(1 + x_S - \sqrt{1 + 2\lambda x_S})\,dx_S \tag{31}$$

Integrating Equation (31) from $(1, 0)$ to (ζ, x_S), we have

$$\frac{1}{\tau_R}\frac{C}{\beta}(1 - \zeta) = \frac{1}{(\lambda V_H)^2}\left[x_S + \frac{1}{2}x_S^2 + \frac{1}{3\lambda} - \frac{1}{3\lambda}(1 + 2\lambda x_S)^{3/2}\right] \tag{32}$$

The above equation defines x_S as an implicit function of the design parameters. Given C, β and τ_R, we can solve for x_S at $\zeta = 0.5$ from above either graphically or by numerical iteration. Since $x_S = v_G(t) - V_{T0}$, from Equation (30), the fall delay can be obtained by first calculating

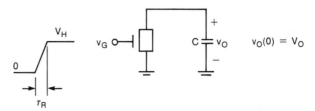

Figure 17(a) A pulldown with a ramp input waveform.

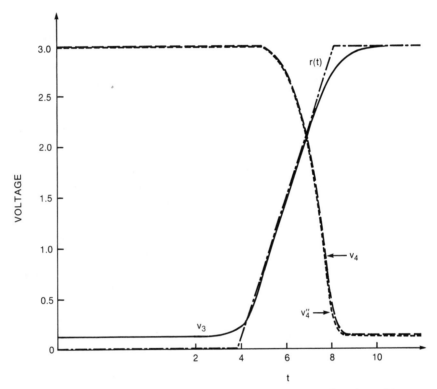

Figure 17(b) Comparison between ramp response v_4'' and v_4 of inverter IV in Figure 14.

$$t_{DHL}' = \tau_R\left[\frac{1}{V_H}(x_S + V_{T0}) - \frac{1}{2}\right] \tag{33}$$

The fall delay $t_{DHL} = t_{DHL}'$ if $t_{DHL}' < \tau_R/2$ so that the input has not yet reached V_H and Equation (31) is valid at $\zeta = 0.5$. In case $t_{DHL}' > \tau_R/2$, the input v_G has already reached V_H before the output falls to the $V_H/2$ point. Let $\zeta = \zeta_r > 0.5$ at $t = \tau_R$ and the rest of the falltime after τ_R can be calculated from Equation (24):

$$T_s = \frac{2\lambda^2}{\xi^2}V_H(\zeta_r - 0.5)\frac{C}{\beta} \tag{34}$$

and $t_{DHL} = \tau_R/2 + T_s$. The value of ζ_r, in turn, can be obtained by substituting $x_S = V_H - V_{T0}$ into Equation (32).

Ramp responses are much more accurate than are step responses for pulldowns. Figure 17(b) approximates the input waveform of v_3 of Figure 16 by a ramp function $r(t)$ and compare the waveforms of ④. As is evident in the figure, the difference in delays is negligibly small. Now, wade through a few examples.

EXAMPLE 11

Estimate the fall delay t_{DHL} of the inverter shown in Figure 18. Assume $v_{10}(0) = 3$ V and the risetime τ_R of e_1 is 2 nS.

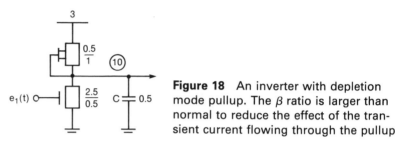

Figure 18 An inverter with depletion mode pullup. The β ratio is larger than normal to reduce the effect of the transient current flowing through the pullup.

Solution

To illustrate the procedure more clearly, this problem will be attacked with a graphical method. Computer programs using numerical iteration are best suited for design automation (DA) in conjunction with logic simulators. They will be the subjects of the next section.

Since the same graph can be used for different problems, plot the function $F(x_S, \lambda)$ defined by

$$F(x_S, \lambda) = \frac{1}{(\lambda V_H)^2} \left\{ x_S + \frac{1}{2} x_S^2 + \frac{1}{3\lambda}[1 - (1 + 2\lambda x_S)^{3/2}] \right\} \tag{35}$$

for $\lambda = 1$ and 0.5 as shown in Figure 19. Equation (32) then becomes

$$\frac{1}{\tau_R} \frac{C}{\beta} (1 - \zeta) = F(x_S, \lambda) \tag{36}$$

In our case, we have

$$\frac{1}{2} \frac{0.5}{(0.082 \times 2.5)/0.5} \times (1 - 0.5) = F(x_S, 1)$$

From Figure 19(a) $x_S = 3.7$. Since $x_S > V_H - V_{T0} = 2.5$, $t'_{DHL} > \tau_R/2$. Indeed

$$t'_{DHL} = 2 + \left(\frac{1}{3} \times (3.7 + 0.5) - \frac{1}{2} \right) = 1.8 > \frac{2}{2} = 1$$

indicating that e_1 has reached V_H at $t = 2$ before v_{10} crosses the $V_H/2$ point. Therefore, we have to find the voltage ratio ζ at $t = 2$. Substituting $x_S = V_H - V_{T0} = 2.5$ into Equation (32), we obtain $\zeta_r = 0.81$. This means $v_{10}(2) = 81\%$ of V_H. After $t = 2$ the input $e_1(t) = V_H$ remains constant. Then apply Equation (34) for the rest of the delay. First, we need the parameter ξ:

$$\xi = \sqrt{1 + 2 \times 1 \times (3 - 0.5)} - 1 = 1.45$$

and then $T_s = \dfrac{2 \times 1^2}{1.45^2} \times 3 \times (0.81 - 0.5) \times \dfrac{0.5}{(0.082 \times 2.5)/0.5} = 1.08.$

The total delay is therefore given by

$$t_{DHL} = 1.08 + 1 = 2.08$$

The actual waveform is shown in Figure 20(a). Since the β ratio of the inverter is large, our calculation agrees very well with that measured directly from the waveforms.

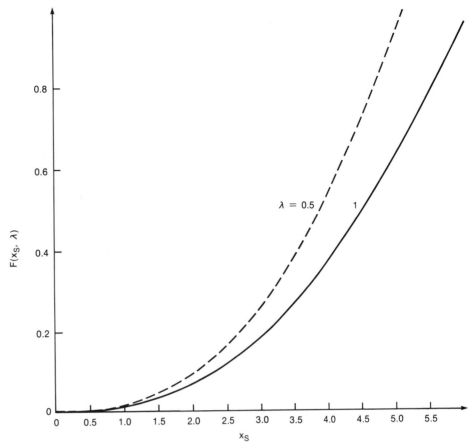

Figure 19(a) The function $F(x_S, \lambda)$ plotted for $x_S \in (0, 5.5)$.

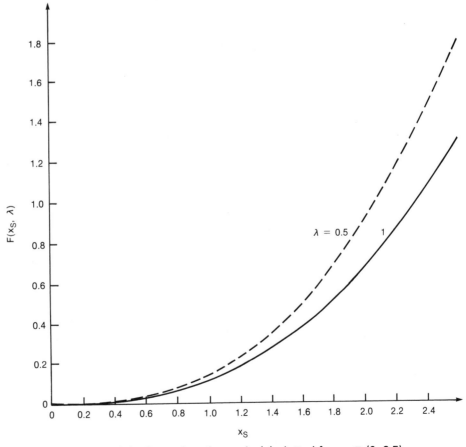

Figure 19(b) Same function as in (a) plotted for $x_S \in (0, 2.5)$.

121

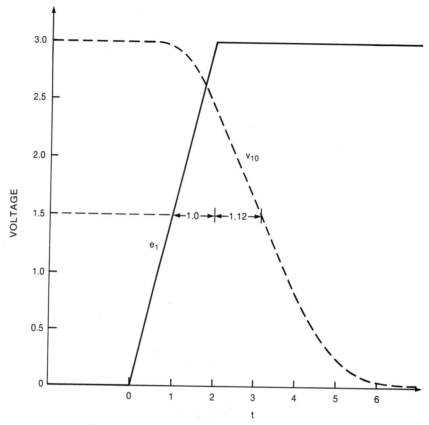

Figure 20(a) Waveform of v_{10} of Example 11.

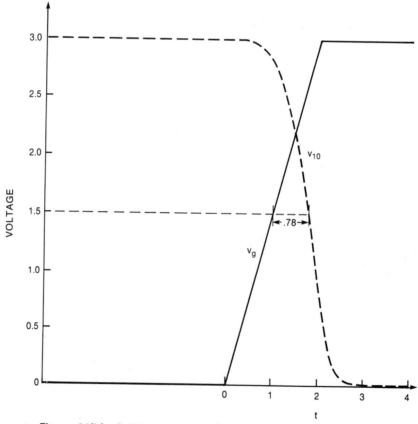

Figure 20(b) Pulldown response of the circuit of Example 12.

EXAMPLE 12

Next consider the same problem with $W/L = 5/0.5$ and $C = 0.25$ pF.

Solution

In this case Equation (35) yields

$$\frac{1}{2} \frac{0.25}{(0.082 \times 5)/0.5} \times (1 - 0.5) = F(x_S, 1)$$

Consulting with Figure 19(b) we obtain

$$x_S = 2.1$$

$$\Rightarrow \qquad t'_{DHL} = 2 \times \left[\frac{1}{3} \times (2.1 + 0.5) - \frac{1}{2} \right] = 0.73 < 1$$

Hence $t_{DHL} = t'_{DHL} = 0.73$.
These waveforms are shown in Figure 20(b).

6. DELAY CALCULATORS

We are now in a position to construct delay calculators for inverters. Since the expressions for pulldowns are more complicated, it makes more sense to write the delay calculator for CMOS inverters first. The inputs to the models are

1. The device sizes.
2. The risetime τ_R or the falltime τ_F of the input ramp functions.
3. The capacitance C_0 that accounts for the internal delay.
4. The external capacitive load C_L.
5. The device parameters $(\gamma, k_1, V_{T0}, \lambda)$ and the parameter α_c that account for the switching "transient currents."
6. Electrical parameters: V_H and the logic uplevel and downlevel.

The outputs of the models are

1. The rise delay t_{DLH} and the risetime T_R of the output waveform approximated as a ramp function if there is a low to high transition.
2. The fall delay t_{DHL} and the falltime T_F of the output waveform approximated as a ramp function if there is a high to low transition.
3. The total delay at each node from the reference input.

Parameters pertaining to n-channel or p-channel devices are subscripted with n and p. Definitions of the parameters are listed in Figure 21.

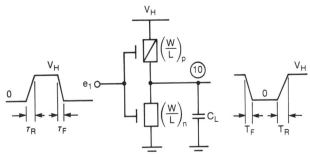

Figure 21 Terminologies used in the CMOS delay calculator.

We use the Newton-Raphson method to solve Equation (32). To calculate the risetime of the output, we determine the times at which the output $\eta = 0.4$ and 0.5 and the risetime $T_R = 10\,[T_R(50\%) - T_R(40\%)]$. Similarly, the falltime is calculated by $T_F = 10[T_F(50\%) - T_F(60\%)]$. The following is a FORTRAN program of the delay calculator. Note that the variable ξ is coded as TSAI instead of XI.

```
C***********************************************************************
C*                                                                    *
C*                         PROGRAM CMOSDLY                             *
C*                                                                    *
C*        CALCULATION OF DELAY TIME OF CMOS INVERTERS                 *
C*                                                                    *
C***********************************************************************
        SUBROUTINE CMOSTD (PWP, PLP, PWN, PLN, TAOR, TAOF, TDLY,
     *                     TDHL, TDLH, TR, TF, XITR, CON, COP, CL)

        IMPLICIT REAL*8(A-H,O-Z)

        COMMON VH, GAMMAN, GAMMAP, PK1N, PK1P, VTON, VTOP, PLMBDN, PLMBDP

C --- GAMMAN = GAMMA OF NMOS
C --- GAMMAP = GAMMA OF PMOS
C --- PK1N = SUBSTRATE SENSITIVITY K1 OF NMOS
C --- PK1P = SUBSTRATE SENSITIVITY K1 OF PMOS
C --- VTON = THRESHOLD VOLTAGE VTO of NMOS
C --- VTOP = THRESHOLD VOLTAGE VTO OF PMOS
C --- PLMBDN = COEF. OF CHANNEL LENGTH DILATION, LAMBDA OF NMOS
C --- PLMBDP = COEF. OF CHANNEL LENGTH DILATION, LAMBDA OF PMOS
C --- XITR = TRANSITION INDICATOR, XITR = 0 IF THE INPUT IS LOW, AND
C ---                              XITR = 1 IF THE INPUT IS HIGH
C --- CON = CAPACITANCE THAT ACCOUNTS FOR INTERNAL FALL DELAY
C --- COP = CAPACITANCE THAT ACCOUNTS FOR INTERNAL RISE DELAY
C --- CL = CAPACITANCE THAT ACCOUNTS FOR EXTERNAL LOADS
C --- ALPHACX, X = N OR P: APPARENT CAPACITIVE LOADING FOR SWITCHING
C ---                      TRANSIENTS
C --- TAOR = RISETIME OF INPUT WAVEFORM
C --- TAOF = FALLTIME OF INPUT WAVEFORM
C --- TDLY = THE TOTAL DELAY THUS FAR FROM THE PRIMARY INPUT
C --- MAX = MAXIMUM NUMBER OF ITERATIONS SPECIFIED BY USER

        IF (XITR .LT. 0.5)  GO TO 100

C *** NMOS PULLDOWN

C --- CALCULATION OF XS

        BETA = GAMMAN*PWN/PLN
        ALPHACN = .05
        C = CON + (1.0+ALPHACN)*CL
        PK = C*((PLMBDN*VH)**2)/(TAOR*BETA)
        PK1 = .5*PK

C --- INITIALIZATION

        XS = VH/2.0
        MAX = 50
        ITER = 1

C --- ITERATION
```

```
   40 IF (ITER .GT. MAX) STOP
      F = XS + (PLMBDN*XS**2)/2.0 - PK1
     *        + (1.0-DSQRT((1.0+2.0*PLMBDN*XS)**3))/(3.0*PLMBDN)

      DF = 1.0 + PLMBDN*XS - DSQRT(1.0+2.0*PLMBDN*XS)

      XSP = XS - F/DF

C --- TEST FOR CONVERGENCE

      DELTA = DABS(XSP-XS)
      IF (DELTA .LT. 1.0D-6) GO TO 50
      XS = XSP
      ITER = ITER + 1
      GO TO 40

C --- CALCULATION OF TDHL

   50 TDHLP = TAOR*((XS+VT0N)/VH-.5)
      HTAOR = TAOR/2.0
      IF (TDHLP .GT. HTAOR) GO TO 60
      TDHL = TDHLP
      GO TO 70
   60 XS = VH - VT0N
      ZETAR = 1.0 - (XS + (PLMBDN*XS**2)/2.0
     *        + (1.0-DSQRT((1.0+2.0*PLMBDN*XS)**3))/(3.0*PLMBDN))/PK
      TSAI = DSQRT(2.0*PLMBDN*XS+1.0) - 1.0
      TS = (2.0*PLMBDN**2/TSAI**2)*VH*(ZETAR-.5)*(C/BETA)
      TDHL = HTAOR + TS

C *** CALCULATION OF TF

   70 PK2 = (1-.6)*PK

      ITER = 1

C --- ITERATION

   42 IF (ITER .GT. MAX) STOP
      F = XS + (PLMBDN*XS**2)/2.0 - PK2
     *        + (1.0-DSQRT((1.0+2.0*PLMBDN*XS)**3))/(3.0*PLMBDN)

      DF = 1.0 + PLMBDN*XS - DSQRT(1.0+2.0*PLMBDN*XS)

      XSP = XS - F/DF

C --- TEST FOR CONVERGENCE

      DELTA = DABS(XSP-XS)
      IF (DELTA .LT. 1.0D-6) GO TO 52
      XS = XSP
      ITER = ITER + 1
      GO TO 42

C --- CALCULATION OF TDHL2

   52 TDHLP = TAOR*((XS+VT0N)/VH-.5)
      HTAOR = TAOR/2.0
      IF (TDHLP .GT. HTAOR) GO TO 62
      TDHL2 = TDHLP
      GO TO 72
```

```
   62 TS = (2.0*PLMBDN**2/TSAI**2)*VH*(ZETAR-.6)*(C/BETA)
      TDHL2 = HTAOR + TS


   72 TF = 10.0*(TDHL-TDHL2)
      XITR = 0.0
      TAOF = TF
      TDLY = TDLY + TDHL

      WRITE(6,1)  PWN, PLN
    1 FORMAT (///' W  =  ',E10.3,'          L  =  'E10.3,'    NMOS')
      WRITE(6,2)  CON, CL
    2 FORMAT (/' CON  =  ',E10.3,'          CL  =  'E10.3,)
      WRITE(6,3)  TAOR
    3 FORMAT (/' INPUT RISE TIME TAOR  =  ',E10.3,)
      WRITE(6,4)  TDHL
    4 FORMAT (/' FALL DELAY TDHL  =  ',E10.3,)
      WRITE(6,5)  TF
    5 FORMAT (/' OUTPUT FALL TIME TF  =  ',E10.3,)
      WRITE(6,6)  XITR
    6 FORMAT (/' TRANSITION XITR  =  ',E10.3,)
      WRITE(6,7) TDLY
    7 FORMAT (/' TOTAL DELAY  =  ',E10.3,//)
      RETURN

C *** PMOS PULLUP

C --- CALCULATION OF XS

  100 BETA = GAMMAP*PWP/PLP
      ALPHACP = 0.05
      C = COP + (1.0+ALPHACP)*CL
      PK = C*((PLMBDP*VH)**2)/(TAOF*BETA)
      PK1 = .5*PK

C --- INITIALIZATION

      XS = -VH/2.0
      MAX = 50
      ITER = 1

C --- ITERATION

  140 IF (ITER .GT. MAX) STOP
      F = XS - (PLMBDP*XS**2)/2.0 + PK1
     *       + (DSQRT((1.0-2.0*PLMBDP*XS)**3)-1.0)/(3.0*PLMBDP)

      DF = 1.0 - PLMBDP*XS - DSQRT(1.0-2.0*PLMBDP*XS)

      XSP = XS - F/DF

C --- TEST FOR CONVERGENCE

      DELTA = DABS(XSP-XS)
      IF (DELTA .LT. 1.0D-6) GO TO 150
      XS = XSP
      ITER = ITER + 1
      GO TO 140

C --- CALCULATION OF TDLH
```

```
    150 TDLHP = TAOF*(.5-(XS-VTOP)/VH)
        HTAOF = TAOF/2.0
        IF (TDLHP .GT. HTAOF) GO TO 160
        TDLH = TDLHP
        GO TO 170
    160 XS = -(VH+VTOP)
        ETAR = -(XS - (PLMBDP*XS**2)/2.0
       *          + (DSQRT((1.0-2.0*PLMBDP*XS)**3)-1.0)/(3.0*PLMBDP))/PK
        TSAI = 1.0 - DSQRT(1.0-2.0*PLMBDP*XS)
        TS = (2.0*PLMBDP**2/TSAI**2)*VH*(.5-ETAR)*(C/BETA)
        TDLH = HTAOF + TS


C *** CALCULATION OF TR

    170 PK2 = .4*PK

        ITER = 1

C --- ITERATION

    142 IF (ITER .GT. MAX) STOP

        F = XS - (PLMBDP*XS**2)/2.0 + PK2
       *          + (DSQRT((1.0-2.0*PLMBDP*XS)**3)-1.0)/(3.0*PLMBDP)

        DF = 1.0 - PLMBDP*XS - DSQRT(1.0-2.0*PLMBDP*XS)

        XSP = XS - F/DF

C --- TEST FOR CONVERGENCE

        DELTA = DABS(XSP-XS)
        IF (DELTA .LT. 1.0D-6) GO TO 152
        XS = XSP
        ITER = ITER + 1
        GO TO 142

C --- CALCULATION OF TDLH2

    152 TDLHP = TAOF*(.5-(XS-VTOP)/VH)
        IF (TDLHP .GT. HTAOF) GO TO 162
        TDLH2 = TDLHP
        GO TO 172
    162 TS = (2.0*PLMBDP**2/TSAI**2)*VH*(.4-ETAR)*(C/BETA)
        TDLH2 = HTAOF + TS

    172 TR = 10.0*(TDLH-TDLH2)
        XITR = 1.0
        TAOR = TR
        TDLY = TDLY + TDLH

        WRITE(6,101)  PWP, PLP
    101 FORMAT (///' W  =  ',E10.3,'         L  =  'E10.3,'    PMOS')
        WRITE(6,102)  COP, CL
    102 FORMAT (/' COP  =  ',E10.3,'         CL  =  'E10.3,)
        WRITE(6,103)  TAOF
    103 FORMAT (/' INPUT FALL TIME TAOF  =  ',E10.3,)
        WRITE(6,104)  TDLH
    104 FORMAT (/' RISE DELAY TDLH  =  ',E10.3,)
        WRITE(6,105)  TR
```

```
 105 FORMAT (/' OUTPUT RISE TIME TR  =   ',E10.3,)
     WRITE(6,106)  XITR
 106 FORMAT (/' TRANSITION XITR  =   ',E10.3,)
     WRITE(6,107)  TDLY
 107 FORMAT (/' TOTAL DELAY  =    ',E10.3,//)

     RETURN
     END
```

As an example, calculate the delay for the inverter chain shown in Figure 22(a) using the previous models. The main program could be coded as follows:

```
     IMPLICIT REAL*8 (A-H,O-Z)

     COMMON VH, GAMMAN, GAMMAP, PK1N, PK1P, VT0N, VT0P, PLMBDN, PLMBDP

     VH = 3.0

     GAMMAN = .082
     GAMMAP = .032

     PK1N = .20
     PK1P = .20

     VT0N = .5
     VT0P = -.5

     PLMBDN = 1.0
     PLMBDP = .5

C --- CIRCUIT PARAMETERS

     PWP = 9.5
     PLP = .5

     PWN = 9.5
     PLN = .5

     TAOR = 0.7
     TAOF = 0.5

     TDHL = 1.0
     TDLH = 1.0

     TR = 1.0
     TF = 1.0

     C0 = .015
     CL = .0

     TDLY = 0.0

     CALL  CMOSTD (PWP, PLP, PWN, PLN, 2.0  , TAOF, TDLY,
    *              TDHL, TDLH, TR, TF, 1.0, C0, C0, .5)

     CALL  CMOSTD (PWP, PLP, PWN, PLN, TAOR, TAOF, TDLY,
    *              TDHL, TDLH, TR, TF, XITR, C0, C0, .25)
```

```
      CALL   CMOSTD  (PWP, PLP, PWN, PLN, TAOR, TAOF, TDLY,
     *                 TDHL, TDLH, TR, TF, XITR, C0, C0, .8)

      CALL   CMOSTD  (PWP, PLP, PWN, PLN, TAOR, TAOF, TDLY,
     *                 TDHL, TDLH, TR, TF, XITR, C0, C0, .8)

      CALL   CMOSTD  (PWP, PLP, PWN, PLN, TAOR, TAOF, TDLY,
     *                 TDHL, TDLH, TR, TF, XITR, C0, C0, .5)

      STOP
      END
```

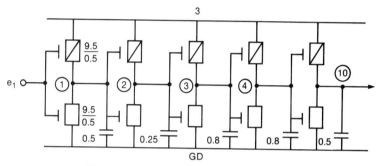

Figure 22(a)   A CMOS inverter chain.

The following is a partial listing of the output:

```
      INPUT RISE TIME TAOR = 0.200D+01
      FALL DELAY TDHL = 0.807D+00
      OUTPUT FALL TIME TF = 0.125D+01
      TOTAL DELAY = 0.807D+00

      INPUT FALL TIME TAOF = 0.125D+01
      RISE DELAY TDLH = 0.677D+00
      OUTPUT RISE TIME TR = 0.903D+00
      TOTAL DELAY = 0.148D+01

      INPUT RISE TIME TAOR = 0.903D+00
      FALL DELAY TDHL = 0.932D+00
      OUTPUT FALL TIME TF = 0.157D+01
      TOTAL DELAY = 0.242D+01

      INPUT FALL TIME TAOF = 0.157D+01
      RISE DELAY TDLH = 0.167D+01
      OUTPUT RISE TIME TR = 0.278D+01
      TOTAL DELAY = 0.409D+01

      INPUT RISE TIME TAOR = 0.278D+01
      FALL DELAY TDHL = 0.870D+00
      OUTPUT FALL TIME TF = 0.152D+01
      TOTAL DELAY = 0.496D+01
```

Figure 22(b) shows the delay for each stage on the actual waveforms. The estimates worked out by the delay calculator are quite accurate.

The delay calculators of pullups are much simpler. Since the risetime T_R (90%) is proportional to the ratio C/β, we can "precalculate" the proportionality constant DK in Equation(29) under specific conditions and then do the real calculation ac-

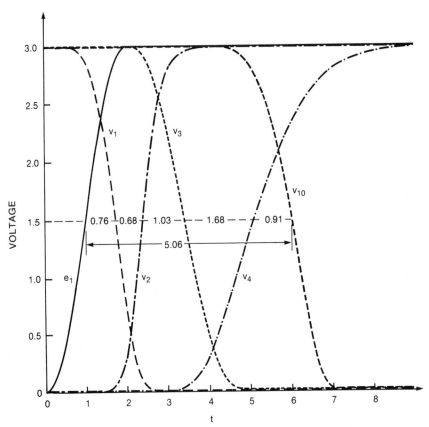

Figure 22(b)   Internal delays of the CMOS inverter chain.

cording to the actual loading and device size. For example, the depletion mode pullup in Example 4 is described by

$$T_R(90\%) = 6.19\frac{C}{\beta} \quad \Rightarrow \quad DK = 6.19$$

A delay calculator for such a pullup can be coded as follows:

```
C *** DEPLETION MODE PULLUP

C --- CALCULATION OF BETA AND C

  100 BETA = GAMMAD*PW/PL
      ALPHACD = 0.05
      C = CO + (1.0+ALPHACD)*CL
      DK = 6.19
      TR = DK*C/BETA
      TDLH = TR/2.0
      XITR = 1.0
      TDLY = TDLY + TDLH
  101 FORMAT (////' W  =  ',E10.3,'        L  =  'E10.3,'      DEP PULLUP')
      WRITE(6,102)  CO, CL
  102 FORMAT (/' CO  =  ',E10.3,'      CL  =  'E10.3,)
      WRITE(6,103)  TAOF
  103 FORMAT (/' INPUT FALL TIME TAOF  =  ',E10.3,)
      WRITE(6,104)  TDLH
```

```
104 FORMAT (/' RISE DELAY TDLH  =   ',E10.3,)
    WRITE(6,105)  TR
105 FORMAT (/' OUTPUT RISE TIME TR  =   ',E10.3,)
    WRITE(6,106)  XITR

106 FORMAT (/' TRANSITION XITR  =   ',E10.3,)
    WRITE(6,107)  TDLY
107 FORMAT (/' TOTAL DELAY  =   ',E10.3,//)

    RETURN
    END
```

More sophisticated delay calculators can be written with all parameters as variables. Such delay calculators can be used for statistical analysis. However, since it will be called constantly in the timing analysis of a complicated logic network, it is advisable to precalculate the constants and keep the delay calculator as simple as possible. Some useful delay calculators written in FORTRAN are listed in the appendices to this chapter.

7. DEVICE CAPACITANCES AND THE MILLER EFFECT

Before we conclude this chapter a discussion about the gate capacitances is in order. The capacitive load of an inverter comes from many sources: the capacitances from the conductor lines, the parasitic capacitances associated with each device, and the thin oxide capacitances of the gates of the driven circuits. Depending on the design style, one component can be more significant than the other. For example, in custom designs where the interconnection lines are very short and all parasitics are minimized, the capacitive loading is caused mostly by gate capacitances. In more structured designs such as gate arrays, wiring capacitances dominate. In this section we shall discuss the gate capacitances of devices.

There are two factors complicating the analysis of gate capacitances. First, the partition of the gate capacitance into C_{gs} and C_{gd} is voltage dependent. In hand calculations we usually assume

$$C_{gs} = \begin{cases} (2/3)C'_o WL, & \text{when the device is in the saturation region} \\ (1/2)C'_o WL, & \text{when the device is in the linear region} \end{cases}$$

and

$$C_{gd} = \begin{cases} 0, & \text{when the device is in the saturation region} \\ (1/2)C'_o WL, & \text{when the device is in the linear region} \end{cases}$$

Their actual values, however, are functions of the terminal voltages. The voltage variation is particularly interesting in CMOS circuits.

The second factor that complicates the analysis is the Miller effect. To understand the instantaneous loading of the gate capacitance to its driver, let us consider the inverter shown in Figure 23. Following the input $v_I(t)$, examine the following events:

1. $v_I(t) < V_{T0}$: Device T_A is off. The capacitance v_I sees is the series combination of the thin oxide capacitance and the depletion region capacitance underneath the gate. It is small.

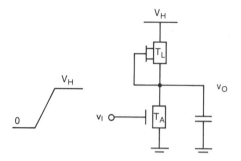

Figure 23 An inverter for the discussion of input capacitance.

2. $V_{T0} < v_I(t)$ and T_A turns on: As $v_I(t)$ increases $v_O(t)$ drops according to the incremental relationship

$$\Delta v_O = A\,\Delta v_I$$

where $A < 0$ is the *small-signal* voltage gain (the slope of the v_I–v_O transfer curve). The input current is, therefore,

$$i_I = C_{gs}\frac{\Delta v_I}{\Delta t} + C_{gd}\frac{(\Delta v_I - \Delta v_O)}{\Delta t}$$

$$= \frac{[C_{gs} + (1 + |A|)C_{gd}]\Delta v_I}{\Delta t}$$

where

$$C_{gd} = \begin{cases} \text{the total parasitic capacitance at the drain of } T_A \\ \text{when } T_A \text{ is in the saturation region} \\ \\ \text{the parasitic capacitances } + C_{gd} \text{ of the device} \\ \text{when } T_A \text{ is in the linear region} \end{cases}$$

In either case notice that the gate-drain capacitance has been amplified by the small-signal voltage gain. The actual amount of charge $Q(t)$ supplies to the input gate depends on the falltime of $v_O(t)$ as well as the risetime of $v_I(t)$ itself. This phenomenon is called *the Miller effect*. It is therefore extremely important to minimize the parasitic capacitance at the drain of a device to achieve high speed.

The Miller effect is most prominent in linear circuits where the capacitance is amplified by the small-signal voltage gain at the DC bias point. In digital inverters, however, its effect is substantially reduced by the various conditions T_A is experiencing. When $|A|$ is at its maximum, T_A is in saturation and C_{gd} is small. When T_A enters the linear region so $C_{gd} \cong (1/2)C'_o WL$ becomes significant, the output v_O is already low and $|A|$ drops considerably. This fact, together with today's self-aligned silicon gate technology which minimizes the gate-drain overlap capacitance, alleviates the problem considerably.

Even though the "instantaneous input capacitance" $C_{in}(t)$ varies during switching, the average input capacitance, C_{IN}, defined as the total charge supplied to the gate by the input divided by the voltage swing, should be constant. For example, assuming the input voltage changes from 0 to V_H, the total amount of charge supplied to the gate is given by $C'_o WL (V_H - V_{T0})$, and the average input capacitance is equal to $C'_o WL (1 - V_{T0}/V_H)$, which is slightly less than the total thin oxide capacitance of the gate by area.

Figure 24(a) and (b) shows the outputs of an E/D inverter with a load $C = 0.25$ and 0.5. Figure 24(c) shows the variations of the gate capacitances and the total charges Q and Q' as functions of time. Notice that both Q and Q' converge to the same final value. For readers who are interested in the loading of the gate capacitance, Figure 24(d) shows the "instantaneous input capacitance" as a function of time. The reader is invited to interpret and compare these curves as an exercise.

Even though the variation of gate capacitances in CMOS circuits is more complicated, as shown in Figure 25, its average input capacitance, C_{IN}, is equal to $C'_o WL(T_A) \times [1 - V_{T0}(T_A)/V_H] + C'_o WL(T_L) \times [1 + V_{T0}(T_L)/V_H]$, which is a constant (assuming the voltage swing is equal to V_H). The corresponding waveforms are shown in Figures 25(b) and (c).

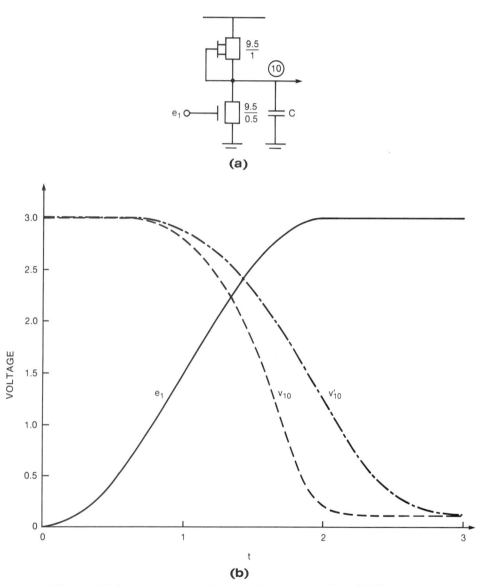

(a)

(b)

Figure 24(a) An inverter with depletion mode pullup. **(b)** The pull-down characteristics of the inverter with $C = 0.25$ and 0.5, respectively. Primed variables pertain to the case of $C = 0.5$. **(c)** The gate capacitances and total charges for $C = 0.25$ and 0.5. **(d)** The input gate capacitances $C_{in}(t)$.

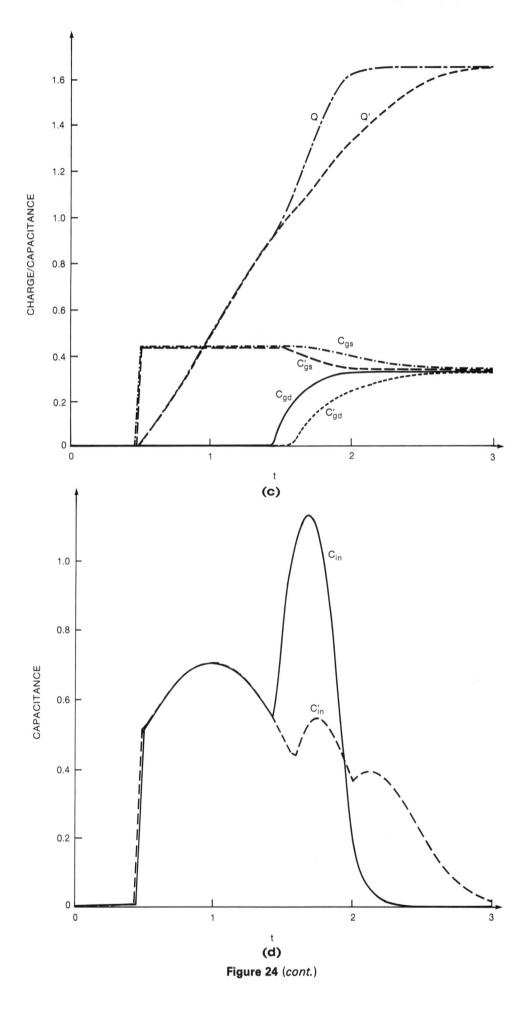

Figure 24 (*cont.*)

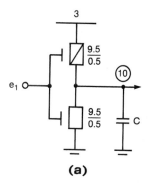

(a)

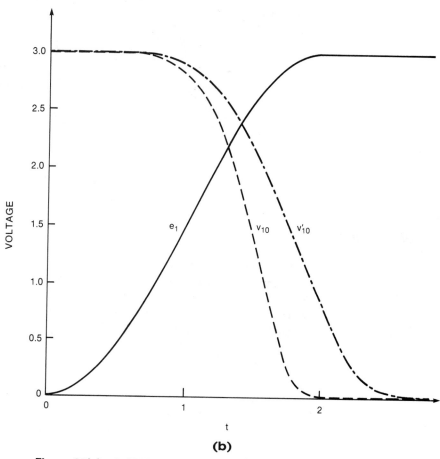

(b)

Figure 25(a) A CMOS inverter. **(b)** The pulldown characteristics of the inverter with $C = 0.25$ and 0.5, respectively. Primed variables pertain to the case of $C = 0.5$. **(c)** The total input gate capacitance to the source (C_s and C_s') and the drain (C_d and C_d') of the devices. Q and Q' are the total charges supplied to the input gate capacitances.

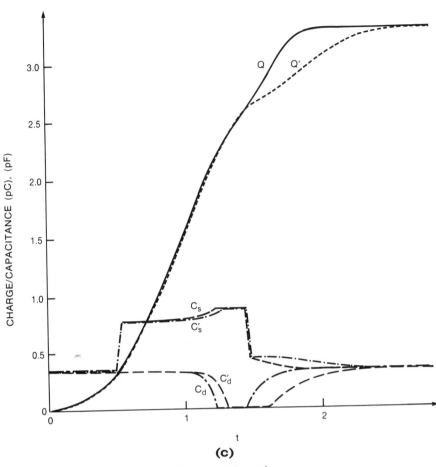

(c)

Figure 25 (*cont.*)

8. CONCLUSION

In this chapter we have discussed the operation of inverters in detail. The response of an inverter is determined by the rise/falltime of its input as well as its capacitive loads. The inverter, in turn, imposes a capacitive load to its driver. The input capacitance of an inverter, although a complicated function of voltage, can usually be estimated by the total gate capacitance by area.

FORTRAN programs of some of the delay calculators are listed in the appendices to this chapter.

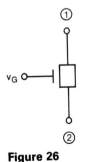

Figure 26

PROBLEMS

1. Consider the device in Figure 26. Show that

$$x_S - x_D = (1 + k_1)(v_1 - v_2)$$

when the device is in the linear region.

2. Show that at the transition time T_s the output

$$v_O = V_H - \frac{\sqrt{2X_D/\lambda}}{1 + k_1} \quad \text{and} \quad \xi_s = \xi(T_s) = \sqrt{2\lambda X_D}$$

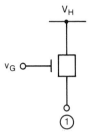

Figure 27

3. Consider the device in Figure 27. Show that

$$\eta_s = 1 - \frac{\xi_s}{(1 + k_1)\lambda V_H}$$

4. The gate capacitance can foster the bootstrap effect even without the extra bootstrap capacitor. Refer to Figure 6. What is the bootstrap efficiency of the circuit when $C_B = 0$?

5. Refer to Figure 9. Identify all possible parasitic capacitances comprising C_{p1} and C_{p2}.

6. Consider the bootstrap circuit in Figure 28.

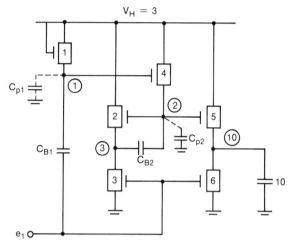

Figure 28

(a) Explain the operation of the circuit.

(b) Determine the device sizes for a $T_R = 2$ nS.

7. Calculate the BC and WC performances of the pullup in Figure 2(a) under the same conditions as in Example 10 except that $V_{T0} \in (0.45, 0.55)$. Which corner of the T_j–V_H rectangle gives the worst case performance?

8. Calculate the risetime $T_R(50\%)$ of the step responses for the circuits in Figures 2, 3, 5, and 8. Express each answer in the form of $T_R = RK \times C/\beta$, where RK is a constant.

9. Interpret and compare the curves in Figure 24(d) according to the different operating regions of the device.

APPENDICES

In the following programs all Greek letters are spelled out such as β = BETA, η = ETA and so on, except ξ is coded as TSAI instead of XI.

1. DELAY CALCULATOR FOR CONSTANT PULLUPS

```
      SUBROUTINE CPUN (PW, PL, DVTO, DLMBD, ALPHAP,
     *                 VG, VO,   CL, TDLH, TR)
C*********************************************************************
C*                                                                 *
C*                     PROGRAM CPUN                                *
C*                                                                 *
C*    CALCULATION OF THE RISETIME AND DELAY FOR CONSTANT PULLUP, NMOS  *
C*                                                                 *
C*********************************************************************
      IMPLICIT REAL*8 (A-H, O-Z)
```

```
         COMMON VH, GAMMAN, GAMMAP, PK1N, PK1P, VT0N, VT0P,
       *        PLMBDN, PLMBDP

C --- VH = POWER SUPPLY
C --- GAMMAN = GAMMA OF NMOS
C --- GAMMAP = GAMMA OF PMOS
C --- PK1N = SUBSTRATE SENSITIVITY OF NMOS
C --- PK1P = SUBSTRATE SENSITIVITY OF PMOS
C --- VT0N = THRESHOLD VOLTAGE OF ENHANCEMENT MODE NMOS
C --- VT0P = THRESHOLD VOLTAGE OF ENHANCEMENT MODE PMOS
C --- PLMBDN = LAMBDA OF NMOS
C --- PLMBDP = LAMBDA OF PMOS

C --- DVT0 = DELTA VT0, MODIFIER OF VT0
C --- DLMBD = DELTA LAMBDA, MODIFIER OF PLMBDN OR PLMBDP
C --- ALPHAP = ALPHA P, A FACTOR FOR PARASITIC CAPACITANCES
C --- VO  = INITIAL VOLTAGE LEVEL OF THE OUTPUT VOLTAGE
C --- VOF = FINAL VOLTAGE LEVEL OF THE OUTPUT VOLTAGE

         VT0 = VT0N + DVT0
         PLMBD = PLMBDN + DLMBD
         PK1 = PK1N
         C = (1.0 + ALPHAP)*CL
         BETA = GAMMAN*PW/PL

         ETA2 = .5

         VOF = (VG-VT0)/(1.0+PK1)
         IF (VOF .GT. VH)  VOF = VH

         WRITE(6,1)  VT0, PLMBD, PK1, C, BETA, ETA2, VOF
       1 FORMAT (1H1,/ 7(F10.5/))

         XD = VG - VT0 -(1.0+PK1)*VH

         XS0 = VG - VT0
         TSAI0 = DSQRT(1.0+2.0*PLMBD*XS0) - 1.0

         XS1 = VG - VT0 -(1.0+PK1)*VO
         TSAI1 = DSQRT(1.0+2.0*PLMBD*XS1) - 1.0

         WRITE(6,2)  XD, XS0, TSAI0, XS1, TSAI1
       2 FORMAT (/ 5(E15.8/))

         IF (XD .GT. 0.0) GO TO 100

C --- DEVICE OPERATES IN SATURATION REGION ONLY

      50 XS2 = VG - VT0 -(1.0+PK1)*(VH/2.0)
         TSAI2  = DSQRT(1.0+2.0*PLMBD*XS2) - 1.0

         TR1 = 2.0*PLMBD*(DLOG(TSAI0/TSAI1) + (TSAI0-TSAI1)/(TSAI0*TSAI1))
       *       *C/((1.0+PK1)*BETA)

         TR2 = 2.0*PLMBD*(DLOG(TSAI0/TSAI2) + (TSAI0-TSAI2)/(TSAI0*TSAI2))
       *       *C/((1.0+PK1)*BETA)

         TDLH = TR2 - TR1
         TR = 2.0*TR2
```

```
            RETURN

       100 Z = DSQRT(2.0*XD/PLMBD)/(1.0+PK1)
           PK = 1.0 + 2.0*XD/((1.0+PK1)*VH)

           IF (VH .GT. Z) GO TO 200

C --- DEVICE OPERATES IN LINEAR REGION ONLY

           ETA1 = VO/VH
           T1 = (DLOG((PK)*(1.0-ETA1)/(PK-ETA1))/XD
          *       +2.0*PLMBD*DLOG((PK-ETA1)/PK))*C/(BETA*(1.0+PK1))
           T2 = (DLOG((PK)*(1.0-ETA2)/(PK-ETA2))/XD
          *       +2.0*PLMBD*DLOG((PK-ETA2)/PK))*C/(BETA*(1.0+PK1))
           TDLH = T2 - T1
           TR = 2.0*T2

           RETURN

C --- DEVICE SWITCHES FROM SATURATION TO LINEAR REGION

       200 TSAIS = DSQRT(2.0*PLMBD*XD)
           ETAS = 1.0 - TSAIS/((1.0+PK1)*PLMBD*VH)

           WRITE(6,3)  TSAIS, ETAS
         3 FORMAT (/ 2(E15.8/))

           IF (ETAS .GE. ETA2)   GO TO 50

           TS1 = 2.0*PLMBD*(DLOG(TSAIO/TSAI1) + (TSAIO-TSAI1)/(TSAIO*TSAI1))
          *      *C/((1.0+PK1)*BETA)

           TS2 = 2.0*PLMBD*(DLOG(TSAIO/TSAIS) + (TSAIO-TSAIS)/(TSAIO*TSAIS))
          *      *C/((1.0+PK1)*BETA)

           TS = TS2 -TS1

           TL = (DLOG((PK-ETA2)*(1.0-ETAS)/((PK-ETAS)*(1.0-ETA2)))/XD
          *       +2.0*PLMBD*DLOG((PK-ETAS)/(PK-ETA2)))*C/(BETA*(1.0+PK1))

           TDLH = TS + TL
           TR = 2.0*(TS2+TL)

           WRITE(6,4)  TS1, TS2, TS, TL, TDLH, TR
         4 FORMAT (//// 6(E15.8/))
           RETURN
           END
```

2. DELAY CALCULATOR FOR DEPLETION MODE PULLUPS

```
       SUBROUTINE DPUN (PW, PL, DVT0, DLMBD, ALPHAP,
      *                    VO,  CL,  TDLH, TR)
C**************************************************************************
C*                                                                      *
C*                          PROGRAM DPUN                                *
C*                                                                      *
C*   CALCULATION OF THE RISETIME AND DELAY FOR DEPLETION PULLUP, NMOS   *
C*                                                                      *
C**************************************************************************
       IMPLICIT REAL*8(A-H,O-Z)
```

```
          COMMON VH, GAMMAN, GAMMAP,  PK1N, PK1P,  VT0N, VT0P,
     *         PLMBDN, PLMBDP
C --- VH = POWER SUPPLY
C --- GAMMAN = GAMMA OF NMOS
C --- GAMMAP = GAMMA OF PMOS
C --- PK1N = SUBSTRATE SENSITIVITY OF NMOS
C --- PK1P = SUBSTRATE SENSITIVITY OF PMOS
C --- VT0N = THRESHOLD VOLTAGE OF ENHANCEMENT MODE NMOS

C --- VT0P = THRESHOLD VOLTAGE OF ENHANCEMENT MODE PMOS
C --- PLMBDN = LAMBDA OF NMOS
C --- PLMBDP = LAMBDA OF PMOS

C --- DVT0 = DELTA VT0, MODIFIER OF VT0
C --- DLMBD = DELTA LAMBDA, MODIFIER OF PLMBDN OR PLMBDP
C --- ALPHAP = ALPHA P, A FACTOR FOR PARASITIC CAPACITANCES

C --- VO = INITIAL OUTPUT VOLTAGE VO(0)
C --- VB0 = PRECHARGE OF THE BOOTSTRAP CAPACITOR
C --- CB = BOOTSTRAP CAPACITOR
C --- CP = PARASITIC CAPACITOR

      ETA = .9

      VT0 = VT0N + DVT0
      PLMBD = PLMBDN + DLMBD
      PK1 = PK1N
      C = (1.0 + ALPHAP)*CL
      BETA = GAMMAN*PW/PL

      XS0 = - VT0 - PK1*VO
      TSAI0 = DSQRT(1.0+2.0*PLMBD*XS0) - 1.0

      WRITE(6,1)  ETA, VT0, PLMBD, PK1, C, BETA, XS0, TSAI0
    1 FORMAT (1H1, / 8(F10.5/))

      X = - VT0 -  PK1*VH
      Z1 = 1.0/(1.0+PK1)

      TSAIS = - Z1 + DSQRT( Z1**2 + 2.0*PLMBD*X )

      WRITE(6,2)  X, Z1, TSAIS
    2 FORMAT (1H1, / 3(F10.5/))

      IF (TSAI0 .GT. TSAIS) GO TO 100

C --- DEVICE OPERATES IN LINEAR REGION ONLY

      ETAS = VO/VH
      TS = 0.D0
      GO TO 105

C --- DEVICE SWITCHES FROM SATURATION REGION TO LINEAR REGION

  100 ETAS = 1.0 - TSAIS/((1.0+PK1)*PLMBD*VH)
      PK = 1.0 - 2.0*X/((1.0-PK1)*VH)

      WRITE(6,3)  ETAS, PK
    3 FORMAT (1H1, / 2(F10.5/))
```

```
      TS = 2.0*PLMBD*(DLOG(TSAI0/TSAIS)+(TSAI0-TSAIS)/(TSAI0*TSAIS))
     *      *C/(PK1*BETA)
105 TL = (DLOG((ETA-PK)*(1.0-ETAS)/((1.0-ETA)*(ETAS-PK)))/((1.0+PK1)
     *      *X)
     *      +2.0*PLMBD*DLOG((ETA-PK)/(ETAS-PK))/(1.0-PK1))
     *      *C/BETA
    WRITE(6,4)  TS, TL
  4 FORMAT (1H1,/ 2(F10.5/))

    TR = TS + TL
    TDLH = TR/2.0

    RETURN
    END
```

RECEIVERS
AND DRIVERS

INTRODUCTION

One special indispensible class of circuits in any application is the receiver and driver. A receiver accepts signals off chip, usually in a different technology such as TTL or ECL, and converts them to internal MOS logic levels (i.e., *GD* and V_H). An offchip receiver, commonly referred to as an OCR, must be able to respond to the input signal with minimum delay while at the same time maintaining a reasonable noise margin. An offchip driver, or OCD, on the other hand, converts internal MOS levels into logic levels of another technology and sends the signal out with adequate power. Unlike receivers, however, drivers are not limited to chip input/output (I/O) applications. Internal drivers of different power levels, from the small output buffer of a logic gate to the huge driver driving a clock line across an entire chip, can be found in the same chip in many applications.

In this chapter we shall concentrate on receiver and driver design. Section 1 is dedicated to receivers. Both differential and single-ended receivers will be considered. In Section 2 the basic principles of driver design are discussed. Many circuit techniques covered in previous chapters, such as bootstrap and pushpull circuits, will be applied to the design of these drivers. Section 3 considers OCDs. As chips become denser and denser, MOS line drivers become more and more important. A detailed discussion of the problems associated with MOS line drivers is thus presented in Section 4. Section 5 discusses the system noises relating to an often neglected area: chip packaging.

1. OFFCHIP RECEIVERS

1.1 Differential Receivers

In many applications signals must travel long distances. Such signals are subject to heavy noise effects. Since most noise appears as a common mode disturbance, these signals are often transmitted in differential form.

A differential signal requires two wires. The polarity of the voltage across these wires carries the digital information 0 or 1. Because of excellent noise immunity, the amplitudes of differential signals are usually much smaller than their single-ended counterparts. As a result, circuits that carry differential signals also switch much faster. The major drawback of differential signals, however, is their cost. The extra wire for each signal doubles the interconnections at all levels, from chips and modules to cards and boards. Therefore, in addition to machine environment considerations the choice between differential and single-ended signals also depends on the availability of I/O pins and the cost of packaging.

1.1.1 Static Differential Receivers. A typical static receiver consisting of a differential amplifier and an output buffer is shown in Figure 1. The reference voltage V_{REF} biases device T_0 in the saturation region so that the current I_0 remains constant, independent of v_0. Devices (T_1, T_2) and (T_3, T_4) are identical pairs and are laid out as symmetrically as possible.

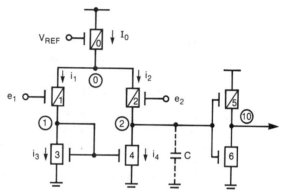

Figure 1. A typical differential receiver. Capacitor C represents all capacitive loading at node ②. Device parameters: $(\gamma, k_1, V_{T0}, \lambda) =$ (0.082, 0.15, 0.5, 0) for n-channel devices, $(\gamma', k_1', V_{T0}', \lambda') = (0.032, 0.20, -0.5, 0)$ for p-channel devices.

Two input voltages, e_1 and e_2 are applied to the differential amplifier. When dealing with differential signals it is convenient to divide inputs into two components: the common-mode input, e_C, and the differential-mode input, Δe, as follows:

$$e_C = \frac{e_1 + e_2}{2} \quad \text{and} \quad \Delta e = e_1 - e_2$$

so that

$$e_1 = e_C + \frac{\Delta e}{2} \quad \text{and} \quad e_2 = e_C - \frac{\Delta e}{2}$$

Thus e_1 and e_2 vary around e_C with equal amplitude. Since nodes ① and ② are the device drains, the common-mode input e_C has no effect on v_1 and v_2 so long as the devices are all in saturation. The symmetry of the circuit then forces $i_1 = i_2 = I_0/2$, which maintains $v_1 = v_2 = $ constant as e_C varies. The differential-mode input Δe, on the other hand, changes i_1 and i_2. While v_1 varies only gradually with i_1 (why?), v_2 is much more responsive to i_2 due to the large output resistances of T_2 and T_4. The gain of the amplifier is designed in such a way so v_2 switches with sufficient amplitude to toggle the inverter buffer (T_5, T_6), which, in turn, drives the output v_{10} between GD and V_H.

For digital applications we shall analyze the circuit from a nonlinear large-signal point of view. Since the analysis is quite involved, simplified square laws with $\lambda = 0$ for device currents will be used. More accurate results can always be obtained through circuit simulation with more elaborate models.

Let us first assume $e_1 = e_2 = e_C$ so that $\Delta e = 0$. The differential amplifier is in a perfectly balanced condition with $i_1 = i_2 = I_0/2$. Equating $i_1 = I_0/2$, we obtain

$$\tfrac{1}{2}\beta_1(e_C - v_0 - V_{T1})^2 = \tfrac{1}{2}I_0$$

$\Rightarrow$
$$v_0 = e_C - V_{T1} + \sqrt{I_0/\beta_1}$$

$$\triangleq V_0$$

Approximating $V_{T1} = V'_{T0} - k'_1(V_H - V_0)$,

$$V_0 = \frac{1}{1 + k'_1}(e_C - V'_{T0} + k'_1 V_H + \sqrt{I_0/\beta_1}) \tag{1}$$

Node voltage V_0 should be low enough to allow device T_0 to remain in saturation so that I_0 is constant for the specified range of input voltages.

Now assume $e_1 \neq e_2$ so that $\Delta e \neq 0$. Let Δe_m be the amplitude of Δe. We will discuss the DC responses of the circuit according to the sign of Δe:

(1) $\Delta e = \Delta e_m > 0$: Since $e_1 > e_2$, device current i_2 tends to increase while i_1 tends to decrease. Node voltage v_2 rises and T_4 remains in saturation. Since T_3 and T_4 have a common gate voltage, however, $i_3 = i_4$, which, in turn, forces $i_1 = i_2$. Consequently both v_0 and v_2 go up; T_2 moves into the linear region, whereas T_1 remains in the saturation region. Now, since $i_1 = I_0/2$

$$v_0 = V_0 + \frac{\Delta e_m}{2} \tag{2}$$

where in Equation (2) we assumed that $\Delta e_m/2 \ll V_0$ so that the change of V_{T1} is negligible. Now, consider the currents of T_1 and T_2:

$$\frac{1}{2}\beta_1\left(e_C + \frac{\Delta e_m}{2} - v_0 - V_{T1}\right)^2 \tag{3}$$

$$= \beta_2(v_2 - v_0)\left(e_C - \frac{\Delta e_m}{2} - v_0 - V_{T2} - \frac{v_2 - v_0}{2}\right) \tag{4}$$

$$= \frac{I_0}{2}$$

Substituting Equation (3) into Equation (4) with the assumption that $\beta_2 = \beta_1$ and $V_{T2} = V_{T1}$, then solving for $(v_2 - v_0)$ and eliminating v_0 by Equation (2), we obtain the uplevel of v_2:

$$v_2 = \frac{1}{1 + k'_1}(e_C - V'_{T0} + k'_1 V_H) - \frac{k'_1}{1 + k'_1}\sqrt{I_0/\beta_1} - \frac{\Delta e_m}{2}$$

$$+ \sqrt{(\sqrt{I_0/\beta_1} + \Delta e_m)^2 - I_0/\beta_1}$$

$$\triangleq \bar{V}_2 \tag{5}$$

(2) $\Delta e = -\Delta e_m < 0$: In this case $i_1 > i_2$ and v_2 falls. Device T_4 enters the linear region while T_1 and T_2 are both in the saturation region. The device currents are

$$i_1 = \frac{1}{2}\beta_1\left(e_C - \frac{\Delta e_m}{2} - v_0 - V_{T1}\right)^2$$

$$= \frac{1}{2}\beta_1\left(v_C - \frac{\Delta e_m}{2}\right)^2 \tag{6}$$

where $v_C \triangleq e_C - v_0 - V_{T1}$ and

$$i_2 = \frac{1}{2}\beta_1\left(v_C + \frac{\Delta e_m}{2}\right)^2 \tag{7}$$

Since

$$i_1 + i_2 = I_0 \tag{8}$$

adding Equations (6) and (7) yields

$$v_C = -\sqrt{I_0/\beta_1 - (\Delta e_m/2)^2} \tag{9}$$

which is a constant. On the other hand, subtracting Equation (7) from Equation (6) gives

$$i_1 - i_2 = -\beta_1 v_C \Delta e_m \tag{10}$$

From Equations (8) and (10) we obtain

$$i_1 = \frac{1}{2}(I_0 - \beta_1 v_C \Delta e_m) \tag{11}$$

and

$$i_2 = \frac{1}{2}(I_0 + \beta_1 v_C \Delta e_m) \tag{12}$$

where v_C is defined in Equation (9).

To solve for the downlevel of v_2, note that

$$i_1 = \frac{1}{2}\beta_3(v_1 - V_{T3})^2 \tag{13}$$

and

$$i_2 = \beta_4 v_2\left(v_1 - V_{T4} - \frac{v_2}{2}\right)$$

With $\beta_4 = \beta_3$ and $V_{T4} = V_{T3}$, substituting i_1 into i_2 yields

$$\beta_3 v_2\left(\sqrt{2i_1/\beta_3} - \frac{v_2}{2}\right) = i_2$$

Solving for v_2, the downlevel of v_2 is

$$v_2 = \sqrt{2/\beta_3}\,(\sqrt{i_1} - \sqrt{i_1 - i_2}) \tag{14}$$

$$\triangleq \underline{V}_2.$$

In the above analysis we have assumed that the differential input $\Delta e_m < 2\sqrt{I_0/\beta_1}$ so that Equation (9) is valid and both i_1 and i_2 are nonzero. In case Δe_m is large device T_2 may be cut off and $i_2 = 0$. $\underline{V}_2$ will then fall to GD.

The following example demonstrates the techniques just presented.

EXAMPLE 1

For the differential receiver shown in Figure 1, let the inputs e_1 and e_2 vary from 0.9 to 1.1. Assume the inverter (T_5, T_6) switches from $\underline{V}_2 = 0.05$ to $\bar{V}_2 = 2.0$. Determine the sizes of T_1 and T_3 for $I_0 = 0.1$.

Solution

The common-mode and differential-mode inputs are

$$e_C = \frac{0.9 + 1.1}{2} = 1.0$$

and

$$\Delta e_m = 1.1 - 0.9 = 0.2$$

Substituting $\bar{V}_2 = 2.0$ into Equation (5), we obtain

$$\frac{1}{1 + 0.2}(1.0 + 0.5 + 0.2 \times 3) - \frac{0.2}{1 + 0.2}\sqrt{I_0/\beta_1} - \frac{0.2}{2}$$
$$+ \sqrt{(\sqrt{I_0/\beta_1} + 0.2)^2 - I_0/\beta_1} = 2.0$$

Solving for $\sqrt{I_0/\beta_1}$, we have

$$\sqrt{I_0/\beta_1} = 0.3 \quad \Rightarrow \quad \frac{I_0}{\beta_1} = 0.09$$

Since $I_0 = 0.1$, $\beta_1 = 0.1/0.09 = 1.11$.

The size of T_1 is therefore

$$\frac{W}{L}(T_1) = \frac{1.11}{0.032} = 34.7$$

For T_3, consider $\underline{V}_2$. From Equation (9),

$$v_C = -\sqrt{0.09 - (0.2/2)^2} = -0.283$$

From Equations (11) and (12)

$$i_1 = \frac{0.1 + 1.11 \times 0.283 \times 0.2}{2} = 0.081$$

$$i_2 = \frac{0.1 - 1.11 \times 0.283 \times 0.2}{2} = 0.019$$

Substituting i_1 and i_2 into Equation (14) yields

$$\sqrt{2/\beta_3}\,(\sqrt{0.081} - \sqrt{0.081 - 0.019}) = 0.05$$

$$\Rightarrow \qquad \beta_3 = 1.014$$

hence the size of T_3:

$$\frac{W}{L}(T_3) = \frac{1.014}{0.082} = 12.4$$

In this example T_2 always conducts current because $\Delta e_m = 0.2$ is quite small. In case Δe_m is so large that T_2 cuts off the size of T_3 and T_4 is determined by the DC bias of v_1 which should be close to the transition voltage V_{tr} of inverter (T_5, T_6).

It is interesting to note that while v_2 varies from 0.05 to 2.0 as the inputs switch, v_1 changes only slightly. From Equation (13) the total DC change of v_1 is given by

$$\Delta v_1 = (V_{T3} + \sqrt{2 \times 0.081/1.014}) - (V_{T3} + \sqrt{2 \times 0.05/1.014})$$
$$= 0.9 - 0.81 = 0.09$$

It is also instructive to analyze the transient response of the differential amplifier. For simplicity the loading effect of the inverter (T_5, T_6) is modeled as a constant capacitance C.

Assume that $e_1 < e_2$ for $t < 0$. Node voltage $v_2 = \underline{V_2}$ and $v_{10} = V_H$. Let e_1 and e_2 switch at $t = 0$ so that $\Delta e = \Delta e_m > 0$. Since $e_2 < e_1$, $i_2 > i_1$ and v_2 starts to move up. There are three regions v_2 travels through before it finally reaches $\overline{V_2}$:

(1) $\underline{V_2} < v_2 < v_1 - V_{T4}$

In this case T_4 is in the linear region while T_1, T_2, and T_3 are all in saturation.

(2) $v_1 - V_{T4} < v_2 < e_2 - V_{T2}$

In this case all devices are in the saturation region.

(3) $e_2 - V_{T2} < v_2$

In this case T_2 is in the linear region while all others are in saturation.

The detailed calculation of the transient response of v_2 is quite involved. Fortunately the expression for the current charging capacitor C is particularly simple in case (2) in which the inverter (T_5, T_6) switches. It is left to the reader to show that in this case

$$C \frac{d}{dt} v_2 = i_2 - i_4 = i_2 - i_1$$

$$= -\beta_1 v_C \Delta e_m$$

where v_C is defined in Equation (9). Integrating the above equation from $(v_1 - V_{T4})$ to $(e_2 - V_{T2})$, we obtain the time duration Δt of case (2):

$$\Delta t = -\frac{C}{\beta_1 v_C \Delta e_m}(e_2 - v_1 - V_{T2} + V_{T4}) \tag{15}$$

where v_1 can be calculated as in Example 1. Similar expressions can be obtained for the falltime of v_2. In general the falltime of v_2 is much shorter than its risetime.

EXERCISE

Express Δt as a function of C for Example 1. Is Δt a reasonable representation of the circuit delay? Why?

In summary, while the downlevel V_2 is fairly low with a reasonable β_3, the uplevel $\overline{V}_2$ is a function of the ratio I_0/β_1. The speed of the amplifier is proportional to the product $\beta_1 v_C$. In case $\Delta e_m \ll \sqrt{I_0/\beta_1}$ so that $v_C \cong - \sqrt{I_0/\beta_1}$ as in some applications, the speed of the amplifier is roughly proportional to $\sqrt{I_0 \beta_1}$.

1.1.2 Dynamic Differential Receivers. The static receiver considered so far draws a constant current I_0 independent of the inputs. Such receivers consume DC power. Static receivers are very popular in logic applications where the inputs change asynchronously. When the inputs are "sampled" with a fixed clocking scheme, the dynamic latch shown in Figure 2 is widely used. This circuit does not consume DC power and is found in many dynamic RAM applications.

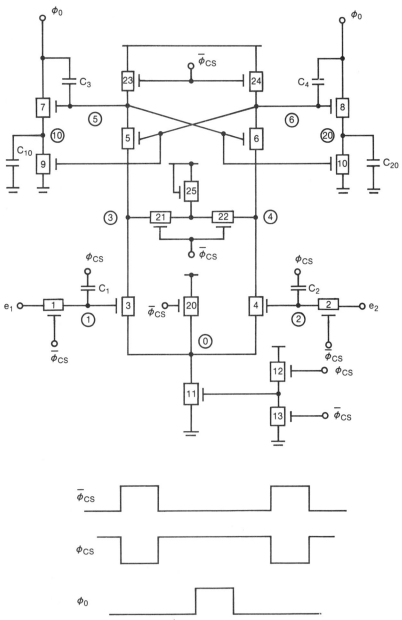

Figure 2. A dynamic differential receiver with typical timing.

Three clock pulses are needed to run the circuit; ϕ_{CS} and $\overline{\phi}_{CS}$ are complementary, nonoverlapping pulses standing for *chip select* and *chip select not*. The third pulse, ϕ_0, drives the outputs.

Before the chip is selected, ϕ_{CS} is low and $\overline{\phi}_{CS}$ is high. Devices ($T_{20}, \ldots, T_{25}$) have precharged nodes ⓪, ③, ④, ⑤, and ⑥ to ($V_H - V_T$). Since T_1 and T_2 are on, $v_1 = e_1$ and $v_2 = e_2$. As soon as the chip is selected ϕ_{CS} goes up and $\overline{\phi}_{CS}$ goes down. Device T_{11} turns on, discharging node ⓪ to ground. The capacitive coupling of C_1 and C_2 pushes v_1 and v_2 up and T_3 and T_4 turn on hard as v_0 falls. For purposes of discussion assume that initially $v_1 > v_2$. Device T_3 will turn on harder than T_4, and v_3 and v_5 will fall much faster than v_4 and v_6. As a result, device T_6 turns off quickly, trapping charges on node ⑥. If the latch operates properly, v_6 falls only slightly from its precharged value as v_5 and v_3 fall to the ground. Consequently capacitor C_4 is only slightly disturbed while C_3 is fully discharged to ground. As ϕ_0 goes up the gate of T_8 is boosted very high and the bootstrap action of T_8 will charge C_{20} all the way to V_H. The voltage at node ⑩, however, remains low because T_7 is off and T_9 is on. Thus at the end of ϕ_0, node voltage v_{20} has gone up while v_{10} remains low. In case $v_1 < v_2$ initially, the polarity of the outputs reverses.

Interesting applications of such receivers will be discussed in the next chapter as input T/C generators of dynamic RAMs.

1.2 Single-Ended Receivers

If the input to the receiver is single ended, the circuit described earlier can be easily turned into a single-ended receiver by applying a constant voltage V_{REF} to one of its inputs. The level of V_{REF} is chosen in the middle of the input voltage swing so the receiver can switch according to the polarity of the input. Such circuits are widely used in dynamic RAMs.

Since the operation of dynamic circuits requires elaborate clocking, the most popular type of single-ended receiver is the Schmitt trigger. A Schmitt trigger is basically a voltage level detector with some hysteresis. The hysteresis offers better noise margin and more stable operation as we will see shortly.

A simple but also very popular Schmitt trigger circuit is shown in Figure 3. The input e_I is applied to the composite pulldown (T_2, T_3) of an otherwise simple inverter: (T_1, (T_2, T_3)). The output of the inverter, v_2, is fed back to the input at node ① via the source follower T_4. We shall now show that due to this positive feedback the TC plot of the Schmitt trigger exhibits a hysteresis loop.

Let the input e_I start at 0. With e_I low, devices T_2 and T_3 are off; v_2 is at V_H and

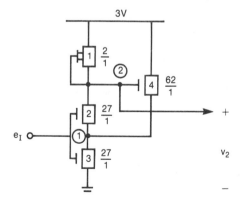

Figure 3. An NMOS Schmitt trigger. Device parameters: $(\gamma, k_1, V_{T0}, \lambda) = (0.082, 0.08, 0.5, 0.5)$ for enhancement mode devices, $(\gamma', k_1', V_{T0}', \lambda') = (0.082, 0.08, -1.3, 0.5)$ for depletion mode devices.

v_1 is at $(V_H - V_{T4})$. Now let e_I go up. Both v_1 and v_2 remain constant until e_I reaches V_{T3} when T_3 turns on, and v_1 starts to fall according to the β ratio between T_3 and T_4. As e_I increases and v_1 decreases the gatedrive of T_2, $(e_I - v_1)$, increases. Eventually $(e_I - v_1)$ reaches V_{T2}, turning on T_2. The turnon of T_2 discharges node ②, which, in turn, helps discharge node ①, and increases $(e_I - v_1)$. Consequently, the output v_2 falls sharply because of the positive feedback and settles to a specified low level V_{2L}.

The input voltage $e_I = E_H$ when the regenerative action begins is called *the upper trigger (trip) point (UTP)*; thus, $E_H = v_1 + V_{T2}$. Further increase of e_I does not have any significant effect on v_2. The waveforms of a typical HL transition are shown in Figure 4(a).

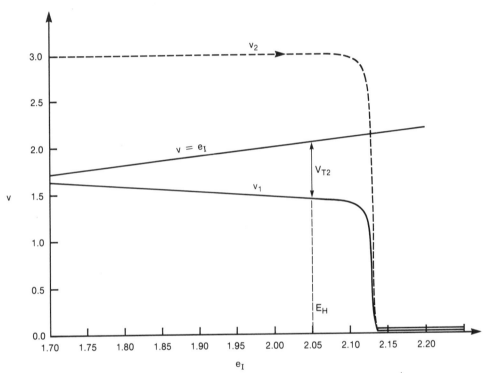

Figure 4(a). Definition of E_H. Data from Example 2 show that v_2 falls completely for $e_I > E_H + 0.15$.

Now, let e_I decrease from V_H. With e_I at V_H, $v_1 < v_2 < V_{2L}$ are both very low. As e_I decreases v_1 and v_2 start to rise according to the β ratios among T_1, T_2, and T_3. Considering T_1 and (T_2, T_3) as an inverter, e_I eventually reaches the V_{ts} of its TC curve and v_2 rises sharply, turning on T_4. As soon as T_4 is on, v_1 rises. This reduces the gatedrive of T_2, allowing v_2 to go up even further. Eventually the output v_2 reaches V_H due to the positive feedback.

The input voltage $e_I = E_L$ when the regenerative action begins is called *the lower trigger (trip) point (LTP)*; thus, $v_2 - v_1 = V_{T4}$ when $e_I = E_L$. Further decrease of e_I does not have significant effect on v_2. The waveforms of a typical LH transition are shown in Figure 4(b).

The trigger points are independently controlled by device sizes. If E_H is made higher than E_L, an hysteresis loop is created in the TC plot of the circuit. Let us now illustrate the design by an example.

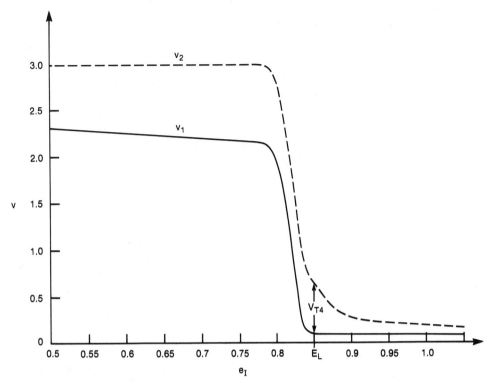

Figure 4(b). Definition of E_L. Data from Example 2 show that v_2 rises to V_H completely for $e_I < E_L - 0.1$.

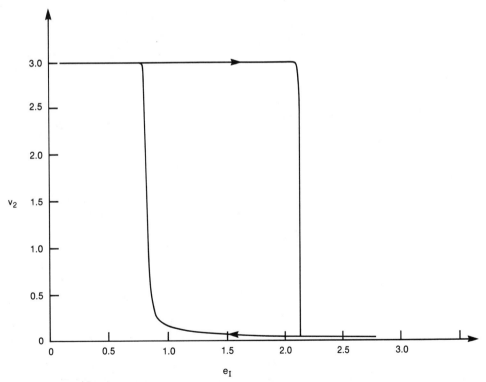

Figure 4(c). The hysteresis loop of the Schmitt trigger in Example 2.

EXAMPLE 2

Consider the Schmitt trigger shown in Figure 3. Assume that from the power and performance requirements the size of T_1 is $W/L(T_1) = 2/1$. Estimate the sizes of T_2, T_3, and T_4 so that $E_L = 1$ V and $E_H = 2$ V.

Solution

At $e_I = E_L$ the inverter $(T_1, (T_2, T_3))$ is in transition, where (T_2, T_3) is a composite pulldown. Let the β ratio of the inverter be β_r. From Problem 4 of Chapter 3 the transition point V_{ts} can be approximated by

$$V_{ts} = V_{T0} - \frac{V'_{T0}}{\sqrt{\beta_r}}$$

$$\Rightarrow \qquad \beta_r = \left(\frac{-V'_{T0}}{V_{ts} - V_{T0}}\right)^2$$

Now let $V_{ts} = E_L = 1$, $V_{T0} = 0.5$ and $V'_{T0} = -1.3$, we obtain

$$\beta_r = \left(\frac{-1.3}{1 - 0.5}\right)^2 = 6.76$$

Letting $\beta_2 = \beta_3$ and neglecting the substrate sensitivity we can approximate

$$\frac{\beta_2}{\beta_1} = \frac{\beta_3}{\beta_1} = 2 \times \beta_r = 13.5$$

Thus,

$$\frac{W}{L}(T_2) = \frac{W}{L}(T_3) = 13.5 \times \frac{2}{1} = \frac{27}{1}$$

The size of T_4, on the other hand, is determined by the inverter (T_4, T_3) and E_H. When $e_I = E_H$, device T_2 turns on, and we have

$$E_H - v_1 = V_{T2} = V_{T4}$$

Since at this point it is still true that $I_3 = I_4$, we have

$$\frac{\beta_4}{2}(V_H - V_{T4} - v_1)^2 = \frac{\beta_3}{2}(E_H - V_{T0})^2$$

Combining the above equations, we obtain

$$\frac{\beta_3}{\beta_4} = \left(\frac{V_H - E_H}{E_H - V_{T0}}\right)^2$$

With $E_H = 2$, $V_H = 3$, and $V_{T0} = 0.5$, we obtain

$$\frac{\beta_3}{\beta_4} = \left(\frac{3 - 2}{2 - 0.5}\right)^2 = 0.44$$

Hence

$$\frac{W}{L}(T_4) = \frac{W}{L}(T_3) \times \frac{1}{0.44} = \frac{62}{1}$$

The TC plot of the circuit is shown in Figure 4(c). The trigger points are close to, but not quite at, the specified values. The calculation with these simple equations, however, serves well in a first-cut design. Fine-tuning of the circuit is left to the reader as an exercise.

A CMOS Schmitt trigger that uses p-channel devices for the pullup is shown in Figure 5. As in the previous example the UTP is determined by the ratio β_4/β_3. The LTP, however, is determined by the ratio β_7/β_6.

Another very popular CMOS Schmitt trigger is shown in Figure 6(a). The UTP is determined by the transition of the composite inverter $(T_2, (T_1, T_3))$ where (T_1, T_3) are in parallel. The LTP is likewise determined by inverter $((T_2, T_4), T_1)$ where (T_2, T_4) are in parallel.

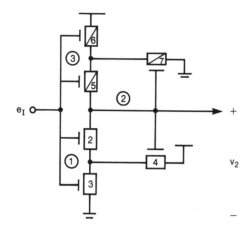

Figure 5.   A CMOS Schmitt trigger.

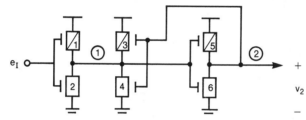

Figure 6(a).   A CMOS Schmitt trigger.

EXAMPLE 3

Determine the β ratios for the circuit in Figure 6(a) for $E_H = 2$ V and $E_L = 1$ V.

Solution

Consider the composite inverter $(T_2, (T_1, T_3))$. At $e_I = E_H = 2$, v_1 starts to fall sharply. First, calculate the device currents at E_H. For T_1,

$$i_1 = -\frac{1}{2}\beta_1 \frac{x_P^2 - x_S^2}{1 + \lambda(x_P - x_S)}$$

where

$$x_S = E_H - V_H - V_{T0}' = 2 - 3 + 0.5 = -0.5$$

$$x_P = x_S - \frac{1 - \sqrt{1 - 2\lambda x_S}}{\lambda}$$

$$= -0.5 - \frac{1 - \sqrt{1 + 2 \times 0.5 \times 0.5}}{0.5} = -0.05$$

Substituting x_S and x_P into i_1 we obtain

$$i_1 = 0.10\beta_1$$

For T_2, we have

$$i_2 = -\frac{1}{2}\beta_2 \frac{x_S^2 - x_P^2}{1 + \lambda(x_S - x_P)}$$

where

$$x_S = E_H - V_{T0} = 2 - 0.5 = 1.5$$

$$x_P = x_S + \frac{1 - \sqrt{1 + 2\lambda x_S}}{\lambda}$$

$$= 1.5 + \frac{1 - \sqrt{1 + 2 \times 0.5 \times 1.5}}{0.5} = 0.34$$

Therefore,

$$i_2 = 0.68\beta_2$$

Finally, for T_3,

$$i_3 = -\frac{1}{2}\beta_3 \frac{x_D^2 - x_S^2}{1 + \lambda(x_D - x_S)}$$

where

$$x_S = v_2 - V_H - V'_{T0} = 0 - 3 + 0.5 = -2.5$$

$$x_D = v_2 - V'_{T0} + k'_1 V_H - (1 + k'_1)v_1 = -0.775$$

and

$$i_3 = 1.52\beta_3$$

Since $i_1 + i_3 = i_2$, we have

$$0.10\beta_1 - 0.68\beta_2 = -1.52\beta_3$$

For $e_I = E_L$ the device currents can be calculated in the same fashion. However, since $|E_H - V_H| = (E_L - 0)$, due to this symmetry of the specification and the fact that $i_2 + i_4 = i_1$ we conclude that at $e_I = E_L$:

$$-0.68\beta_1 + 0.10\beta_2 = -1.52\beta_4$$

If we let $\beta_3 = \beta_4$ then $\beta_1 = \beta_2$. The above equation yields

$$\beta_1 = 2.62\beta_3$$

which is the required β ratio. For example, let $W/L(T_4) = 4/1$. Since $\beta_3 = \beta_4$, we have

$$\frac{W}{L}(T_3) = \frac{0.082}{0.032} \times \frac{4}{1} \times \frac{0.5}{0.5} = \frac{5}{0.5}$$

The device sizes of T_2 and T_1 are then given by

$$\frac{W}{L}(T_2) = 2.62 \times \frac{4}{1} = \frac{10.5}{1.0}$$

and

$$\frac{W}{L}(T_1) = 2.62 \times \frac{5}{0.5} = \frac{13}{0.5}$$

The TC plot of this circuit is shown in Figure 6(b).

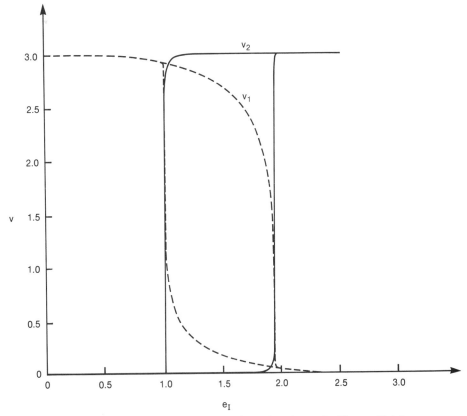

Figure 6(b). The TC plot of the Schmitt trigger in Figure 6(a) for Example 3.

In all these examples our first-order hand calculations of the device sizes served well as a design starting point. Since the incoming logic levels do not necessarily track with the power supply and temperature of the receiver, statistical runs are generally required to determine the BC and WC TCs of the circuit to ensure that both the UTP and LTP are within the specified tolerance around their nominal values. Such tolerances define the noise margins of the receiver.

1.3 Noise Margins of a Receiver

The noise margins of a receiver are defined as the maximum amount of noise on top of the minimum logic levels (i.e., LPUL and MPDL) of the receiver input before it switches to the wrong state. Since the receiver responds to the input with finite speed, both the amplitude and the duration of the noise must be considered when analyzing the noise margins. A noise pulse with a large amplitude and a short pulsewidth may not be able to affect the circuit whereas a noise pulse with a lower amplitude and a long lifespan may "eventually" toggle the receiver.

The receiver's response to noise is best summarized by its *noise tolerance curves*. A pair of typical noise tolerance curves are shown in Figure 7. The maximum noise pulse amplitudes are plotted as functions of noise pulsewidth. Two curves are generated in the figure, one for each logic state. Each point on the curve corresponds to a noise condition beyond which the receiver switches to the wrong state. Again, since the receiver is subjected to external as well as internal noises, these curves are generated by circuit simulation with statistical runs.

As shown in the figure, the maximum noise amplitude varies the most for short pulsewidths. As the pulsewidth increases the receiver switches at a lower noise level. The curves level off completely as the noise pulsewidth becomes longer than the internal delays of the receiver; beyond these points the maximum magnitude of the noise margin is simply the DC margin that triggers the receiver. For a Schmitt trigger, for example, these margins are (LPUL − E_L) and (E_H − MPDL) for the

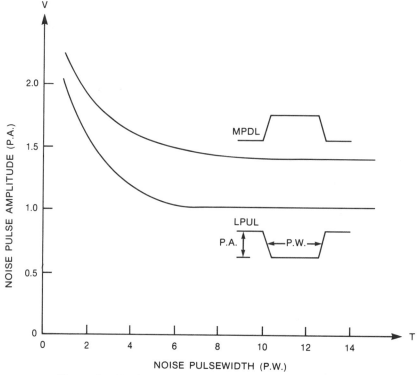

Figure 7. Typical noise tolerance curves of a receiver.

two logic states. For a differential receiver without hysteresis the DC noise margins are simply (LPUL − MPDL) for both logic states.

Noises come from many sources. The ground shift and power supply drop are notable DC noises. Reflections and crosstalks due to electric and magnetic coupling are noise pulses with finite pulsewidths (see Section 4). Since most of these noises come offchip, noise analysis is one of the major responsibilities of module (chip packaging) and card designers. Since the faster the receiver the smaller the noise margins in a given environment, a tradeoff between receiver speed and noise tolerance must be carefully assessed for a particular application.

2. HIGH-POWER DRIVERS

2.1 The Differential Drivers

Differential drivers, like differential receivers, offer better noise margins with smaller voltage swings. A typical differential driver is shown in Figure 8. The basic circuit of the driver is a low-gain, high-speed differential amplifier consisting of devices (T_0, T_1, T_2) and resistors (R_1, R_2). The differential inputs come from the DC bias circuit. The polarity of the differential input is, in turn, controlled by the input pulse e_I and its complement $\bar{e}_I$ via the pass devices (T_7, T_8) and (T_9, T_{10}). Amplifier outputs are further buffered by the large source followers (T_3, T_5) and (T_4, T_6), which then drive the external loads with high power.

To reduce transient disturbance to the bias circuit during switching, low resistance paths are provided to nodes Ⓗ, and Ⓛ, through T_{11} and T_{12}, respectively. To generate true differential waveforms that cross each other at the 50% point, it is important to be able to maintain equal output resistances for the pullup and pulldown at the driver's outputs. This can be accomplished by adjusting the sizes and bias points of T_3 (T_4) and T_5 (T_6).

In practice only one input, e_I, is available, and $\bar{e}_I$ is generated by an inverter associated with the driver. The symmetry between (T_7, T_9) and (T_8, T_{10}) reduces the

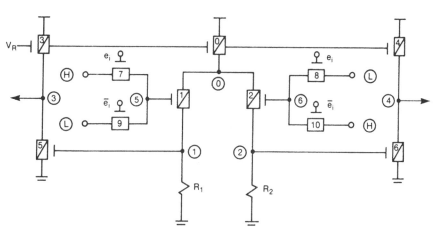

Figure 8(a). A high-speed differential driver. [14]

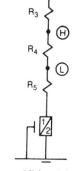

Figure 8(b). DC bias circuit for the differential inputs.

effect of skew between e_I and $\bar{e}_I$ and produces a differential output with 50% crosspoint. Thus this driver converts single-ended input into complementary differential outputs symmetric with respect to their common mode voltage.

2.2 Single-Ended Drivers

As we saw in the previous chapter simple inverters that drive large capacitive loads with reasonable performance require huge output pullup and pulldown. Since the performance of an inverter is proportional to its size, simple small inverters are seldom used to drive heavy loads directly. Instead, they are used as predrives in a large driver to drive the gates of the output pullup and pulldown, which, in turn, drive the external loads.

Reducing power and improving performance as well the operating conditions of output devices are always complementary in the sense that whenever the pullup is on, the pulldown is off, and vice versa. These drivers are called *pushpull drivers*. Since, in practice, all MOS high-power single-ended drivers are pushpull drivers, in this subsection we shall discuss the pushpull drivers in detail.

2.2.1 Basic Concept of Pushpull Drivers. Consider the typical pushpull driver shown in Figure 9(a). Here devices T_1, T_2, and T_3 form a bootstrap predrive stage and the pair (T_4, T_5) is the output stage. Prior to $t = 0$, the input is high and T_3 is on, allowing T_1 to charge C_B to $(V_H - V_T)$. Since T_5 is also on strongly and T_4 is off, the output v_{10} is at GD. Now at $t = 0$ input e_1 goes down, turning off T_3 and T_5. Since the load of T_2 is mainly the small gate capacitance of T_4, v_2 moves up quickly. As v_2 moves up, T_4 turns on and starts to charge up C_L. If the speed of v_2 is much faster than of v_{10}, T_4 will charge C_L with its gate potential at V_H most of the time. The speed of T_4 is therefore comparable to that of a simple inverter with the same pullup as shown in Figure 9(b). The advantages of the pushpull driver over the simple inverter are:

1. Since T_4 and T_5 are driven by complementary signals, there is no DC current flowing through the output devices. DC power consumption comes from only the predriver and is much smaller than the power consumed by the simple inverter of Figure 9(b).

2. The complementary operation eliminates the β ratio requirement between T_4 and T_5. The size of T_5 is determined by the falltime alone, which can be much smaller than T_5' in Figure 9(b).

Now, for a practical example.

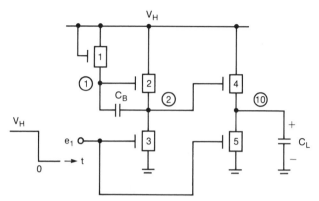

Figure 9(a). A bootstrap pushpull driver.

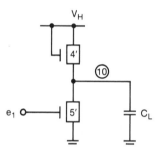

Figure 9(b). An equivalent simple inverter. Devices T_4 and T_4' are of the same size.

EXAMPLE 4

Consider the pushpull driver shown in Figure 9(a). Let the maximum DC power be 1 mW. Determine the device sizes so that the driver pulls up the load $C_L = 10$ pF with reasonably good performance. The process variations are as follows:

(a) $V_{T0} \in (0.4, 0.6)$

(b) $\Delta W \in (-0.25, 0.25)$

(c) $\Delta L \in (-0.15, 0.15)$

Solution

Maximum DC power imposes a restriction to the size of T_2, which is the only device that burns DC power. Since the worst power conditions occur at high temperature and power supply, we shall calculate the DC power $P_{DC}(T_2)$ at $T_j = 70°C$ and $V_H = 3.3$ V:

$$P_{DC}(T_2) = V_H \times \frac{1}{2}\beta\frac{X_S^2 - X_P^2}{1 + \lambda(X_S - X_P)}$$

where

$$\beta = \gamma\frac{W}{L}$$

Since the channel width $W(T_2)$ is to be determined, a Monte Carlo simulation is performed for the above equation with $W = 5$ and $L = 1$ nominal. As shown in Figure 10(a), the 3σ point WC power is 1.79 mW. The proper channel width of T_2 can then be obtained by scaling: $5 \times (1/1.79) = 2.8$. With $W = 2.8$ nominal, the Monte Carlo simulation now yields 3σ point WC power of 1.01 mW as required. The distribution of the DC power in this case is shown in Figure 10(b).

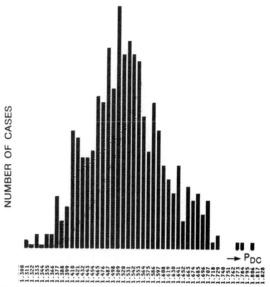

Figure 10(a). Distribution of DC power for a device with nominal channel width $W = 5$ and channel length $L = 1$.

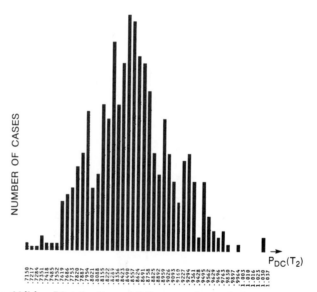

Figure 10(b). Distribution of DC power for a device with nominal channel width $W = 2.8$ and channel length $L = 1$. The 3σ point WC power is 1.01 mW.

With $W/L\,(T_2) = 2.8/1$ we can now determine the size of T_4. In Figure 11(a) the rise delay t_{DLH} of the circuit is shown for different channel widths of T_4. These results are obtained from circuit simulations. As shown in the figure, t_{DLH} declines monotonically as $W(T_4)$ increases. The rate of decrease, however, diminishes quickly for $W(T_4) > 50$. We therefore choose $W(T_4) = 50$, with $t_{DLH} = 5$ nS.

For the rest of the circuit the size of T_1 should be as small as possible and the size of T_3 is determined by the β ratio requirement to ensure a proper downlevel. The reader is invited to calculate the size of T_5 for a comparable $t_{DHL} = 5$ nS for an input with a reasonable risetime.

The response of the inverter is shown in Figure 11(b).

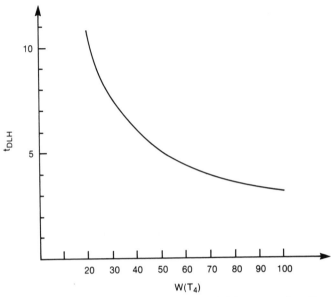

Figure 11(a). Rise delay of the pushpull driver as a function of the channel width $W(T_4)$.

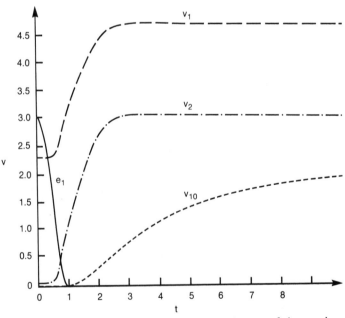

Figure 11(b). Waveforms of the node voltages of the pushpull driver for $W(T_4) = 50$.

In this particular example the final uplevel of v_{10} is $(V_H - V_T)$. If this is unacceptable because of performance or noise problems, the circuit shown in Figure 12 can be used. As we have pointed out in Chapter 4, however, the bootstrap capacitor C_B in Figure 12 is larger than that in Figure 9(a) because C_B has to share charges with the gate capacitance of T_4 as v_2 moves up. Notice also that T_4 burns DC power when the output is low. However, since the gatedrive of T_4 is much larger, the size of T_4 is much smaller than T_4' in Figure 9(b) and hence so is the DC power.

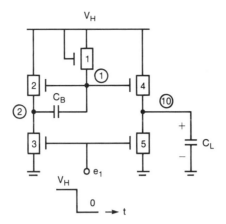

Figure 12. A pushpull driver whose output reaches V_H.

Bootstrap pushpull circuits are widely used in memory chips with fixed clocking schemes so the bootstrap capacitors can be "refreshed" regularly. In case the input patterns are totally random the outputs may have to stay high for a long period of time. Refer to Figure 9(a). The potential at node ① of C_B will eventually reduce to $(V_H - V_T)$ due to junction leakage. The output at node ⑩ can then be as low as $(V_H - 3V_T)$, where the logic level degrades and the noise margin diminishes. Sometimes a small holdup device is used to compensate the leakage. For an example, see Section 1.2(7) of Chapter 6. Bootstrap drivers are generally not used in random logic circuits. The pushpull circuits discussed next use depletion mode pullups and are much more popular among NMOS logic circuit designers.

2.2.2 Pushpull Drivers with Depletion Mode Pullups. Both inverting and noninverting drivers can be implemented with depletion mode pullups as shown in Figure 13. Since the threshold voltage of a depletion mode device is negative, the

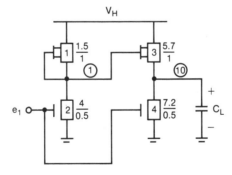

Figure 13(a). An inverting pushpull driver with depletion mode pullup. Device parameters: $(\gamma, k_1, V_{T0}, \lambda) = (0.082, 0.08, 0.5, 1)$ for enhancement mode devices, $(\gamma', k_1', V_{T0}', \lambda') = (0.082, 0.08, -1.3, 0.5)$ for depletion mode devices.

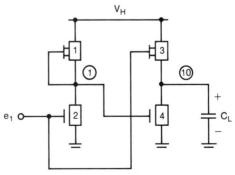

Figure 13(b). A noninverting pushpull driver with depletion mode pullup.

output can be held at V_H indefinitely as long as the pulldowns are cut off. For the same reason, however, both T_1 and T_3 burn power when the output is low. To save power, weak depletion mode and enhancement mode devices are sometimes used for T_3 at the expense of a larger device size and/or degradation of the uplevel. The normal depletion mode pullups are, however, most popular in high speed and moderate loading (0.5–5 pF) applications.

The circuit design in Figure 13 is a little different from previous designs. Since both T_1 and T_3 burn power, it is important to divide the total power between them properly to optimize performance. Let us illustrate the procedure by an example.

EXAMPLE 5

Determine the device sizes of the circuit in Figure 13 for a load $C_L = 1$ pF and WC power of 1 mW. The process parameter variations are as follows:

(a) $V'_{T0} \in (-1.2, -1.4)$

(b) $\Delta W \in (-0.25, 0.25)$

(c) $\Delta L \in (-0.15, 0.15)$

Solution

Again, since the sizes of T_1 and T_3 are unknown, let us run a Monte Carlo simulation of the DC power:

$$P_{DC}(W) = V_H \times \frac{1}{2}\beta \frac{X_S^2 - X_P^2}{1 + \lambda(X_S - X_P)}$$

where $\beta = \gamma W/L$ for a $W = 5$ nominal under the condition $V_H = 3.3$ V and $T_j = 70°C$. The WC power is found to be 0.69 mW. By scaling, the proper value of W is given by $5 \times (1/0.69) = 7.2$. A rerun of the Monte Carlo simulation with $W = 7.2$ nominal then gives a WC power of 1.0 mW.

Now, with $W(T_1) + W(T_3) = 7.2$ make a few test runs to find the most efficient power division between T_1 and T_3. Figure 14 displays a plot of t_{DLH} of the circuit as a function of $W(T_1)$. Notice that t_{DLH} is a relatively shallow convex function of the partition. The minimum occurs around $W(T_1) = 1.25–1.5$ and $W(T_3) = 5.95–5.7$, where it is approximately 1.7 nS. We therefore chose $W(T_1) = 1.5$ and $W(T_3) = 5.7$. $W(T_1) = 1.5$ was chosen because it may be

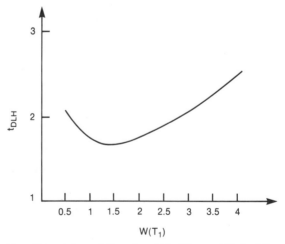

Figure 14. Rise delay of the circuit in Figure 13(a) of Example 5 as a function of $W(T_1)$.

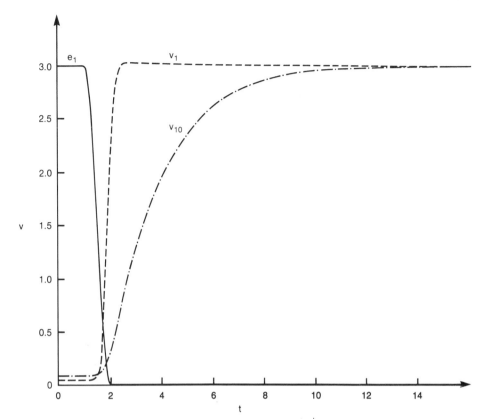

Figure 15. Waveforms of the node voltages in Example 5.

very close to the minimum device width for layout already. The pullup response of this circuit is shown in Figure 15.

The sizes of T_2 and T_4 are determined either by the downlevel or by stability considerations as discussed in detail in Chapter 3. The acute reader may have noticed, however, that the β ratio of (T_1, T_2) is larger than that of (T_3, T_4). When the output is low both T_1 and T_3 burn DC power. Whereas the DC current of T_1 is independent of its downlevel, the DC current of T_3 is a strong function of its gate voltage. To minimize the DC power it is important to keep the gate voltage of T_3 as close to the ground as possible; hence the larger β ratio of (T_1, T_2).

The DC power control is particularly critical with weak depletion mode pullups. Since the threshold voltage of these devices is close to zero, a downlevel of a few hundred mVs can cause a significant DC power increase.

2.2.3 Calculation of Delay of Pushpull Drivers. To facilitate the DA (design automation) process, let us construct a delay calculator for the inverting pushpull driver in Figure 13(a). There are two components of t_{DLH}: t_{DLH1}, which is the rise delay of v_1 relative to e_1, and t_{DLH10}, which accounts for the delay from v_1 to v_{10}. Since T_1 is a depletion mode device driving a light load, $t_{DLH1} = T_R (90\%)/2$ of the step response of the inverter (T_1, T_2), which can be easily obtained by calling subroutine DPUN. The calculation of t_{DLH10}, however, is much more complicated. Since v_{10} follows v_1, t_{DLH10} depends on the risetime of v_1. Thus, for t_{DLH10} we must calculate the transient response of a source follower with a ramp input. Since it is of general interest we shall derive the solution of such a problem here and apply it to t_{DLH10}.

Consider the circuit shown in Figure 16(a) where the input e_G is a ramp function described by

$$e_G(t) = \begin{cases} \dfrac{E_G}{\tau_R} t, & t < \tau_R \\ E_G, & \tau_R < t \end{cases}$$

Since $e_G(t) = E_G$ is a constant for $t > \tau_R$, which is the case for delay calculator CPUN, all we have to do is to calculate the output v_O at $t = \tau_R$ and add the individual delays as shown in Figure 16(b):

$$t_{DLH} = \frac{\tau_R}{2}\frac{V_H}{E_G} + \left[T_R\left(V_H/2\right) - T_R(V_O) \right]$$

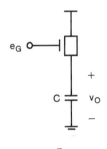

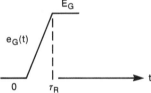

Figure 16(a). A source follower with a ramp input.

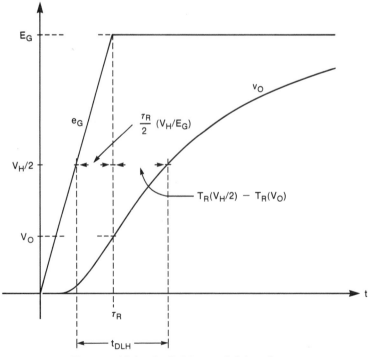

Figure 16(b). Definitions of delay times.

where $V_O = v_O(\tau_R)$. Assume that $E_G < (V_H + V_T)$ so that the device operates in the saturation region. We have, from Chapter 4,

$$C\frac{d}{dt}v_O = \frac{1}{2}\beta\frac{\xi^2}{\lambda^2} \tag{16}$$

where

$$(\xi + 1)^2 = 1 + 2\lambda[e_G - V_{T0} - (1 + k_1)v_O] \tag{17}$$

Differentiating Equation (17) with respect to t and substituting the results into Equation (16), we have

$$\frac{\xi + 1}{K^2 - \xi^2}d\xi = \frac{1 + k_1}{2\lambda}\frac{\beta}{C}dt \tag{18}$$

where

$$K^2 = \frac{2\lambda^2}{1 + k_1}\frac{E_G}{\tau_R}\frac{C}{\beta}$$

is a constant.

By integrating Equation (18) from 0 to t, we obtain

$$\left(\frac{1}{K} - 1\right)\ln\left|\frac{K + \xi}{K + \xi_0}\right| - \left(\frac{1}{K} + 1\right)\ln\left|\frac{K - \xi}{K - \xi_0}\right| = \frac{1 + k_1}{2\lambda}\frac{\beta}{C}t \tag{19}$$

where $\xi = \xi(t)$ and $\xi_0 = \xi(0)$. We are specifically interested in the value of ξ at $t = \tau_R$ at which e_G reaches E_G. After $\xi(\tau_R)$ has been determined, V_O can be calculated from Equation (17) and passed to subroutine CPUN. Since Equation (19) defines ξ as an implicit function of t, a numerical method has to be used to solve for $\xi(\tau_R)$. A FORTRAN subroutine called RPUN has been written that calculates $\xi(\tau_R)$ and V_O for different E_G, τ_R, and device parameters. A listing of RPUN is in Appendix 1 to this chapter.

With subroutine RPUN, writing delay calculators for the pushpull drivers is possible. For the circuit in Example 5, for example, the following program is assembled from the previous subroutines:

```
IMPLICIT REAL*8 (A-H, O-Z)
COMMON VH, GAMMAN, GAMMAP, PK1N, PK1P, VTON, VTOP,
*       PLMBDN, PLMBDP

VH = 3.0
GAMMAN = .082
GAMMAP = .032
PK1N = .08
PK1P = .20
VTON = .5
VTOP = -.5
PLMBDN = .5
PLMBDP = .5

EG = 3.0
```

```
      CALL DPUN (1.5, 1., -1.8, 0.D0, .64,
     *                   0.1D0,  .00787,  TDLH, TR)
      DELAY = TDLH + .5*TR
      TAOR = TR/.9
      WRITE(6,2) DELAY, TR
    2 FORMAT (1H1, / 2(' ',F10.5/))
      CALL RPUN (5.7, 1., -1.8, 0.D0, .01,
     *           EG, TAOR,  VO,  1.)
      CALL CPUN (5.7, 1., -1.8, 0.D0, .01,
     *           EG, VO,  1.,  TDLH, TR)
      DELAY = DELAY + TDLH
      WRITE(6,1) EG, TAOR, VO, TDLH, DELAY, TR
    1 FORMAT (1H1, / 6('   ',F10.5/))

      STOP
      END
```

The first subroutine, DPUN, calculates the delay of v_1. Its output, rise delay t_{DLH1}, is added to the accumulative delay DELAY, and its 90% risetime T_R is fed to the next subroutine RPUN. The input risetime for RPUN is then $\tau_R = T_R/0.9$. Subroutine RPUN then calculates the output voltage V_O at $t = \tau_R$. V_O will then be used by CPUN as the initial voltage for the calculation of the delay from τ_R to the time when v_{10} reaches $V_H/2$. The total delay, then, is the sum of the delay calculated by CPUN and the previous delay t_{DLH1}.

The following is a partial list of the program printout.

```
      EG     =  3.00000
      TAOR   =  0.68590
      VO     =  0.53787
      TDLH   =  0.96187
      DELAY  =  1.61347
      TR     =  2.68431
```

The estimated delay is 1.61 nS. Compared with the delay measured directly from waveforms in Figure 15: 1.7 nS, the agreement is within 6%.

2.2.4 Dynamic Drivers—The Principle of Invert and Delay. The push-pull drivers discussed so far use predrivers for large output devices. The power savings from the output stage are significant, and the predriver does consume DC power. This is because the predriver is a simple inverter driven by a single input, e_1. In many applications both e_1 and $\overline{e}_1$ are available, so the predriver can be replaced by a pass device as shown in Figure 17. We then have a "dynamic driver" that does not consume any DC power. There is always an inverter somewhere on the chip to provide the complementary signals, but the driver itself is totally dynamic in the

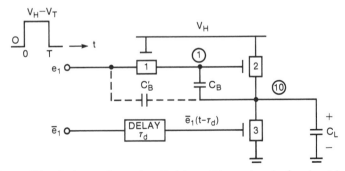

Figure 17. A dynamic pushpull driver. The uplevel of e_1 is at least $(V_H - V_T)$.

sense that the only power it consumes is associated with charging and discharging the load and some of its internal capacitances.

The dynamic driver in Figure 17 operates on the principle of "invert and delay." Prior to $t = 0$ the true input e_1 is low and its complement $\bar{e}_1$ is high. This ensures that capacitor C_B is fully discharged and device T_2 is off. Then at $t = 0$ the input e_1 starts to move up and its complement $\bar{e}_1$ goes down. Since $\bar{e}_1$ is delayed, however, device T_3 remains unaffected until $t = \tau_d$. Meanwhile the output v_{10} will be held to GD, providing a conductive path for e_1 to charge up C_B. If τ_d is long enough, C_B will be charged up to $(V_H - V_T)$ before $\bar{e}_1(t - \tau_d)$ finally reaches T_3. After $t = \tau_d$ device T_3 starts to turn off allowing T_2 to pull up the output with a large gatedrive provided by the bootstrap capacitor C_B. This is the pullup. To see how pulldown works, first assume that at $t = T$, the input e_1 goes down. Both C_B and C_L will discharge through T_1 by charge redistribution. Then, at $t = T + \tau_d$ device T_3 starts to turn on, further discharging and holding the output to ground. Since the gate of T_2 is charged up only when needed and remains low when the output is low, there is no DC current flowing through T_2 and T_3, and hence no DC power.

Sometimes the uplevel of e_1 may be slightly lower than $(V_H - V_{T1})$ because of noise. In this event C_B will discharge through T_1 as v_{10} starts to bootstrap. A feedback capacitor C_B' is thus added to the input from node ⑩. As v_{10} rises the input is pushed up slightly by the capacitive coupling of C_B', preventing the discharge of C_B.

One important factor that must be optimized is the internal delay τ_d. If it is too short, device T_3 will turn off before C_B is fully charged up, reducing the gatedrive of T_2. On the other hand, if τ_d is too long, both T_2 and T_3 will be on hard "for a while," dissipating high power. In practice, because of difficulties in controlling the timing between e_1 and $\bar{e}_1$, a small "delay and invert" circuit is put at the front end of the driver to generate $\bar{e}_1(t - \tau_d)$. The DC power is then back, but it is small while allowing much better control of the internal timing.

For a real heavy load another output stage can be attached to the circuit as shown in Figure 18. From $t = 0$ to $t = \tau_d$, both C_B and the gate capacitance $C_{gs}(T_4)$ are charged up, turning on T_2 and T_4. At $t = \tau_d$ devices T_3 and T_5 are turning off, allowing v_2 and v_{10} to move up. Since the loads at node ② and ⑩ are vastly different, v_2 moves up much faster than v_{10}. The potential at node ① will be boosted very high through C_B. Device T_4 then charges up C_L with a large gatedrive. This is the pullup. The pulldown operation is straightforward, and its explanation is left to the reader as a simple exercise.

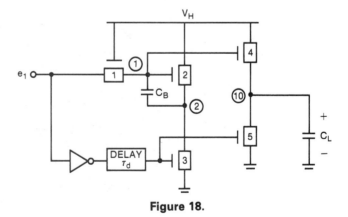

Figure 18.

There is an interesting relationship between the sizes of C_B and T_4. The speed of T_4 is determined by two factors: the precharge of $C_{gs}(T_4)$ and the loss of gatedrive through C_B when v_{10} moves up. Assuming that v_{10} does not move before (T_2, T_3) has completed its bootstrap action, the precharge of $C_{gs}(T_4)$ before v_{10} moves up can be calculated by charge conservation:

$$C_B(V_H - V_T) + \frac{2}{3}C_o(V_H - 2V_T) = C_B(V_1' - V_H) + C_o\left(V_1' - V_T - \frac{V_H}{2}\right),$$

where $C_o = C_o' WL(\text{T}_4)$ and V_1' is the highest potential v_1 reaches. In this equation the last term is the sum of the charges stored in T_4 at the source and drain:

$$\frac{1}{2}C_o(V_1' - V_T) + \frac{1}{2}C_o(V_1' - V_H - V_T) = C_o\left(V_1' - V_T - \frac{V_H}{2}\right)$$

Notice that T_4 is now in the linear region and $C_{gs}(\text{T}_4) = C_{gd}(\text{T}_4) = C_o/2$. Since v_{10} is still at GD, V_1' is also the precharged potential of $C_{gs}(\text{T}_4)$. Solving for V_1', we have

$$V_1' = \frac{2\alpha + 7/6}{\alpha + 1}V_H - \frac{\alpha + 1/3}{\alpha + 1}V_T \tag{20}$$

where $\alpha \triangleq C_B/C_o$. Thus, a large value of α implies higher precharge of $C_{gs}(\text{T}_4)$ and $V_1' = (2V_H - V_T)$ as $\alpha \to \infty$, as expected.

As v_{10} moves up, however, $C_{gs}(\text{T}_4)$ dumps charge back to C_B through T_2, losing its gatedrive along the way. The bootstrap efficiency of $C_{gs}(\text{T}_4)$ is thus inversely proportional to the size of C_B. The net effect of this charge sharing can again be calculated by charge conservation for $t \to \infty$:

$$C_B(V_1'' - V_H) + C_o(V_1'' - V_H - V_T) = C_B(V_H - V_T) + \frac{2}{3}C_o(V_H - 2V_T)$$

$$\Rightarrow \qquad V_1'' - V_H = \frac{\alpha + 2/3}{\alpha + 1}V_H - \frac{\alpha + 1/3}{\alpha + 1}V_T \tag{21}$$

where V_1'' is the final uplevel of v_1 and $(V_1'' - V_H)$ is the final value of the v_{GS} of T_4. The above equation indicates that $(V_1'' - V_H)$ is a moderate increasing function of α.

Earlier, in Chapter 4, the coefficient of the precharge loss, k_2, was defined as

$$v_B(t) - v_B(0) = k_2[v_O(0) - v_O(t)]$$

where v_B is the v_{GS} of the device and v_O is the output voltage at the source of the device. Applying this information to the question at hand, $v_B = v_{C_{gs}}(\text{T}_4)$ and $v_O = v_{10}$. The initial voltage at node ⑩ is $v_O(0) = v_{10}(0) = 0$. Taking the limit as $t \to \infty$, we obtain

$$(V_1'' - V_H) - V_1' = -k_2 V_H$$

Substituting Equations (20) and (21) into the above, we obtain

$$k_2 = \frac{V_1' - (V_1'' - V_H)}{V_H} = \frac{\alpha + 1/2}{\alpha + 1}$$

The bootstrap efficiency is therefore given by the expression

$$1 - k_2 = \frac{1/2}{\alpha + 1}$$

Bootstrap efficiency, then, decreases as α increases. It is also interesting to note that bootstrap efficiency is always less than 50%. Generally speaking, the speed of the driver improves with V_1' as α increases, but with diminishing returns due to lower bootstrap efficiency. In practice, the size of C_B is determined more by the area available and some other considerations than performance, as we shall see in the following example.

EXAMPLE 6

Consider the dynamic driver shown in Figure 19. Determine the device sizes for the driver to drive $C_L = 100$ pF with a rise/falltime $T_R = T_F = 10$ nS.

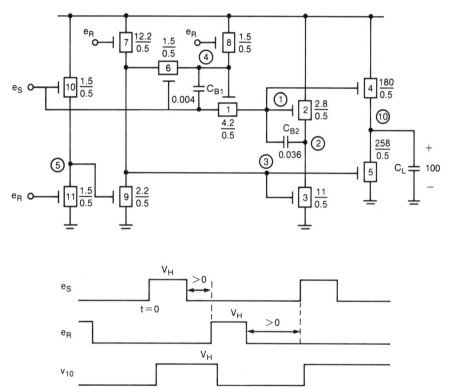

Figure 19. A high-power dynamic driver. Device parameters: $(\gamma, k_1, V_{T0}, \lambda) = (0.082, 0.08, 0.5, 1)$.

Solution

Since this is a fairly complex circuit illustrating many of the fine points encountered in practical design, the analysis will be carried out in detail. In particular, the equations derived in previous chapters will be applied to each device according to its particular operating condition. To make the problem mathematically tractible, however, simplifying assumptions about input will be made, and all parasitic capacitances will be neglected for the first design cut. Other parameters will be considered in subsequent refinements after the circuit has been laid out.

There are two complementary input signals: e_S for set and e_R for reset. When e_S goes up the output goes up. The output stays up even though e_S goes down at some later time. It is pulled down only when e_R comes up. After the output has been pulled down, it stays down even though e_R goes down later. It is pulled up only when e_S goes up again. In short, the circuit is set by e_S and reset by e_R, maintaining the status quo in between. Such a circuit is said to possess memory. Obviously the inputs e_S and e_R must be nonoverlapping for proper operation.

To understand this circuit start with the reset condition. Here, e_S is low and e_R is high. Node ⑤ is at GD and T_9 is off. Nodes ③ and ④ are charged up to $(V_H - V_T)$ by T_7 and T_8, respectively. The bootstrap capacitor C_{B1} is precharged to $(V_H - V_T)$ and C_{B2} is completely discharged to zero; e_R then goes down sometime later, leaving all internal nodes floating. At $t = 0$, e_S moves up. Since C_{B1} is precharged, v_4 will be boosted very high, turning on T_1 very hard and allowing e_S to charge C_{B2} and $C_{gs}(T_4)$ up to V_H quickly. At the same time v_5 is also pulled up by T_{10}, discharging node ③. The discharge rate of node ③, however, is designed to be slow to allow e_S to charge up C_{B2} and $C_{gs}(T_4)$ before T_3 and T_5 cut off. As soon as T_3 and T_5 turn off, v_2 and v_{10} move up. Since the load of T_2 is very small, v_2 will move up to V_H quickly, boosting v_1 real high. This precharges the gate capacitance $C_{gs}(T_4)$ to a large initial voltage. As v_{10} moves up, v_1 will be boosted up even higher due to the bootstrap action of T_4. The bootstrap then provides a large gatedrive for T_4 to pull up the heavy load C_L.

An interesting timing effect around T_1, T_6, and C_{B1} is worth noting. When e_S goes up it turns on T_6. Since T_9 is slow, nothing happens until it is turned on hard. This allows C_{B1} to bootstrap v_4. Now, as soon as T_9 starts to pull down v_3, capacitor C_{B1} will discharge through T_9. In fact, eventually, it will be charged in the opposite direction. This discharge of node ④ is important because T_1 has to be completely shut off when v_2 starts to move up and boosts v_1 via C_{B2}.

For our design we shall allow 0.5 nS for T_1 to charge up C_{B2} and $C_{gs}(T_4)$, 1 nS for T_9 to discharge node ③ and 0.5 nS for v_2 to bootstrap itself. The internal delay of the driver is therefore about 1.5–2 nS, which is the price we have to pay for a total dynamic operation.

Start with the output stage and work out each subcircuit in reverse order toward the inputs.

(1) Size of T_4: Since C_{B2} and $C_{gs}(T_4)$ are precharged all the way up to V_H, V_1' and V_1'' are higher than those in the previous case for the same value of α. By charge conservation it is easy to show that

$$V_1' = \frac{2\alpha + 7/6}{\alpha + 1} V_H - \frac{1/3}{\alpha + 1} V_T$$

and that

$$V_1'' - V_H = \frac{\alpha + 2/3}{\alpha + 1} V_H - \frac{1/3}{\alpha + 1} V_T$$

The coefficient of precharge loss, k_2, is easily proved to be

$$k_2 = \frac{\alpha + 1/2}{\alpha + 1}$$

Suppose due to area consideration we choose $\alpha = 0.3$. From the above equations we then have $k_2 = 0.62$, $V_1' = 3.90$ and $(V_1'' - V_H) = 2.05$. Following the procedure in Chapter 4, we have

$$k = k_1 + k_2 = 0.08 + 0.62 = 0.70$$

$$V_{B0} = V_{B0}' = 3.90$$

$$\Rightarrow \qquad X = V_{B0}' - V_{T0}' - kV_H = 3.90 - 0.5 - 0.7 \times 3 = 1.30$$

and

$$K = 1 - \frac{2X}{(1 - k_1 - 2k_2)V_H} = 1 - \frac{2 \times 1.30}{-0.32 \times 3} = 3.71$$

At $t = 0$

$$x_S(0) = V_1' - V_{T0} = 3.90 - 0.5 = 3.40$$

$$\xi_0 = \sqrt{1 + 2 \times 1 \times 3.40} - 1 = 1.79$$

At $t = T_s$

$$\xi_s = -\frac{1 - 0.62}{1.08} + \sqrt{\left(\frac{1 - 0.62}{1.08}\right)^2 + 2 \times 1 \times 1.30} = 1.30$$

$$\Rightarrow \quad \eta_s = 1 - \frac{1.30}{1.08 \times 3} = 0.60$$

We then have

$$T_s = \frac{2}{0.7}\left(\ln \frac{1.79}{1.30} + \frac{1.79 - 1.30}{1.79 \times 1.30}\right)\frac{C}{\beta} = 1.52\frac{C}{\beta}$$

and

$$T_l = \left[\frac{1}{1.08 \times 1.30} \ln \frac{(0.9 - 3.71) \times (0.60 - 1)}{(0.9 - 1) \times (0.60 - 3.71)}\right.$$

$$\left. + \frac{2}{(-0.32)} \ln \left(\frac{0.9 - 3.71}{0.60 - 3.71}\right)\right]\frac{C}{\beta}$$

$$= 1.55\frac{C}{\beta}$$

and the total risetime

$$T_R = (1.52 + 1.55)\frac{C}{\beta} = 3.07\frac{C}{\beta}$$

Substituting $C = C_L = 100$ and $T_R = 10$ into the above equation yields

$$\beta = \frac{3.07 \times 100}{10} = 30.7 \quad \text{and} \quad \frac{W}{L}(T_4) = \frac{30.7}{0.082} = \frac{187}{0.5}$$

The total gate capacitance of T_4 is given by

$$C_o(T_4) = 1.38 \times 10^{-3} \times 187 \times 0.5 = 0.129$$

$$\Rightarrow \quad C_{B2} = 0.3 \times 0.129 = 0.039$$

(2) *Size of T_5:* The size of T_5 is determined by a comparable falltime $T_F = 10$ nS.

For simplicity's sake, assume the input of T_5 is a step function of 2.32 V, because v_3 is pulled up by T_7 to $(V_H - V_T)$. Following the same procedure used in Example 8 of Chapter 4, it can be easily shown

$$T_F = 4.23\frac{C}{\beta} \quad \Rightarrow \quad \frac{W}{L}(T_5) = \frac{4.23 \times 100}{10 \times 0.082} = \frac{258}{0.5}$$

The total gate capacitance of T_5 is

$$C_o(T_5) = 1.38 \times 10^{-3} \times 258 \times 0.5 = 0.18$$

(3) Size of T_2: Device T_2 is a bootstrap pullup with bootstrap capacitor C_{B2}. Since v_1 is coupled to v_{10} via the gate capacitance $C_{gs}(T_4) = (2/3)C_o(T_4) = 0.086$, the capacitive load of T_4 is the series combination of C_{B2} and $C_{gs}(T_4)$:

$$C = \frac{0.039 \times 0.086}{0.039 + 0.086} = 0.027$$

The coefficient of precharge loss k_2 is given by

$$k_2 = \frac{0.086}{0.086 + 0.039} = 0.69$$

With a precharge of 3 V, the reader should be able to show that the risetime T_R is given by

$$T_R = 9.14\frac{C}{\beta} \quad \text{and} \quad \frac{W}{L}(T_2) = \frac{9.14 \times 0.027}{0.5 \times 0.082} = \frac{3.0}{0.5}$$

Here we allow 0.5 nS for T_2 to charge up node ② to 90% of its final value. The gate capacitance of T_2 is given by

$$C_o(T_2) = 1.38 \times 10^{-3} \times 3.0 \times 0.5 = 0.002$$

(4) Size of T_3: Device T_3 should be big enough to hold node ③ down when the gate of T_2 is being charged up to V_H. There is a β ratio requirement here. The reader is challenged to show that a β ratio of 4 is a good choice. Hence the size of T_3 is

$$\frac{W}{L}(T_3) = 4\frac{W}{L}(T_2) = \frac{12}{0.5}$$

and

$$C_o(T_3) = 1.38 \times 10^{-3} \times 12 \times 0.5 = 0.008$$

(5) Size of T_1: To determine the size of T_1 refer to Figure 20. The bootstrap capacitor C_{B1} is precharged to $(3 - 0.5)/1.08 = 2.32$ V before e_s goes up. When e_s moves up the gate voltage of T_1 will be boosted high. Assume $e_s(t)$ is a step function of 3 V; then the gate voltage of T_1 (node ④) is given by $V_G = [2.32 + (1 - k_2) \times 3]$ V where k_2 is the coefficient of the precharge loss. Assume $k_2 = 0.2$ so that $V_G = 4.72$. The capacitive load C of T_1 is the parallel combination of C_{B2} and $(2/3)[C_o(T_4) + C_o(T_2)] = 0.13$. Since e_s is a step function of 3 V, and we allow 0.5 nS for T_1 to charge C up to the 90% point, we have

$$X_D = v_G - V_{T0} - (1 + k_1)V_H$$

$$= 4.72 - 0.5 - 1.08 \times 3 = 0.98$$

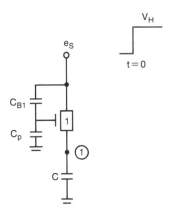

Figure 20. A pushpull configuration.

and

$$K = 1 + \frac{2X_D}{(1 + k_1) \times V_H}$$

$$= 1 + \frac{2 \times 0.98}{1.08 \times 3} = 1.60$$

It can be shown that the device starts from the saturation region at $t = 0$ with $e_s(0) = 3$ and enters the linear region at $t = T_s$. Now,

At $t = 0$

$$x_s(0) = 4.72 - 0.5 = 4.22$$

$$\Rightarrow \qquad \xi_0 = \sqrt{1 + 2 \times 1 \times 4.22} - 1 = 2.07$$

At $t = T_s$

$$\xi_s = \sqrt{2 \times 1 \times 0.98} = 1.40$$

$$\Rightarrow \qquad \eta_s = 1 - \frac{1.40}{1.08 \times 1 \times 3} = 0.57$$

Therefore,

$$T_s = \frac{2}{1.08}\left(\ln\frac{2.07}{1.40} + \frac{2.07 - 1.4}{2.07 \times 1.4}\right)\frac{C}{\beta}$$

$$= 1.15\frac{C}{\beta}$$

and

$$T_l = \frac{1}{1.08}\left[\frac{1}{0.98}\ln\frac{(1.60 - 0.9) \times (1 - 0.57)}{(1 - 0.9) \times (1.60 - 0.57)} + 2 \times 1 \times \ln\left(\frac{1.60 - 0.57}{1.60 - 0.9}\right)\right]\frac{C}{\beta}$$

$$= 1.72\frac{C}{\beta}$$

and the total risetime

$$T_R = (1.15 + 1.72)\frac{C}{\beta} = 2.87\frac{C}{\beta}$$

With $C = 0.13$ and $T_R = 0.5$, the size of T_1 is

$$\frac{W}{L}(T_1) = \frac{2.87 \times 0.13}{0.5 \times 0.082} = \frac{4.6}{0.5}$$

(6) Size of T_6: Device T_6 serves as a path to discharge C_{B1}. Since C_{B1} is fairly small, we can choose a small dimension for T_6:

$$\frac{W}{L}(T_6) = \frac{1.5}{0.5} \qquad \Rightarrow \qquad C_o(T_6) = 1.38 \times 10^{-3} \times 1.5 \times 0.5 = 0.001$$

Since $k_2 = 0.2$ in the previous case,

$$C_{B1} = 4 \times 0.001 = 0.004$$

(7) Size of T_9: Since device T_9 is the "invert-delay" element, we shall allow 1 nS for T_9 to discharge node ③ from its precharge level $(3 - 0.5)/1.08 = 2.32$ V to 0.5 V at which both T_3 and T_5 are cut off. The load for T_9 can be estimated as

$$C = C_o(T_5) + C_o(T_3) + 2 \times C_{B1} = 0.20$$

The reader is challenged to show that in this case

$$T_F = 1.78\frac{C}{\beta} \quad \Rightarrow \quad \frac{W}{L}(T_9) = \frac{1.78 \times 0.20}{1 \times 0.082} = \frac{2.2}{0.5}$$

(8) Size of T_7: The load of T_7 is approximately $C = C_o(T_3) + C_o(T_5) = 0.008 + 0.18 = 0.188$. Assuming e_R is a step function, it is left to the reader to show that Equation (3) of Chapter 4 gives

$$T_R(90\%) = 10.63\frac{C}{\beta}$$

Since the falltime is 10 nS, we can assume $T_R(90\%) = 1$ nS to validate the step input assumption to T_5. The above equation then gives

$$\beta = \frac{10.63 \times 0.188}{1} = 2.00$$

$$\Rightarrow \qquad \frac{W}{L}(T_7) = \frac{2.00}{0.082} = 24.37 = \frac{12.2}{0.5}$$

(9) Sizes of T_8, T_{10}, and T_{11}: Since the loads of these devices are small, they can be small devices:

$$\frac{W}{L}(T_8) = \frac{W}{L}(T_{10}) = \frac{W}{L}(T_{11}) = \frac{1.5}{0.5}$$

This concludes our "first cut" or "preliminary" design.

The next step is to simulate this design on ASTAP. The waveforms of the node voltages are shown in Figure 21. Despite all the simplifying assumptions it's clear that the results are reasonably close to our expectations. When all

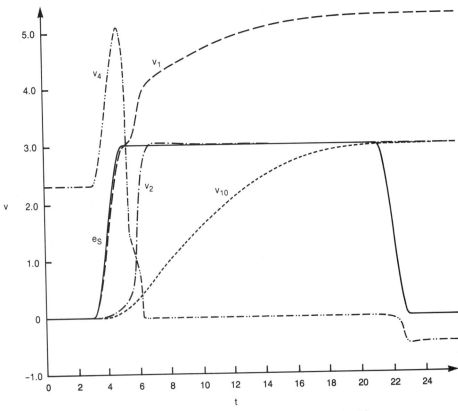

Figure 21(a). Pullup waveforms of the dynamic driver.

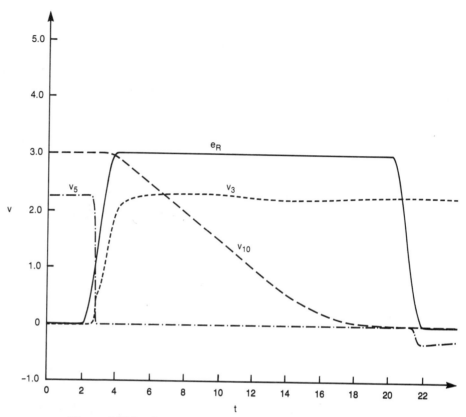

Figure 21(b). Pulldown waveforms of the dynamic driver.

the device sizes are determined, proceed with the layout of the circuit. With physical layout it is easier to estimate the parasitic capacitances more accurately and reevaluate the device sizes accordingly. This procedure needs to be repeated once more for the new design. For complex circuits like this example, several iterations may be needed for fine-tuning.

2.2.5 CMOS Drivers. Power consumption with CMOS drivers is less restrictive because the DC power can be zero. Since the n-channel and p-channel devices operate in a complementary fashion, the pullup and pulldown can share the same input. When the input goes up three things happen: n-channel device turns on, the p-channel device turns off, and the output is pulled down to ground. When the input goes down the p-channel device turns on, the n-channel device turns off, and the output goes up to V_H. Since the output always reaches either GD or V_H, bootstrapping is usually not needed and the noise margins are better. The main disadvantage of CMOS circuits, in addition to their process complexity, is their large area. Because of the larger input capacitances, they consume higher transient power, as well.

As with NMOS, predrivers are used to drive large output devices for heavy loads. Figure 22 shows a CMOS driver chain consisting of $N - 1$ stages of predrivers of increasing sizes toward the output. For simplicity, the p-channel and n-channel devices in each inverter have the same device size. The number of predrivers depends on both the input capacitance restriction and the load, as we shall see shortly.

In the previous chapter we proved that the average delay t_D of an inverter is given by

$$t_D = \frac{1}{2}(t_{DLH} + t_{DHL}) \tag{22}$$

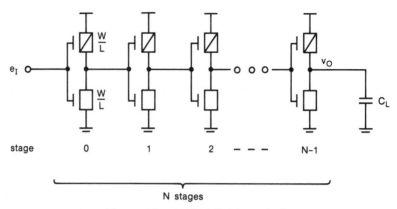

Figure 22. A CMOS driver chain.

with

$$t_{DLH} = K_P \frac{C}{\beta_P} + B_P$$

and

$$t_{DHL} = K_N \frac{C}{\beta_N} + B_N$$

where K_X and B_X are constants and $\beta_X = \gamma_X W/L$. The subscript $X = P$ for p-channel device and $X = N$ for n-channel device. Here the inherent delay of devices and the effects of parasitic capacitances are grouped into the constants B_P and B_N. The average input capacitance C_{IN} of each inverter is equal to

$$C_{IN} = C_o' WL \left(2 - \frac{V_{T0} - V_{T0}'}{V_H}\right)$$

where W and L are device width and length, respectively, of both n-channel and p-channel devices. Substituting the above into Equation (22) yields

$$t_D = A \frac{C}{C_{IN}} + B$$

where $A = C_o' L^2 [2 - (V_{T0} - V_{T0}')/V_H](K_p/\gamma_P + K_N/\gamma_N)/2$ and $B = (B_P + B_N)/2$ are constants.

Since the input capacitance of one stage in a driver chain is the load capacitance of its previous stage, the total delay is given by

$$T_D = A \left(\frac{C_1}{C_0} + \frac{C_2}{C_1} + \cdots + \frac{C_L}{C_{N-1}}\right) + NB$$

where C_k is the input capacitance of stage k, $k = 0, \ldots, N - 1$. To find the minimum of T_D, differentiate the above with respect to C_k:

$$\frac{\partial}{\partial C_k} T_D = A \left(\frac{1}{C_{k-1}} - \frac{C_{k+1}}{C_k^2}\right) = 0 \qquad \Rightarrow \qquad \frac{C_k}{C_{k-1}} = \frac{C_{k+1}}{C_k}$$

$$k = 1, \ldots, N - 2$$

and

$$\frac{\partial}{\partial C_{N-1}} T_D = A\left(\frac{1}{C_{N-2}} - \frac{C_L}{C_{N-1}^2}\right) = 0 \quad \Rightarrow \quad \frac{C_{N-1}}{C_{N-2}} = \frac{C_L}{C_{N-1}}$$

Equating the product of all left-hand terms to the product of all right-hand terms in these equations, we obtain

$$\frac{C_{N-1}}{C_0} = \frac{C_L}{C_1}$$

Substituting the above into each equation then yields

$$\frac{C_k}{C_{k-1}} = \frac{C_L}{C_{N-1}}, \qquad k = 1, \ldots, N-1$$

And, by multiplying all terms together, we have

$$\frac{C_k}{C_{k-1}} = \frac{C_L}{C_{N-1}} = \left(\frac{C_L}{C_0}\right)^{1/N}, \qquad k = 1, \ldots, N-1 \tag{23}$$

The total delay is then equal to

$$T_D(N) = N\left[A\left(\frac{C_L}{C_0}\right)^{1/N} + B\right] \tag{24}$$

Thus, the total delay is minimized when each stage contributes equally to the delay with an I/O capacitance ratio given by $(C_0/C_L)^{1/N}$

In a chain driver the input capacitance $C_{IN} = C_0$ is limited by the driving capability of its previous driver and the load C_L is determined by the task the circuit must accomplish. Both are given conditions. To find the proper number of stages for the chain driver, differentiate Equation (24) with respect to N:

$$\frac{\partial}{\partial N} T_D(N) = A\left(\frac{C_L}{C_{IN}}\right)^{1/N}\left[1 - \frac{\ln(C_L/C_{IN})}{N}\right] + B = 0 \tag{25}$$

The solution of the above equation gives the number of stages N_o in the chain driver required to minimize the delay. If the constant B is very small, the solution takes a simple form:

$$N_o = \ln\left(\frac{C_L}{C_{IN}}\right)$$

Substituting N_o into Equation (23), we then obtain

$$\frac{C_1}{C_{IN}} = \frac{C_k}{C_{k-1}} = \frac{C_L}{C_{N-1}} = e, \qquad k = 2, \ldots, N-1$$

where e is the base of the natural logarithm. If B is not negligible, Equation (25) can be solved by numerical means. Since the constant B represents the inherent delay of the inverters it is affected by many (including unknown) factors such as layout, device parameters, and parasitics. The value of B is best determined by circuit simulation.

This analysis applies to all kinds of "chain drivers." It is, however, particularly popular with CMOS circuits because of the zero DC power in CMOS design.

EXAMPLE 7

Let the input capacitance of the driver be limited to 0.011 pF, design a CMOS chain driver to drive a load $C_L = 10$ pF. Assume an equal size for both n-channel and p-channel devices.

Solution

The average input capacitance of 0.011 pF corresponds to the inverter shown in Figure 23. Finding the inherent delay of the inverter requires a simulation of the circuit with different loads C_L. By measuring the slope and extrapolating as shown in the figure, we obtain

$$A = 0.014 \quad \text{and} \quad B = 0.098$$

The capacitance ratio $C_L/C_{IN} = 10/0.011 = 909$. Substituting these values into Equation (25) and solving for N, we have

$$N = 3.45 \quad \text{and} \quad \frac{C_k}{C_{k-1}} = 909^{1/3.45} = 7.20$$

Use $N_o = 3$ or 4, with the corresponding capacitance ratio. Since the device sizes form a geometric sequence, the total device area can be estimated by a geometric series:

For $N = 3$

$$\left(\frac{C_L}{C_{IN}}\right)^{1/3} = 9.69 \quad \Rightarrow \quad T_D = 3 \times (0.014 \times 9.69 + 0.098) = 0.70$$

with an area $= (2 \times 9.5 \times 0.5) \times (909 - 1)/(909^{1/3} - 1) = 993 \ \mu^2$. For $N = 4$

$$\left(\frac{C_L}{C_{IN}}\right)^{1/4} = 5.49 \quad \Rightarrow \quad T_D = 4 \times (0.014 \times 5.49 + 0.098) = 0.699$$

with an area $= (2 \times 9.5 \times 0.5) \times (909 - 1)/(909^{1/4} - 1) = 1920 \ \mu^2$!

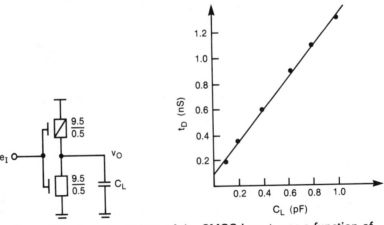

Figure 23. Average delays of the CMOS inverter as a function of load.

Since the difference in area is huge and the performance improvement is negligible, take $N_o = 3$. The small difference in performance indicates that the T_D versus N relation is a relatively shallow convex curve around the optimal point which is, as we saw in Section 2.2.2, fairly characteristic of many cases in MOS circuit design. Due to the area restriction, the number of stages in a chain driver seldom exceeds 4.

Due to the difference in mobility and possible threshold voltage unsymmetry, a p-channel device is slower than an n-channel device of the same size. As a result designing different β ratios between p-channel and n-channel devices is suggested to make the risetime and falltime equal. The improvement is most noticeable for single-stage drivers where the input voltage comes from an ideal voltage source with equal rise and falltimes. Since a p-channel device always drives an n-channel device in internal circuits, however, the difference in rise and fall delay is automatically compensated after a few stages. It is therefore more important to minimize the area and line delay than artificially equalizing the rise- and falltime by adjusting the β ratios for each individual inverter. Interested readers are referred to the pertinent literature.[11]

3. OFFCHIP DRIVERS

In addition to delivering high power to its load, an offchip driver, or OCD, must also satisfy specific system requirements. First, the output voltage levels must comply with the technology of the load it drives. The output of a TTL driver, for example, switches between an MPDL = 0.6 V and an LPUL = 2.4 V. Second, almost all OCDs are tristate drivers, and most cable drivers are equipped with short-circuit protection. Furthermore, drivers that drive long lines must be terminated properly. Although all these features are not needed in all applications, most OCDs are equipped with most of them.

3.1 TTL Compatibility

The output logic levels of most OCDs are specified with sink and source currents such as MPDL = 0.6 V at I (sink) = 8 mA and LPUL = 2.4 V at I (source) = 4 mA. This is because TTL receivers draw finite input currents, and the output devices of the MOS driver must be large enough to accommodate the currents while still being able to maintain output voltage levels. If a driver drives MOS receivers, the size of the output devices is only determined by the speed required to charge or discharge the load. Then, the current specifications do not apply.

Since the output devices are very large, their threshold voltages are lower. Lower threshold voltage means larger leakage current. For TTL applications, the channel length of the output devices is usually designed slightly longer than the minimum to reduce the leakage current when the device is off.

3.2 Tristate Drivers

In many applications, a heavy bus is used to allow more than one chip to be attached to a single line. Since the line can only be occupied by a single signal at any given time, it is necessary to deactivate all "standby" OCDs. In this case the output of each OCD is controlled by two signals: the normal "data signal" driving the pushpull output in a complementary fashion; and an "enable signal" that can turn off both output devices simultaneously, thereby detaching the OCD from the bus. Such OCDs, therefore, have three "states": output high, output low, and output floating; hence the name tristate.

The design of a tristate driver is the same as the design of an ordinary driver except some simple logic has to be incorporated to control the output devices separately. Two such circuits are shown in Figure 24. In Figure 24(a) the input $A0$ is the data signal that drives the output either high or low. The control signals $B0$ and $B1$ are low when the driver is in the active mode. In standby mode, one of these signals goes high, turning off both output devices. The output node ⑩ is then left floating and is driven by other circuits on the bus.

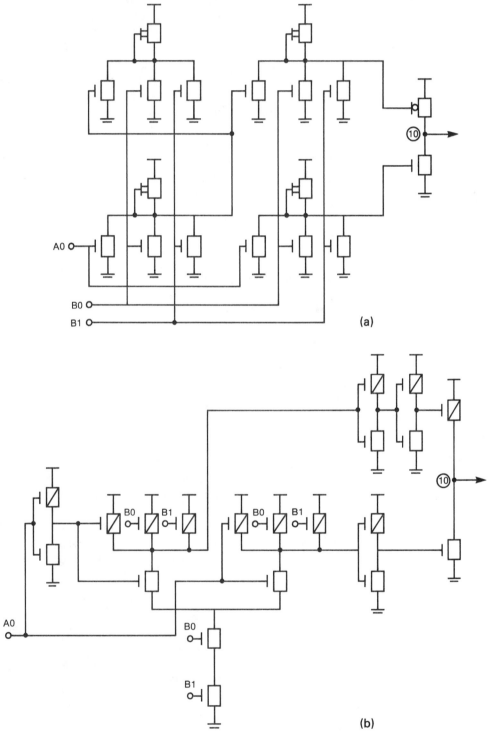

Figure 24. (a) A tristate NMOS OCD. (b) A tristate CMOS OCD.

Two control signals are used for the driver. Input $B0$ tristates the driver in normal operations and hence is controlled by internal logic. Input $B1$ is for testing. The $B1$ inputs of all drivers on the same chip are tied together to a chip I/O pad. During its operation, the tester can detach the chip from the bus by holding $B1$ high while testing some other chip on the same bus.

The CMOS driver shown in Figure 24(b) operates in the same fashion except that $v_{10} = \overline{A0}$ and both $B0$ and $B1$ are high when the driver is in the active mode.

3.3 Short-Circuit Protection

The short-circuit protection feature is found in OCDs that drive lines connecting different "boxes" of a machine. The output of such a driver is subject to being short circuited (by mistake) during either installation or service in the field. Since the duration of short circuiting is unpredictable, special circuitry must be provided to protect the driver from burning out by a large output current.

Simple short-circuit protection circuit is shown in Figure 25. The inverters (T_n, T'_n), n = 1, 2, 3, and 4, form a driver chain with the output v_{10} in phase with the input e_1. The positive feedback from v_{10} to node ① via devices T'_7 and T_8 is controlled by the buffer (T_5, T'_5), (T_6, T'_6) with appropriate delay. To see how the circuit works, assume input $e_1(t)$ rises from 0 to V_H at time $t = 0$. For $t < 0$ both v_{10} and v_5 are low. Since the n-channel devices T_7 and T_8 are off, the feedback circuit is cut off from the driver chain. Now the rise of e_1 propagates through the chain, and the output v_{10} moves up and eventually cuts off device T'_7. If the delay of v_5 is such that T_7 turns on after T'_7 has turned off, the feedback circuitry will not interfere with normal operation and v_{10} will stay high thereafter. Now, suppose v_{10} is shorted to ground by mistake. Since T'_4 is on, a large current flows through T'_4 to GD; however, as soon as v_{10} goes down device T'_7 turns on. Since T_7 is also on, v_1 starts to move up. If the β ratio between (T_7, T'_7) and T_1 is such that v_1 rises beyond $\overline{V}_{ts}$ of (T_2, T'_2), v_2 falls and v_3 rises. The rise of v_3 turns off T'_4 and shuts off the output current. The only power dissipation of the circuit will be associated with the current through (T_7, T'_7, T_1), which can be controlled by restricting the size of T_1.

Operations in the case where e_1 switches from V_H to GD can be analyzed in the same fashion.

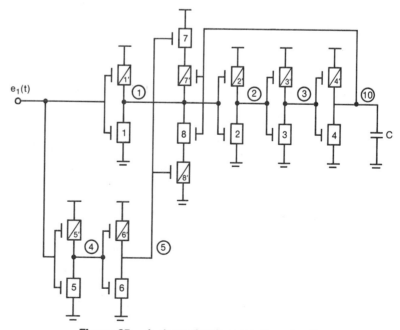

Figure 25.   A short-circuit protection circuit.

Because of the race condition between v_{10} and v_5, the delay of v_5 should be longer than the delay from e_1 to v_{10}, but shorter than the cycle time of the OCD.

4. MOS LINE DRIVERS

In previous sections we have used lumped capacitors to model the driver loads. Such capacitors can be obtained from the circuit layout. While these capacitances seem to be "real" parameters directly related to "physical" elements, the concept of lumped circuits, and, for that matter, the associated Kirchhoff's laws, are nevertheless approximations to the more general Maxwell equations. In a lumped circuit, signals are assumed to travel between circuit elements without delay. This assumption holds true when the physical dimension of the circuit is small compared with the wavelength of the signal. In reality, since signals propagate with finite velocity, a wire becomes a waveguide if the rise/falltime of the signal is much shorter than the propagation delay from one end of the wire to the other. In such cases, the lumped model becomes inadequate, and the wire must be modeled as a transmission line.

In this section, the transmission line theory is applied to the design of MOS line drivers and receivers. A brief review of the basic transmission line theory can be found in Appendix 2 to this chapter.

4.1 Analysis of Ideal Transmission Lines

An ideal transmission line has three main parameters: its inductance per unit length L, its capacitance per unit length C, and the line length ℓ. Given L and C, the characteristic resistance (impedance) Z_0 and the propagation velocity u along the line are then given by

$$Z_0 = \sqrt{\frac{L}{C}} \quad \text{and} \quad u = \frac{1}{\sqrt{LC}}$$

The propagation delay of the line, T_D, defined as the time taken by a signal to travel from one end of the line to the other, is given by

$$T_D = \frac{\ell}{u} = \ell\sqrt{LC}$$

Either ℓ or T_D is sometimes called the "time of flight" of the line.

Thus, given the length, a transmission line is completely characterized by two parameters: Z_0 and T_D.

Consider the simple case of a linear driver/receiver pair shown in Figure 26. At the source, in the left end, is the output characteristic of the driver modeled by its Thevenin's equivalent circuit ($e_S(t)$, R_S). At the load, in the right end, is the input characteristic of the receiver modeled by a linear resistor R_L. Let the input $e_S(t) = E_S u(t)$ be a step function of magnitude E_S. At $t = 0$, a wave is generated, continuing down the line. The wavefronts of the voltage and current are given by

$$E = \frac{Z_0}{Z_0 + R_S}E_S \quad \text{and} \quad I = \frac{E}{Z_0} \tag{26}$$

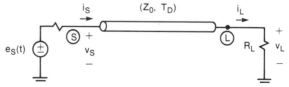

Figure 26. A typical transmission line configuration.

which travel along the line with velocity u. The wavefronts reach the load at $t = T_D$. Since $R_L \neq Z_0$ in general, a reflected wave of magnitude $(\Delta E, \Delta I)$ is generated at R_L such that

$$\frac{E + \Delta E}{I + \Delta I} = R_L \tag{27}$$

The reflected wave propagates from the load toward the source. Since the direction of the current is reversed,

$$\frac{\Delta E}{\Delta I} = -Z_0 \tag{28}$$

Substituting Equations (26) and (28) into Equation (27), we obtain

$$\frac{\Delta E}{E} = \rho_L \quad \text{and} \quad \frac{\Delta I}{I} = -\rho_L$$

where

$$\rho_L \triangleq \frac{R_L - Z_0}{R_L + Z_0}$$

is called the *reflection coefficient* at R_L. This equation characterizes the effect of transmission line discontinuity at the load. It is worthwhile to take a close look at the meaning of the reflection coefficient. First, $|\rho_L| < 1$. $\rho_L > 0$ if $R_L > Z_0$ and the reflected voltage reinforces the incident. In the limiting case where $R_L = \infty$, $\Delta E = E$ results in a total positive voltage reflection. On the other hand, $\rho_L < 0$ if $R_L < Z_0$, and the incident voltage is reduced by the reflection. In case $R_L = 0$ and the line is shorted to the ground at the load, $\Delta E = -E$ and the incident voltage is totally annihilated at R_L.

Reflections of currents can be interpreted in the same fashion except that the current reflection coefficient is always equal in magnitude, but opposite in sign, to that of voltage.

Thus, at $t = T_D$, the voltage and current at R_L are given by

$$v_L = E + \Delta E = (1 + \rho_L)E$$

and

$$i_L = I + \Delta I = (1 - \rho_L)I$$

The reflected wave reaches the source at $t = 2T_D$. Again, in case $R_S \neq Z_0$, a second reflection occurs. Defining

$$\rho_S \triangleq \frac{R_S - Z_0}{R_S + Z_0}$$

it is easy to show that the magnitudes of the second reflection due to $(\Delta E, \Delta I)$ are given by $(\rho_S \Delta E, -\rho_S \Delta I)$. By superposition, the voltage and current at node $\circledS$ are equal to

$$v_S = E + (1 + \rho_S)\Delta E = (1 + \rho_L + \rho_L \rho_S)E$$

and

$$i_S = I + (1 - \rho_S)\Delta I = (1 - \rho_L + \rho_L\rho_S)I$$

As this reflected wave reaches R_L at $t = 3T_D$, another reflection is generated at R_L and travels back toward the source. In other words, multiple reflections take place. As an exercise, show that at time t,

$$v_S(t) = Eu(t) + (1 + \rho_S)\rho_L E \sum_{n=0}^{\infty} \rho_L^n \rho_S^n u[t - 2(n + 1)T_D] \qquad (29)$$

and

$$v_L(t) = (1 + \rho_L) E \sum_{n=0}^{\infty} \rho_L^n \rho_S^n u[t - 2(n + 1)T_D] \qquad (30)$$

where E is defined in Equation (26).

EXERCISE

Derive the corresponding expressions for the currents $i_S(t)$ and $i_L(t)$.

The multiple reflections can also be analyzed with a graphical method. In Figure 27(a) the curve

$$\mathcal{L}_S : v = E_S - R_S i$$

represents the output characteristic of the driver and

$$\mathcal{L}_L : v = R_L i$$

represents the input characteristic of the receiver. Prior to $t = 0$, $i_S = i_L = 0$. At $t = 0$, v_S jumps from 0 to E and i_S jumps from 0 to I and a wave of (E, I) starts to travel from the source to the load along the line. Since E and I must satisfy $E = Z_0 I$, the value of (E, I) is the intersection of $\mathcal{L}_S$ and the line $v = Z_0 i$, which is shown in Figure 27(a) as point $\mathbf{S}^0$. Both v_S and i_S will stay at $\mathbf{S}^0$ for the period $t \in [0, 2T_D)$ until the reflection from the load reaches the source at $t = 2T_D$. Now, at $t = T_D$, the wave reaches R_L and generates a reflection $(\Delta E, \Delta I)$. Since $\Delta E/\Delta I = -Z_0$, the voltage and current of R_L at $t = T_D$ can be obtained by drawing a line segment from $\mathbf{S}^0$ with slope $-Z_0$ toward $\mathcal{L}_L$. The intersection, $\mathbf{L}^1$, then, represents the value of v_L and i_L at $t = T_D$. The values of the voltage and current at this point will stay constant for $t \in [T_D, 3T_D)$. By the same construction and argument, it is easy to see that in Figure 27(a), point $\mathbf{S}^2$ represents (v_S, i_S) for $t \in [2T_D, 4T_D)$ and $\mathbf{L}^3$ represents (v_L, i_L) for $t \in [3T_D, 5T_D)$, and so on. The sequences $\mathbf{S}^{2n}$ and $\mathbf{L}^{2n+1}$ for $n = 0, 1, 2, \ldots$ eventually converges to the intersection of $\mathcal{L}_L$ and $\mathcal{L}_S$, which is, of course, the DC solution of the circuit.

The waveforms of v_S and v_L are also shown in Figure 27(b) as functions of time. Since $\rho_L > 0$ and $\rho_S < 0$, the waveforms "ring." For example, v_L overshoots for $t \in [T_D, 3T_D)$ and undershoots for $t \in [3T_D, 5T_D)$. However, the amplitude of the oscillation decreases monotonically with time t.

In all previous analyses, we have assumed that the load was purely resistive as was represented by R_L. In reality, however, there is always a certain amount of ca-

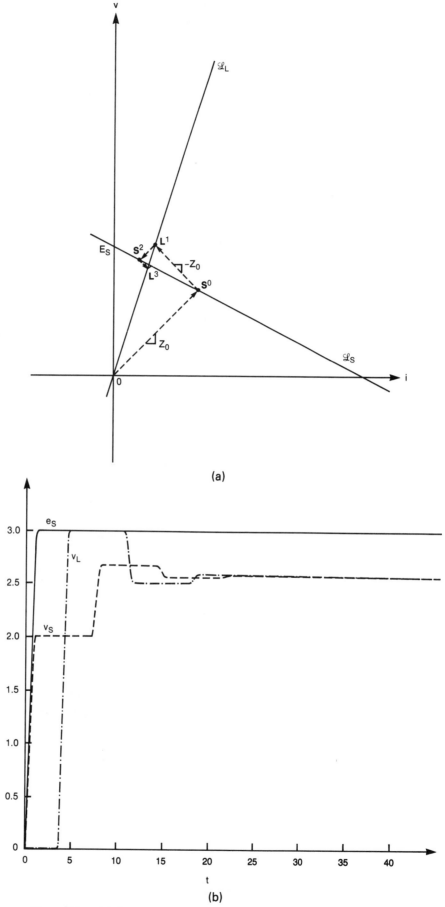

(a)

(b)

Figure 27. (a) Graphical construction of reflections on a transmission line. (b) Waveforms of v_S and v_L for a transmission line with $(Z_0, T_D) = (0.080, 3.5)$. (c) Waveforms in (b) with 5 pF pad capacitance.

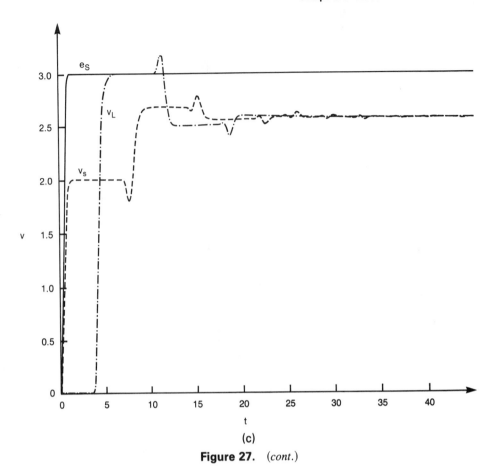

Figure 27. (*cont.*)

pacitance and inductance associated with the input pad of a chip. Figure 27(c) shows the waveforms under the same conditions as Figure 27(b), except that a 5 pF capacitance is added to the load. The "extra bumps" on the waveforms are a result of the capacitor trying to maintain a continuous voltage jump. Transmission line analysis with reactive elements is similar to the above analysis except that the load impedance Z_L is now a complex number. We will now apply this method to analyze MOS driver/receiver pairs.

4.2 Analysis of MOS Driver/Receiver Pairs

The graphical method can be easily applied to the analysis of MOS driver/receivers. The output characteristic of an MOS driver is nonlinear, and the input characteristic of an MOS receiver is purely capacitive and so is practically an open circuit. The procedure of analysis is best illustrated by an example.

EXAMPLE 8

Consider the driver/receiver pair shown in Figure 28. Assume the LTP = 1.0 V and UTP = 2.0 V for the receiver. Estimate the delays t_{DLH} and t_{DHL} for the receiver.

Solution

Let us consider the up transition first. The v–i characteristic of the driver's pullup device with its gate tied to *GD* is shown in Figure 29(a) as $\mathcal{L}_S$, and the vertical axis $i = 0$ is $\mathcal{L}_L$. Starting at the DC solution $(v,i) = (0, 0)$, the line

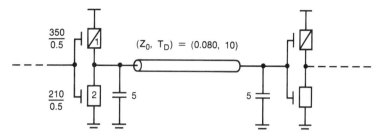

Figure 28.

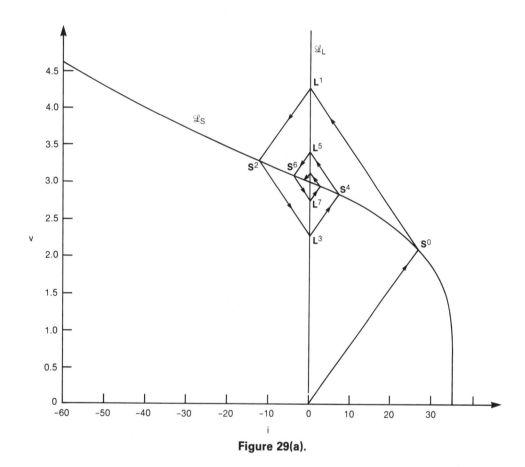

Figure 29(a).

segment of slope Z_0 intersects $\mathcal{L}_S$ at $\mathbf{S}^0$. This gives the output voltage and current of the driver for $t \in [0, 2T_D)$. By drawing a line segment of slope $-Z_0$ from $\mathbf{S}^0$ to intersect $\mathcal{L}_L$, we obtain $\mathbf{L}^1$, which gives the voltage and current at the receiver for $t \in [T_D, 3T_D)$. Since the load is an open circuit, the incident voltage doubles, and the receiver switches shortly after $t = T_D$. However, the voltage rings thereafter, with no effect on the receiver. The delay t_{DLH} is therefore approximately T_D.

For the down transition the output characteristic of the driver is obtained by plotting the v–i curve of the pulldown with its gate tied to V_H. The DC solution prior to time 0 is $(v, i) = (V_H, 0)$. Following the same procedure, we obtain the voltage and current at the receiver, $\mathbf{L}^i$, $i = 1, 3, \ldots$. Since $\mathbf{L}^3$ is below 2.0 V, the delay t_{DHL}, is also approximately T_D.

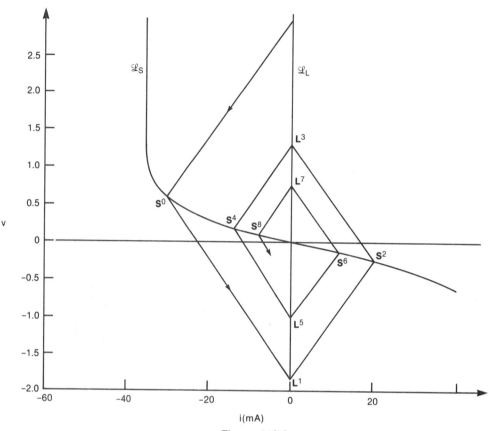

Figure 29(b).

The waveforms of both transitions are shown in Figure 30. One particular concern is the large negative overshoot in the down transition of more than 1.5 V. A large overshoot below the ground turns on the ESD protective diode of the receiver, resulting in large transient currents. The cumulative effect of these currents causes electromigration wearout of metallization as well as slow degradation of the ESD diode. [13]. Since the overshoot in our example lasts as long as $2T_D$, some means of reducing it must be considered.

4.3 Termination of Transmission Lines

Both the overshoots and the ringing can be reduced by proper line termination. In practice there are three methods:

(1) Parallel termination: To eliminate reflection *from the receiver*, a pair of resistors R_1 and R_2 are tied to the load as shown in Figure 31. The parallel combination of $R_1 \parallel R_2$ equals Z_0. The input characteristic of the receiver, $\mathscr{L}_L$, now becomes

$$i = \left(\frac{1}{R_2} + \frac{1}{R_1}\right)v - \frac{V_H}{R_1}$$

$$= \frac{v}{Z_0} - \frac{V_H}{R_1}$$

The intersection of $\mathscr{L}_L$ and the v–i curve of the pullup forms the LPUL, $\overline{\mathbf{L}}$, of the driver, and the intersection of $\mathscr{L}_L$ and the v–i curve of the pulldown forms the

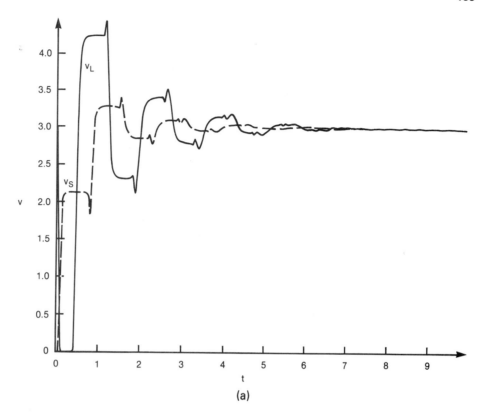

(a)

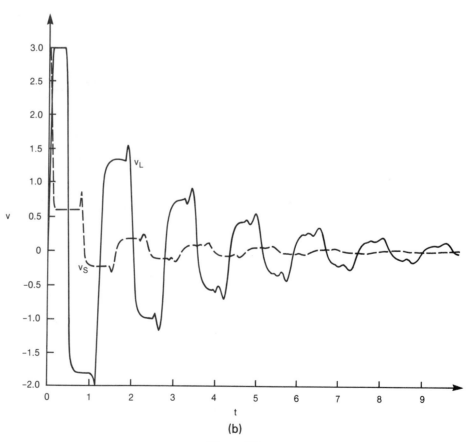

(b)

Figure 30.

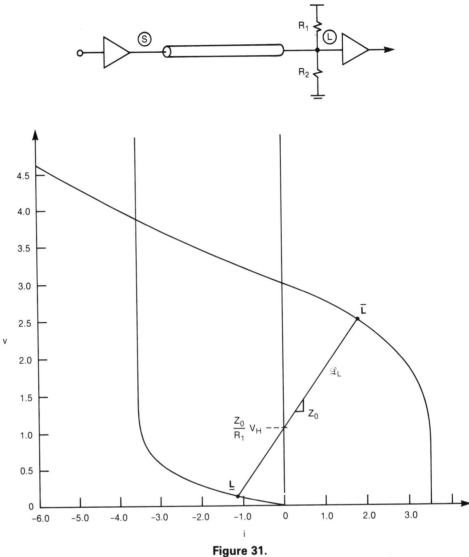

Figure 31.

MPDL, $\underline{L}$, of the driver. The levels can be adjusted by choosing the value of R_1. In operation, both the output of the driver and the input of the receiver switch back and forth between $\overline{L}$ and $\underline{L}$ along $\mathscr{L}_L$ without disturbances. The delays $t_{DLH} = t_{DHL} = T_D$.

The main drawback of parallel termination is high power consumption. If the data rate is less than $1/2T_D$ so that a single reflection from the load can be tolerated, the series termination shown in Figure 32 is more preferable.

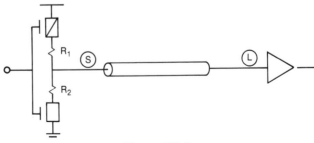

Figure 32(a).

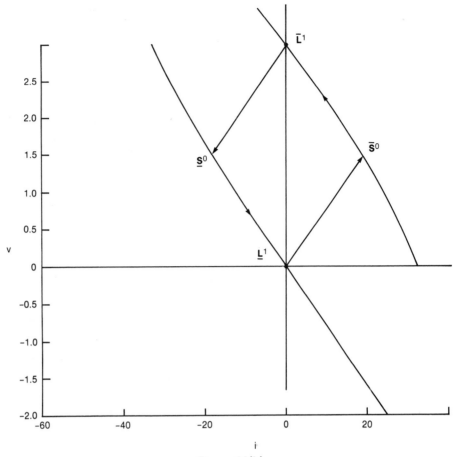

Figure 32(b).

(2) Series termination: In series termination, two resistors R_1 and R_2 are connected to the output devices so that the combination of the resistor and the incremental output resistance of the device is roughly Z_0. See Figure 32(a). Since the output resistances of the device are very small, the series resistances can effectively compensate, though not totally eliminate, the reflections from the source.

The switching characteristics of an example is shown in Figure 32(b). If perfectly compensated the up transition consists of only two points: $\overline{\mathbf{S}}^0$ and $\overline{\mathbf{L}}^1$. The load reaches $\overline{\mathbf{L}}^1$ at $t = T_D$ and the source stabilizes at $\overline{\mathbf{S}}^2 = \overline{\mathbf{L}}^1$ at $t = 2T_D$. The same analysis applies to the down transition.

(3) Circuit clamping: Another approach for reducing reflection in high data rate applications uses a clamp circuit to limit the voltage excursions at the receiver. As shown in Figure 33, device T_1 is biased at $V_{RN} \cong V_{T1}$ above GD so it is off as long as v_L is positive. In a down transition, v_L may overshoot below the ground. As soon as the overshoot happens, device T_1 turns on. If the resistance of R_1 is large enough, node voltage v_1 falls almost to the ground, turning on T_2 strongly. The large current supplied by T_2 will then clamp v_L close to the ground.

The operation of (T_{11}, R_{10}, T_{12}) for positive overshoot control is the same as above except that $V_{RP} \cong V_H + V_{T11}$ so that device T_{11} is normally off when $v_L < V_H$.

The generation of V_{RN} and V_{RP} is straightforward. For example, as long as device T_6 is biased in the saturation region, the output V_{RN} of the T_1 bias circuit can be made equal to the threshold voltage of the n-channel devices. Design details are left to the reader. The analysis of Problem 7 in Chapter 3 may be helpful.

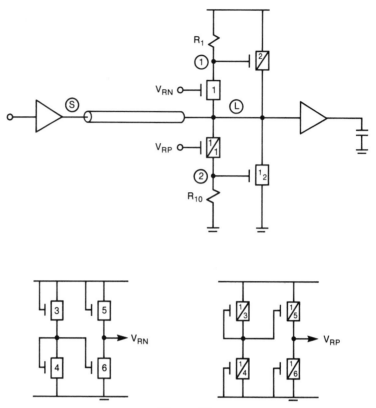

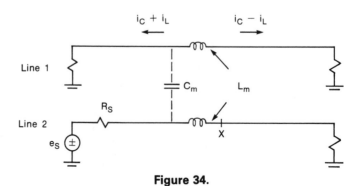

Figure 33.

5. *SYSTEM NOISES*

In addition to the reflections, a driver/receiver pair is also subjected to different kinds of system noises. By system noises we mean the spurious signals generated by the interactions among various parts of the system, which, unlike the random noises picked up from outside, are fairly predictable and controllable through preventative design. Two major sources of such noises are the crosstalks between two lines and the voltage spikes generated by the large transient currents of the drivers.

5.1 Crosstalk

Crosstalk is the noise caused by electric and magnetic coupling and switching activities between two long, parallel lines. A typical example is illustrated in Figure 34,

Figure 34.

where the parameters L_m and C_m are the mutual inductance and capacitance per unit length between the lines, respectively. A voltage source $e_S(t)$ with finite rise/falltime T_R/T_F is attached to line 2, called the active line. Line 1, on the other hand, is not controlled by any active sources, and it is therefore called the quiet line. When the active line switches, the quiet line is also disturbed via electric and magnetic coupling. Thus the two lines "talk" to each other across the space; hence the name crosstalk.

As $e_S(t)$ switches, a wave starts to propagate down line 2. At a particular point x, currents I_C are induced on line 1 through C_m as the wave passes by. These currents flow from x to both ends of line 1.

On the other hand, a current I_L is also induced on line 1 through L_m. Since the magnetic flux increases toward the right, I_L flows to the left; thus, on line 1 a wave $(I_C + I_L)$ is induced that propagates to the left (the near end) of the line, and a wave $(I_C - I_L)$ is induced that propagates to the right (the far end) of the line. The wave that travels to the near end is called the *backward crosstalk,* and the one traveling to the far end, the *forward crosstalk.*

The backward crosstalk is generated at the near end as soon as $e_S(t)$ switches. It lasts for a period of $2T_D$ until the wave generated at the far end of line 1 finally arrives at the near end and is absorbed by the proper termination. In Appendix 2 to this chapter, it is shown that the backward crosstalk at the near end is given by

$$v_b = K_b[e(t) - e(t - 2T_D)]$$

where

$$e(t) = \frac{Z_0}{R_S + Z_0} e_S(t)$$

and

$$K_b \triangleq \frac{\ell}{4T_D}\left(\frac{L_m}{Z_0} + Z_0 C_m\right)$$

is called the *backward crosstalk constant.* A receiver attached to the near end of line 1 will therefore experience a noise voltage pulse whose magnitude is proportional to $e_S(t)$ with a pulsewidth $2T_D$ plus T_R or T_F. A simplified waveform of a typical case is shown in Figure 35(a).

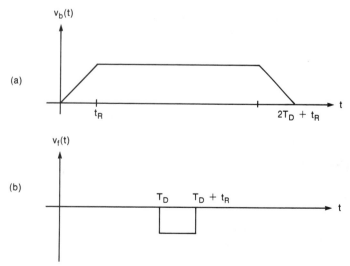

Figure 35. (a) Near-end backward crosstalk. (b) Far-end forward crosstalk.

As shown in Appendix 2 to this chapter, the forward crosstalk at the load is given by

$$v_f = K_f \ell \frac{d}{dt} e(t - T_D)$$

where

$$K_f \triangleq -\frac{1}{2}\left(\frac{L_m}{Z_0} - Z_o C_m\right)$$

is the *forward crosstalk constant*. A receiver attached to the far end of line 1 will therefore experience a noise voltage pulse whose magnitude is proportional to T_R or T_F of $e_S(t)$ with a pulsewidth equal to T_R or T_F. A simplified waveform of a typical case is shown in Figure 35(b).

In many cases, crosstalk problems are more serious than reflections due to signal pattern randomness and long noise duration. In practice grounded lines can be inserted between the signal lines to reduce couplings. Since these grounded lines affect the packaging density, a tradeoff must be made between maximum signal-to-ground line ratio and adequate noise margin.

5.2 Simultaneous Switching of OCDs

When a high-power driver switches, a large transient current is drawn from V_H or GD, and a voltage spike is induced on the power buses according to the relation $v = L \, (di/dt)$ where L is the self-inductance of the power line. Since a driver's output impedance is very low, a quiet driver on the same power bus will propagate the noise spike to its receiver almost without attenuation. This noise, called Δi noise, is most serious when a large number of drivers on the same power bus all switch simultaneously in a short period of time. Since the input impedance of a CMOS receiver is an open circuit, these large, wide-noise pulses double in magnitude when they reach the input of the receiver. It is possible therefore for receivers without adequate noise margins to switch to the wrong state.

The inductances of the power buses not only depend on the layouts of the power lines themselves but also the power pads and the power distribution lines on the module. A complete analysis of the noise margins of all driver/receiver pairs must take both the chip and the packaging into account. After careful analysis, a chip design rule should be issued limiting the maximum number of drivers on each power bus. This rule must be adhered to by chip designers to ensure adequate noise margins.

Unless LPUL = V_H is required, n-channel pullups should be used at the OCD output to reduce the effect of V_H bus disturbance (Why?).

6. CONCLUSION

This chapter has focused on the design of MOS receivers and drivers. Our discussions, though fairly detailed, are necessarily simplified because of the complexity of the problems. All circuits are sensitive to their layouts and no transmission line is ideal in practice. These discussions were meant to serve as a foundation for "practical" designs. In the next two chapters we shall apply these concepts to the design of some "real" products, namely, the MOS memory and logic chips.

REFERENCES

1. Jarvis, D. B. "The Effect of Interconnections on High-Speed Logic Circuits." *IEEE Trans. on Computers,* October 1963, pp. 476–487.

2. Singleton, R. S. "No Need to Juggle Equations to Find Reflection—Just Draw Three Lines." *Electronics,* October 28, 1968, pp. 93–99.

3. Matik, R. E. *Transmission Line for Digital and Communication Networks.* New York: McGraw-Hill, 1969.

4. Chua, L. O. *Nonlinear Network Theory.* New York: McGraw-Hill, 1969.

5. Barna, A. *High Speed Pulse and Digital Techniques.* New York: John Wiley, 1980.

6. DeFalco, J. A. "Reflection and Crosstalk in Logic Circuit Interconnections." *IEEE Spectrum,* July 1970.

7. Carr, W. N., and J. P. Mize. *MOS LSI Design and Application.* New York: McGraw-Hill, 1972.

8. Penney, W. M., and L. Law. *MOS Integrated Circuits.* Huntington, N.Y., Robert E. Krieger, 1979.

9. Blood, W. R. Jr. *MECL System Design Handbook,* 3rd ed., Phoenix, Az.: Motorola Semiconductor Products Inc., 1980.

10. McCarthy, O. J. *MOS Device and Circuit Design.* New York: John Wiley, 1982.

11. Kanuma, A. "CMOS Circuit Optimization." *Solid State Electronics,* Vol. 26, no.1, 1983, pp. 47–58.

12. Annaratone, M. *Digital CMOS Circuit Design.* Hingham, MA: Kluwer, 1986.

13. Dillinger, T. E. *VLSI Engineering.* Englewood Cliffs, NJ: Prentice Hall, 1988.

14. Wang, N. N., and P. W. Chung. "A CMOS Differential Driver." IBM Corporation, US Patent pending.

APPENDICES

1. FORTRAN LISTING OF SUBROUTINE RPUN

This subroutine calculates the output voltage at $t = \tau_R$ with a ramp input $v_I = (V_G/\tau_R)t$. NMOS device is assumed. Greek letter ξ is coded as TSAI instead of XI.

```
      SUBROUTINE RPUN (PW, PL, DVT0, DLMBD, ALPHAP,
     *                 VG,  TAOR,   VO,   CL)


      IMPLICIT REAL*8 (A-H,O-Z)


      COMMON VH, GAMMAN, GAMMAP,  PK1N,  PK1P,  VT0N, VT0P,
     *        PLMBDN, PLMBDP


C --- VH = POWER SUPPLY
C --- GAMMAN = GAMMA OF NMOS
C --- GAMMAP = GAMMA OF PMOS
C --- PK1N = SUBSTRATE SENSITIVITY OF NMOS
C --- PK1P = SUBSTRATE SENSITIVITY OF PMOS
C --- VT0N = THRESHOLD VOLTAGE OF ENHANCEMENT MODE NMOS
C --- VT0P = THRESHOLD VOLTAGE OF ENHANCEMENT MODE PMOS
C --- PLMBDN = LAMBDA OF NMOS
C --- PLMBDP = LAMBDA OF PMOS
```

```
C --- DVT0 = DELTA VT0, MODIFIER OF VT0
C --- DLMBD = DELTA LAMBDA, MODIFIER OF PLMBDN OR PLMBDP
C --- ALPHAP = ALPHA P, A FACTOR FOR PARASITIC CAPACITANCES

C --- VO = OUTPUT VOLTAGE
C --- TAOR = RISETIME OF THE INPUT WAVEFORM
C --- VG = UP LEVEL OF THE INPUT WAVEFORM
C --- MAX = MAXIMUM NUMBER OF ITERATIONS, DEFAULT 20
C --- SS = STEP SIZE FOR CALCULATION OF DERIVATIVES, DEFAULT 1.0D-6

      MAX = 20
      SS = 1.0D-6

      VT0 = VT0N + DVT0
      PLMBD = PLMBDN + DLMBD
      PK1 = PK1N
      C = (1.0 + ALPHAP)*CL
      BETA = GAMMAN*PW/PL

C --- CONSTANTS

      PK = DSQRT(2.0*(PLMBD**2)*C*VG/(BETA*(1.0 + PK1)*TAOR))
      PKI = 1.0/PK

C --- CALCULATION OF THE CONSTANTS OF FUNCTION F

      F0 = -2.0*DLOG(PK)
      FK = (1.0 + PK1)*BETA*(1.0-VT0/VH)*TAOR/(PLMBD*C)

      WRITE(6,3)  F0, FK
    3 FORMAT (1H1,/ 2(F10.5/))

C --- START ITERATION

      ITER = 0

C --- INITIAL GUESS

  100 DPK = 5.0D-3
      TSAI = PK - DPK

  150 F =(-1.0+PKI)*DLOG(DABS(TSAI+PK))-(1.0+PKI)*DLOG(DABS(TSAI-PK))
     *    -F0 -FK

      IF (F .GT. 0.D0)  GO TO 200

      DPK = DPK/2.0
      IF (DPK .LE. 1.D-5) GO TO 400
      GO TO 100

  200 TSAIP = TSAI + SS
      TSAIM = TSAI - SS

      WRITE(6,4)  TSAI, TSAIP, TSAIM
    4 FORMAT (1H1,/ 3(F10.5/))

      FP=(-1.0+PKI)*DLOG(DABS(TSAIP+PK))-(1.0+PKI)*DLOG(DABS(TSAIP-PK))
     *    -F0-FK
```

```
      FM=(-1.0+PKI)*DLOG(DABS(TSAIM+PK))-(1.0+PKI)*DLOG(DABS(TSAIM-PK))
     *    -F0-FK
      WRITE(6,5)  F, FP, FM
    5 FORMAT (1H1,/ 3(F10.5/))

C --- DERIVATIVE OF F

      DF = (FP-FM)/(2.0*SS)

      WRITE(6,6)  DF
    6 FORMAT (1H1,/ 1(E15.7/))

      IF (DABS(DF).GE. 1.0D-8) GO TO 250
      DF = 1.0D-8

C --- ITERATION ON TSAI

      IF (ITER .GE. MAX) GO TO 500

  250 TSAIP = TSAI - F/DF

      WRITE(6,7)  TSAIP
    7 FORMAT (1H1,/ 1(F10.5/))

C --- TEST FOR CONVERGENCE

      DTSAI = DABS(TSAIP - TSAI)

      IF (DTSAI .LE. 1.0D-5) GO TO 300

      TSAI = TSAIP

      ITER = ITER + 1

      GO TO 150

  300 VO = (VG - VTO -((TSAI+1.0)**2-1.0)/(2.0*PLMBD))/(1.0+PK1)

      WRITE(6,1)  TSAI, VO
    1 FORMAT (1H1,/ 2(F10.5/))

      RETURN

  400 WRITE(6,8)
    8 FORMAT (' IMPROPER ASSUMPTION' )

  500 WRITE(6,9)
    9 FORMAT (' DIVERGENCE')

      RETURN
      END
```

2. IDEAL TRANSMISSION LINE THEORY

An ideal transmission line system consists of two straight, parallel lines with uniform surroundings. The term "lines" means conductors. In practice one of the lines may be a wire while the other "line" may be another wire or just a flat ground plane. By uniform surroundings, we mean that both the material and the cross-sectional geometry of the whole system must be uniform throughout the length of the lines.

The analysis of the general cases of transmission lines involves the Maxwell's equations in their most general forms. The results are extremely complicated. Fortunately in digital applications, the mode of the wave is mostly TEM (transverse electromagnetic) in the sense that the electric and magnetic fields are always orthogonal to each other. Under this condition, the currents and voltages on the lines represent the fields (why?), and the transmission line system can be simply analyzed as a distributed RLC network. Let us consider the distributed RLC network shown in Figure 36 where the parameters L and C are the inductance and capacitance per unit length of the line, and R and G are the resistances per unit length accounting for the power loss and leakage current. To derive the equations for the voltage and current propagation, consider a small section of the line of length Δx at point x. Summing the voltages around the loop $\circled{x}-\boxed{x + \Delta x}-GD$, we obtain

$$v(t, x + \Delta x) - v(t, x) = -i(t, x)R\Delta x - L\Delta x \frac{\partial}{\partial t} i(t, x)$$

Notice that v and i are functions of both time t and position x. Divide both sides by Δx and let $\Delta x \to 0$, we obtain a partial differential equation:

$$\frac{\partial}{\partial x} v(t, x) = -i(t, x)R - L\frac{\partial}{\partial t} i(t, x)$$

Similarly, by summing the currents at node $\boxed{x + \Delta x}$, it can be shown

$$\frac{\partial}{\partial x} i(t, x) = -v(t, x)G - C\frac{\partial}{\partial t} v(t, x)$$

In applications where the operating frequencies are moderately high (say, less than 100 MHz), the effects of the resistances are generally negligible. By setting $R = G = 0$ in the above, we then obtain the standard wave equations:

$$\frac{\partial}{\partial x} v(t, x) = -L\frac{\partial}{\partial t} i(t, x)$$

$$\frac{\partial}{\partial x} i(t, x) = -C\frac{\partial}{\partial t} v(t, x)$$

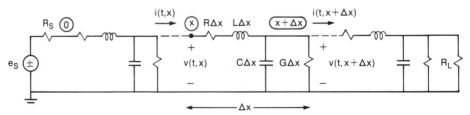

Figure 36. The RLC model of a typical transmission line.

Taking the Laplace transform of the wave equations with respect to the time t, we have

$$\frac{d}{dx}\hat{v}(x) = -sL\hat{i}(x)$$

$$\frac{d}{dx}\hat{i}(x) = -sC\hat{v}(x)$$

where $\hat{v}(x) = \hat{v}(s, x) = \mathcal{L}\{v(t, x)\}$ and $\hat{i}(x) = \hat{i}(s, x) = \mathcal{L}\{i(t, x)\}$.

Rewriting the above in state equation form, we obtain

$$\frac{d}{dx}\begin{bmatrix} \hat{v} \\ \hat{i} \end{bmatrix} = s\begin{bmatrix} 0 & -L \\ -C & 0 \end{bmatrix}\begin{bmatrix} \hat{v} \\ \hat{i} \end{bmatrix} \tag{A1}$$

(1) Solution of the Wave Equations: It is left to the reader to show that the state transition matrix of Equation (A1) is given by

$$\Phi(x - x_0) = \begin{bmatrix} \cosh \Omega(x - x_0) & -Z_0 \sinh \Omega(x - x_0) \\ -Z_0^{-1} \sinh \Omega(x - x_0) & \cosh \Omega(x - x_0) \end{bmatrix}$$

and the solution of Equation (A1) is therefore given by

$$\begin{bmatrix} \hat{v}(x) \\ \hat{i}(x) \end{bmatrix} = \begin{bmatrix} \cosh \Omega(x - x_0) & -Z_0 \sinh \Omega(x - x_0) \\ -Z_0^{-1} \sinh \Omega(x - x_0) & \cosh \Omega(x - x_0) \end{bmatrix}\begin{bmatrix} \hat{v}(x_0) \\ \hat{i}(x_0) \end{bmatrix}$$

where $\Omega \triangleq s\sqrt{LC}$ and $Z_0 \triangleq \sqrt{L/C}$ is called the characteristic impedance (resistance).

Since it is a linear system, let us assume $x_0 = 0$. Expanding the hyperbolic functions into exponentials, we have

$$\begin{bmatrix} \hat{v}(x) \\ \hat{i}(x) \end{bmatrix} = \frac{e^{\Omega x}}{2}\begin{bmatrix} 1 & -Z_0 \\ -Z_0^{-1} & 1 \end{bmatrix}\begin{bmatrix} \hat{v}(0) \\ \hat{i}(0) \end{bmatrix} + \frac{e^{-\Omega x}}{2}\begin{bmatrix} 1 & Z_0 \\ Z_0^{-1} & 1 \end{bmatrix}\begin{bmatrix} \hat{v}(0) \\ \hat{i}(0) \end{bmatrix}$$

$$\triangleq e^{\Omega x}\mathbf{T}_1\begin{bmatrix} \hat{v}(0) \\ \hat{i}(0) \end{bmatrix} + e^{-\Omega x}\mathbf{T}_2\begin{bmatrix} \hat{v}(0) \\ \hat{i}(0) \end{bmatrix} \tag{A2}$$

where

$$\mathbf{T}_1 = \frac{1}{2}\begin{bmatrix} 1 & -Z_0 \\ -Z_0^{-1} & 1 \end{bmatrix} \quad \text{and} \quad \mathbf{T}_2 = \frac{1}{2}\begin{bmatrix} 1 & Z_0 \\ Z_0^{-1} & 1 \end{bmatrix}$$

EXERCISE

Prove that both $\mathbf{T}_1$ and $\mathbf{T}_2$ are singular, $\mathbf{T}_1 + \mathbf{T}_2 = \mathbf{I}$, $\mathbf{T}_1\mathbf{T}_2 = \mathbf{T}_2\mathbf{T}_1 = 0$, and that $\mathbf{T}_1^2 = \mathbf{T}_1$ and $\mathbf{T}_2^2 = \mathbf{T}_2$.

Since $\mathbf{T}_1$ and $\mathbf{T}_2$ have the same null space, the matrix equation (A2) yields only one linearly independent equation. The reader is invited to verify this fact by multiplying out the equations and checking for their linear dependency.

In Section 4.1 we assumed the propagation of voltages and currents along a transmission line as waves. To see the wave nature of the solution, let us consider the semiinfinite line shown in Figure 37. Since $\hat{v}(x)$ and $\hat{i}(x)$ must be finite as $x \to \infty$, the coefficient of the positive exponential in Equation (A2) must be zero. That is,

$$\frac{e^{\Omega x}}{2} \begin{bmatrix} 1 & -Z_0 \\ -Z_0^{-1} & 1 \end{bmatrix} \begin{bmatrix} \hat{v}(0) \\ \hat{i}(0) \end{bmatrix} = \begin{bmatrix} 0 \\ 0 \end{bmatrix}$$

$$\Rightarrow \qquad\qquad \hat{v}(0) = Z_0 \hat{i}(0) \qquad\qquad\qquad (A3)$$

Substitution of the above relation back to Equation (A2) then yields

$$\begin{bmatrix} \hat{v}(x) \\ \hat{i}(x) \end{bmatrix} = e^{-\Omega x} \begin{bmatrix} \hat{v}(0) \\ \hat{i}(0) \end{bmatrix} = e^{-s\sqrt{LC}\,x} \begin{bmatrix} \hat{v}(0) \\ \hat{i}(0) \end{bmatrix}$$

In time domain this means

$$\begin{bmatrix} v(t, x) \\ i(t, x) \end{bmatrix} = \begin{bmatrix} v(t - \sqrt{LC}\,x, 0) \\ i(t - \sqrt{LC}\,x, 0) \end{bmatrix}$$

Thus the waveforms of $v(t, x)$ and $i(t, x)$ are replica of the waveforms of $v(t, 0)$ and $i(t, 0)$, respectively, with a time delay of $\sqrt{LC}\,x$. In other words, a wave is propagated along the line. The propagation velocity of the wave can be obtained by keeping the phase $(t - \sqrt{LC}\,x)$ constant and calculate the derivative of the position x with respect to time t

$$u = \frac{d}{dt}x = \frac{1}{\sqrt{LC}}$$

It is also clear from Equation (A3) that the voltages and currents at any point x are related by the characteristic impedance Z_0:

$$v(t, x) = Z_0 i(t, x)$$

Both u and Z_0 have been mentioned in Section 4.1 as parameters that characterize a transmission line.

Figure 37.

(2) Termination of a Transmission Line of Finite Length: In our application the length of the transmission line is finite. Let the length of the line be ℓ. Substituting $x = \ell$ in Equation (A2) and premultiplying both sides of the equation by $\mathbf{T}_1$, we obtain

$$e^{-\Omega \ell}\,\mathbf{T}_1 \begin{bmatrix} \hat{v}(\ell) \\ \hat{i}(\ell) \end{bmatrix} = \mathbf{T}_1 \begin{bmatrix} \hat{v}(0) \\ \hat{i}(0) \end{bmatrix}$$

Substitution of the above equation into Equation (A2) then gives

$$\begin{bmatrix} \hat{v}(x) \\ \hat{i}(x) \end{bmatrix} = e^{-\Omega x}\,\mathbf{T}_2 \begin{bmatrix} \hat{v}(0) \\ \hat{i}(0) \end{bmatrix} + e^{-\Omega(\ell - x)}\,\mathbf{T}_1 \begin{bmatrix} \hat{v}(\ell) \\ \hat{i}(\ell) \end{bmatrix} \qquad (A4)$$

The above equation states that the complete solution of the wave equation at any point x is the superposition of two waves: a wave traveling from the source to the load as is represented by the first term and a wave traveling from the load to the source as is represented by the second term.

The exact waveforms of the voltages and currents are determined by the input and the termination conditions at the ends of the transmission line. For example, in a linear system the termination conditions can be generally specified by their Thevenin's equivalent circuits

$$\hat{v}(0) = \hat{e}_S - R_S \hat{i}(0)$$

and

$$\hat{v}(\ell) = R_L \hat{i}(\ell) + \hat{e}_L \tag{A5}$$

as shown in Figure 38. The procedure is best illustrated by an example shown below.

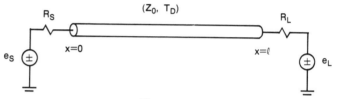

Figure 38.

Before we delve into the details, however, let us define two more important parameters we have mentioned in Section 4.1: the reflection coefficients. Multiplying out Equation (A3), we obtain

$$\hat{v}(x) = \frac{1}{2} e^{\Omega x} [\hat{v}(0) - Z_0 \hat{i}(0)] + \frac{1}{2} e^{-\Omega x} [\hat{v}(0) + Z_0 \hat{i}(0)]$$

Thus, if we stand at point x and look into the source, we see a wave of $[\hat{v}(0) - Z_0 \hat{i}(0)]$ traveling toward the source and a wave of $[\hat{v}(0) + Z_0 \hat{i}(0)]$ traveling out of the source. In the limiting case when $x \to 0$, the wave that is incident to the source is given by $[\hat{v}(0) - Z_0 \hat{i}(0)]$ and that reflected from the source is given by $[\hat{v}(0) + Z_0 \hat{i}(0)]$. By definition of the reflection coefficient

$$\rho_S = \frac{\hat{v}(0) + Z_0 \hat{i}(0)}{\hat{v}(0) - Z_0 \hat{i}(0)}$$

$$= \frac{\hat{v}(0) + Z_0(-\hat{v}(0)/R_S)}{\hat{v}(0) - Z_0(-\hat{v}(0)/R_S)}$$

$$= \frac{R_S - Z_0}{R_S + Z_0}$$

Here we have used $\hat{v}(0) = -R_S \hat{i}(0)$ by setting $\hat{e}_S = 0$ in Equation (A5) because the reflection coefficient is an incremental relationship. Similarly, it is left to the reader to show that

$$\begin{bmatrix} \hat{v}(x) \\ \hat{i}(x) \end{bmatrix} = e^{\Omega(\ell-x)} \mathbf{T}_2 \begin{bmatrix} \hat{v}(\ell) \\ \hat{i}(\ell) \end{bmatrix} + e^{-\Omega(\ell-x)} \mathbf{T}_1 \begin{bmatrix} \hat{v}(\ell) \\ \hat{i}(\ell) \end{bmatrix} \tag{A6}$$

and the reflection coefficient at the load

$$\rho_L = \frac{R_L - Z_0}{R_L + Z_0}$$

Let us now see the example.

EXAMPLE

Consider the transmission line shown in Figure 38. The source and load of the line are terminated by the following conditions:

$$\hat{v}(0) = \hat{e}_S - R_S \hat{i}(0) \tag{A7}$$

and

$$\hat{v}(\ell) = R_L \hat{i}(\ell) \tag{A8}$$

where $R_S = 0.2 Z_0$ and $R_L = 4 Z_0$. Express the voltage at the load, $v(t, \ell)$, in terms of the input voltage $e_S(t)$.

Solution

From Equation (A4) we have

$$\hat{v}(x) = \frac{1}{2} e^{-\Omega x} [\hat{v}(0) + Z_0 \hat{i}(0)] + \frac{1}{2} e^{-\Omega(\ell - x)} [\hat{v}(\ell) - Z_0 \hat{i}(\ell)] \tag{A9}$$

For $x = 0$ the above equation becomes

$$\hat{v}(0) = \frac{1}{2} [\hat{v}(0) + Z_0 \hat{i}(0)] + \frac{1}{2} e^{-\Omega \ell} [\hat{v}(\ell) - Z_0 \hat{i}(\ell)]$$

Substituting Equations (A7) and (A8) into above, we obtain

$$\hat{v}(0) = \frac{Z_0}{R_S + Z_0} \hat{e}_S + e^{-\Omega \ell} \frac{R_S}{R_L} \frac{R_L - Z_0}{R_S + Z_0} \hat{v}(\ell) \tag{A10}$$

Similarly, for $x = \ell$ Equation (A9) gives

$$\hat{v}(\ell) = \frac{1}{2} e^{-\Omega \ell} [\hat{v}(0) + Z_0 \hat{i}(0)] + \frac{1}{2} [\hat{v}(\ell) - Z_0 \hat{i}(\ell)]$$

Substituting Equations (A7) and (A8) into above, we obtain

$$\frac{R_L + Z_0}{R_L} \hat{v}(\ell) = e^{-\Omega \ell} \frac{R_S - Z_0}{R_S} \hat{v}(0) + e^{-\Omega \ell} \frac{Z_0}{R_S} \hat{e}_S \tag{A11}$$

Substitution of Equation (A10) into (A11) then yields

$$\hat{v}(\ell) = \frac{Z_0}{R_S + Z_0} (1 + \rho_L) \frac{e^{-\Omega \ell}}{1 - \rho_S \rho_L e^{-2\Omega \ell}} \hat{e}_S$$

$$= \frac{Z_0}{R_S + Z_0} (1 + \rho_L)(e^{-\Omega \ell} + \rho_S \rho_L e^{-3\Omega \ell} + \rho_S^2 \rho_L^2 e^{-5\Omega \ell} + \cdots) \hat{e}_S$$

With the numerical values we then have

$$\hat{v}(\ell) = \frac{4}{3} \left(e^{-\Omega \ell} - \frac{2}{5} e^{-3\Omega \ell} + \left(\frac{2}{5}\right)^2 e^{-5\Omega \ell} - \left(\frac{2}{5}\right)^3 e^{-7\Omega \ell} + \cdots \right) \hat{e}_S$$

Thus in time domain

$$v(t, \ell) = \frac{4}{3} e_S(t - T_D) - \frac{4}{3} \times \frac{2}{5} e_S(t - 3T_D) + \frac{4}{3} \times \left(\frac{2}{5}\right)^2 e_S(t - 5T_D)$$

$$- \frac{4}{3} \times \left(\frac{2}{5}\right)^3 e_S(t - 7T_D) + \cdots$$

where $T_D = \ell \sqrt{LC} = \ell/u$ is the line delay.

The reader is invited to compare above result with Equation (30), which was "derived" by pure reasoning based on the wave nature of the solutions.

EXERCISE

Express the voltage at the source $v(t, 0)$ as a function of $e_S(t)$ and compare the result with Equation (29).

(3) Crosstalk Between Transmission Lines: When two transmission lines are laid out in proximity the signal on one line interferes with the signal on the other due to electric and magnetic coupling and they "talk." This interaction, called crosstalk, is one of the important factors that affects the choice of transmission lines in application.

To analyze the crosstalk between two lines, consider the situation shown in Figure 39. The mutual inductance L_m represents the magnetic coupling per unit length and the mutual capacitance C_m represents the electric coupling per unit length between the lines. Following the same procedure as we did for Equation (A1), we can derive the wave equations with coupling:

$$\frac{\partial}{\partial x} v_1 = -L_1 \frac{\partial}{\partial t} i_1 + L_m \frac{\partial}{\partial t} i_2$$

$$\frac{\partial}{\partial x} i_1 = -C_1 \frac{\partial}{\partial t} v_1 - C_m \frac{\partial}{\partial t} v_2$$

$$\frac{\partial}{\partial x} v_2 = L_m \frac{\partial}{\partial t} i_1 - L_2 \frac{\partial}{\partial t} i_2$$

$$\frac{\partial}{\partial x} i_2 = -C_m \frac{\partial}{\partial t} v_1 - C_2 \frac{\partial}{\partial t} v_2$$

here both C_m and L_m are positive numbers. The reader should convince himself that these coupling terms are of the appropriate signs for the configuration shown in Figure 39.

Taking the Laplace transform, the above equations become

$$\frac{d}{dx} \begin{bmatrix} \hat{v}_1 \\ \hat{i}_1 \\ \hat{v}_2 \\ \hat{i}_2 \end{bmatrix} = s \begin{bmatrix} 0 & -L_1 & 0 & L_m \\ -C_1 & 0 & -C_m & 0 \\ 0 & L_m & 0 & -L_2 \\ -C_m & 0 & -C_2 & 0 \end{bmatrix} \begin{bmatrix} \hat{v}_1 \\ \hat{i}_1 \\ \hat{v}_2 \\ \hat{i}_2 \end{bmatrix} \tag{A12}$$

The complete solution of Equation (A12) can be obtained in the same fashion as we did for Equation (A2). The results are, however, much more complicated because of

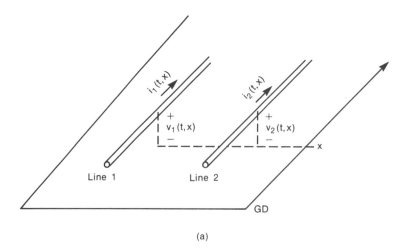

(a)

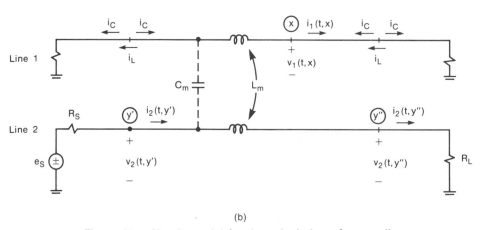

(b)

Figure 39. Circuit model for the calculation of crosstalks.

the cross-coupling between (v_1, i_1) and (v_2, i_2). To make the problem more tractable, we shall analyze the problem under the following simplifying assumptions:

1. Let line 1 be the quiet line and line 2 be the active line that switches. When the quiet line bounces due to crosstalk, the active line is essentially undisturbed. In other words we consider the following differential equations for (v_1, i_1) with (v_2, i_2) as external inputs

$$\frac{d}{dx}\begin{bmatrix} \hat{v}_1 \\ \hat{i}_1 \end{bmatrix} = s\begin{bmatrix} 0 & -L_1 \\ -C_1 & 0 \end{bmatrix}\begin{bmatrix} \hat{v}_1 \\ \hat{i}_1 \end{bmatrix} + s\begin{bmatrix} 0 & L_m \\ -C_m & 0 \end{bmatrix}\begin{bmatrix} \hat{v}_2 \\ \hat{i}_2 \end{bmatrix} \qquad (A13)$$

2. Proper terminations are assumed on the lines so that the effect of reflections on line 1 can be neglected.

In Figure 39 the input $e_S(t)$ is applied to the source of line 2. At $t = 0$ a wave (v_2, i_2) starts to propagate down line 2. As the wavefront of (v_2, i_2) reaches a particular point y the electric coupling induces a current i_C on line 1 flowing from y to both ends of line 1. On the other hand, the magnetic coupling induces a current i_L flowing from y toward the source because the magnetic flux increases as the wavefront moves toward the load. The effect of coupling to a particular point x on line 1 therefore depends on the relative position between x and y. In Figure 39(b) two points of coupling y' and y'' are shown such that $y' < x < y''$. Since $y' < x$ a coupling current $[i_C(y') - i_L(y')]$ is induced and propagated toward x on line 1. This is called forward crosstalk from y'. The total crosstalk to x is therefore the integration of the coupling from all y' such that $0 < y' < x$. On the other hand, since $y'' > x$ a coupling current $[i_C(y'') + i_L(y'')]$ is induced and propagated toward x. This is called backward crosstalk from y''. This total backward crosstalk is likewise the integration of the coupling from all y'' such that $x < y'' < \ell$. The partition of crosstalk according to their directions is particularly convenient if we realize that at $x = 0$ of line 1 there is only backward crosstalk whereas at $x = \ell$ there is only forward crosstalk. Let us now calculate the crosstalks according to their directions.

In Equation (A4) we have derived the following

$$\begin{bmatrix} \hat{v}(x) \\ \hat{\imath}(x) \end{bmatrix} = e^{-\Omega x}\,\mathbf{T}_2 \begin{bmatrix} \hat{e}(0) \\ \hat{\jmath}(0) \end{bmatrix} + e^{-\Omega(\ell-x)}\,\mathbf{T}_1 \begin{bmatrix} \hat{e}'(\ell) \\ \hat{\jmath}'(\ell) \end{bmatrix} \tag{A14}$$

here we have replaced the $\hat{v}$ and $\hat{\imath}$ in the right side by $\hat{e}$ and $\hat{\jmath}$ to emphasize the fact that these voltages and currents can be "external" inputs. Now, the first term

$$\begin{bmatrix} \hat{v}(x) \\ \hat{\imath}(x) \end{bmatrix}_f = e^{-\Omega x}\,\mathbf{T}_2 \begin{bmatrix} \hat{e}(0) \\ \hat{\jmath}(0) \end{bmatrix}$$

represents the effect of an input applied at $y' = 0$ to the point x. Since the forward crosstalk comes from all points $0 < y' < x$, the total effect is the superposition of all individual contributions

$$\begin{bmatrix} \hat{v}(x) \\ \hat{\imath}(x) \end{bmatrix}_f = \int_0^x e^{-\Omega(x-y')}\,\mathbf{T}_2 \begin{bmatrix} \hat{e}(y') \\ \hat{\jmath}(y') \end{bmatrix} dy'$$

In our case the effect to $[\hat{v}_1(x),\ \hat{\imath}_1(x)]$ due to coupling from line 2 at y' is given by Equation (A13):

$$\begin{bmatrix} \hat{e}(y') \\ \hat{\jmath}(y') \end{bmatrix} = s \begin{bmatrix} 0 & L_m \\ -C_m & 0 \end{bmatrix} \begin{bmatrix} \hat{v}_2(y') \\ \hat{\imath}_2(y') \end{bmatrix}$$

and the total forward crosstalk at x is therefore equal to

$$\begin{bmatrix} \hat{v}_1(x) \\ \hat{\imath}_1(x) \end{bmatrix}_f = \int_0^x e^{-\Omega(x-y')}\,\mathbf{T}_2 s \begin{bmatrix} 0 & L_m \\ -C_m & 0 \end{bmatrix} \begin{bmatrix} \hat{v}_2(y') \\ \hat{\imath}_2(y') \end{bmatrix} dy'$$

Since $\hat{v}_2(y') = e^{-\Omega y'}\hat{v}_2(0)$, the above equation gives

$$\hat{v}_{1f}(x) = -\frac{1}{2}\left(\frac{L_m}{Z_0} - C_m Z_0\right)(sx)e^{-\Omega x}\hat{v}_2(0)$$

where we have attached a subscript f for v_1 to designate the forward crosstalk. In time domain we then have

$$v_1(t, x) = -\frac{1}{2}\left(\frac{L_m}{Z_0} - C_m Z_0\right)x\frac{d}{dt}v_2(t - \sqrt{LC}\,x,\, 0)$$

$$\triangleq K_f\, x\, \frac{d}{dt}\, v_2\left(t - \frac{x}{\ell}\,T_D,\, 0\right)$$

where $K_f = -1/2(L_m/Z_0 - C_m Z_0)$ is called the constant of forward crosstalk. Note that $v_1(t, x)$ is proportional to x. It is zero at $x = 0$. It reaches maximum at $x = \ell$.

The backward crosstalk can be calculated in the same fashion. Refer to Equation (A14) again, the second term now represents the effect due to an input at $y' \neq \ell$ to point x:

$$\begin{bmatrix} \hat{v}(x) \\ \hat{\imath}(x) \end{bmatrix}_b = e^{-\Omega(\ell-x)} \mathbf{T}_1 \begin{bmatrix} \hat{e}\,'(\ell) \\ \hat{\jmath}\,'(\ell) \end{bmatrix}$$

Following the same argument, we conclude, for the backward crosstalk at point x:

$$\begin{bmatrix} \hat{v}_1(x) \\ \hat{\imath}_1(x) \end{bmatrix}_b = \int_x^\ell e^{-\Omega(y''-x)} \mathbf{T}_1 s \begin{bmatrix} 0 & L_m \\ -C_m & 0 \end{bmatrix} \begin{bmatrix} \hat{v}_2(y'') \\ \hat{\imath}_2(y'') \end{bmatrix} dy''$$

With $\hat{v}_2(y'') = e^{-\Omega y''}\hat{v}_2(0)$, it can be shown that the backward crosstalk is equal to

$$\hat{v}_{1b}(x) = -\frac{1}{4}\frac{\ell}{T_D}\left(\frac{L_m}{Z_0} + C_m Z_0\right)[e^{-\Omega(2\ell-x)} - e^{-\Omega x}]\hat{v}_2(0)$$

or, in time domain

$$v_{1b}(t, x) = \frac{1}{4}\frac{\ell}{T_D}\left(\frac{L_m}{Z_0} + C_m Z_0\right)\left[v_2\left(t - \frac{x}{\ell}T_D, 0\right) - v_2\left(t - \frac{2\ell-x}{\ell}T_D, 0\right)\right]$$

$$\triangleq K_b\left[v_2\left(t - \frac{x}{\ell}T_D, 0\right) - v_2\left(t - \frac{2\ell-x}{\ell}T_D, 0\right)\right]$$

where

$$K_b = \frac{1}{4}\frac{\ell}{T_D}\left(\frac{L_m}{Z_0} + C_m Z_0\right)$$

is likewise called the constant of backward crosstalk.

In the analysis we have neglected the effects of reflections on line 1. In case the line is not properly terminated and the reflections are significant, the above results can be used as initial waves on line 1. All subsequent reflections can then be calculated either analytically as in this appendix or by pure reasoning based on the wave nature of the solutions as in Section 4.1.

3. SUBSTRATE BIAS GENERATION

In many applications the substrate of the chip is biased at some voltage level other than GD or V_H. In NMOS or n-well CMOS technology the substrate is biased negative relative to GD, whereas in PMOS or p-well CMOS technology the substrate is biased more positive than V_H. In this appendix we shall discuss the basic design of substrate bias generators for NMOS technology. While the devices will be different, the circuit concept applies to other technologies as well.

Biasing the substrate more negative than ground offers many circuit advantages:

(1) Minimization of Substrate Sensitivity: The threshold voltage of an n-channel device is given by Equation (6) in Chapter 1.

$$V_T = V_{FB} + 2|\phi_F| + \frac{\sqrt{2\varepsilon_s q N_A(2|\phi_F| + V_{SB})}}{C_o'}$$

Thus the substrate sensitivity, $\partial V_T / \partial V_S$, decreases as V_{SB} increases. Operating the devices in a low-sensitivity region with a reasonably low threshold voltage optimizes the performance.

(2) Enhancement of Isolation: Large substrate bias increases the threshold voltage of the thick field oxide regions which enhances device isolation.

(3) Reduction of the S/D junction capacitances: The S/D junction capacitances decrease as the S/D depletion regions widen with negative substrate bias.

(4) Improvement of Noise Immunity: Perhaps the most important reason of negative substrate bias on memory chips is noise margin. When I/O circuits switch, large transient voltage pulses are coupled to the floating nodes of internal dynamic circuits. If the substrate is tied to ground these disturbances may forward bias some pn junctions which then inject carriers into substrate. Biasing the substrate to a negative voltage ensures that all pn junctions are reverse biased at all times.

A substrate bias generator consists of an oscillator and a charge pump. For easy start the oscillator is usually a ring oscillator containing odd number of inverting stages such as that shown in Figure 40. Each stage may be simply an inverter or some more elaborate design like a Schmitt trigger. The frequency of the oscillator can be controlled by adjusting the load of the inverting stages. The output of the oscillator, v_O and $\bar{v}_O$, drives the charge pump.

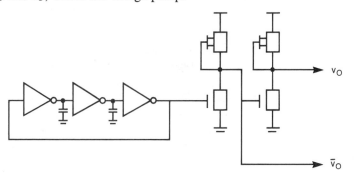

Figure 40.

A simple charge pump circuit is shown in Figure 41 where the substrate is modeled by three elements. Diode D models the intrinsic pn junction of the substrate, C_S is the substrate junction capacitance and R_S represents substrate leakage current. Let us ignore device T_4 for a moment. Assume the substrate voltage $v_B = 0$ when the oscillator starts with v_O going up. Device T_1 will charge capacitor C up to $V_H - V_{T1} - V_{T3}$ through device T_3. As v_O goes down and $\bar{v}_O$ goes up in the second half of the cycle, T_2 turns on. Node ① is discharged to ground and node ② falls below ground through coupling of C. Diode D turns on, discharging the substrate and v_B falls slightly below ground. This process continues as the oscillator moves on. In each cycle a small amount of charge is pumped out of the substrate and v_B goes slightly lower.

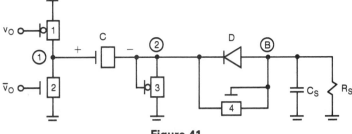

Figure 41.

The substrate voltage will eventually stabilize around some level V_B. Let I_l be the constant leakage current through the pn junction of the substrate. In the half cycle when v_O is high, diode D is reverse biased and the total charge leaked into the substrate is given by

$$Q_1 = I_l t_1$$

where t_1 is the time period in which v_O stays high.

In the half cycle in which $\overline{v}_O$ is high, diode D turns on slightly and charge is pumped out of substrate. This charge is equal to the charge lost by the capacitor during this half cycle:

$$Q_2 = C[V_H - V_{T1} - V_{T3} + (V_B - V_D)]$$

where V_D is the voltage drop of diode D when it is on. Since $Q_1 = Q_2$ in steady state, we have

$$V_B = -(V_H - V_{T1} - V_{T3}) + V_D + \frac{I_l t_1}{C} \tag{A15}$$

The instantaneous value of the substrate bias, $v_B(t)$, of course, fluctuates around V_B slightly in each cycle due to leakage.

One undesirable feature of the above substrate generator is the injection of carriers into substrate by the charge pump. To reduce substrate current, an MOS diode T_4 is connected across the pn junction of substrate with ohmic contact. Since the threshold voltage of T_4 is very low when v_2 falls below ground, T_4 turns on before v_D reaches V_D. Charges are then pumped out of the substrate through T_4 on the surface. The final voltage, V_B, is then given by Equation (A15) with V_D replaced by $V_{DS}(T_4) \ll V_D$. In practice C_S is of the order of several hundred pF's with I_l on the order of several μA's at 85°C. The capacitor C of the charge pump is about 2–3 pF, and the frequency of the oscillator is on the order of 5–20 MHz in most practical designs. The value of V_B depends on the technology and application. For the NMOS process of this book with $V_H = 3$ V and $t_{ox} = 250$ Å, for example, V_B ranges from -2 V to -1 V. For more information, see [13].

MOS MEMORY CHIP DESIGNS

INTRODUCTION

After laying down the ground work of circuit design in the previous chapters, we are now ready to discuss IC *chip designs*. Due to different technology and design methodology requirements, digital IC chip designs are generally classified into two categories: logic and memory. The number and variety of logic functions that can be implemented in a single chip are the main concerns in logic designs. Because of the large number of random logic gates, the interconnections among circuits consume a considerable portion of the chip area. As a result, the number of wiring planes—such as diffusion, polysilicon, M1 and M2, and so on—plays an important role in chip productivity. The layout ground rules, though they certainly affect the performance of any design, are usually less aggressive in logic chips. In addition, the tremendous complexity of the logic also forces logic design methodology to be more structured, more uniform, and, above all, more automated. A logic chip designer is concerned with the wirability of the whole chip as a function of its floor plan, the stability of circuits under different noise conditions, power distribution across the chip, and proper testing for different circuits. She is less interested in layout details. In contrast, a memory chip designer has to watch the "real estate" constantly and try to minimize area consumption by exercising the minimum ground rules wherever possible. While circuit design does affect performance (especially if it is a poor one), access time and other performance gauges are basically limited by technology. Memory chips are highly customized. Design methodologies are less structured, and chip performances are highly sensitive to physical layouts.

The focus of this chapter will be on memory chip designs. We shall concentrate on random access memory designs (the RAMs) and leave other kinds such as programmable logic array (PLA) and read only memory (ROM) to Chapter 7, where logic chip designs will be discussed.

There are two kinds of RAMs: the static RAM (SRAM) and the dynamic

RAM (DRAM). In SRAM the memory cells can hold information indefinitely as long as the power is continuous. In DRAM the memory cells must be refreshed regularly to retain data. As we shall see, typical SRAM memory cells are generally much larger than DRAM cells in the same technology, but the performance of an SRAM is considerably faster than is its dynamic counterpart. SRAM is best suited for high-speed applications such as the cache and buffer memory of a CPU. DRAM is used in large computer systems as the main memory where, because of high volume, cost is of primary concern.

In this chapter both static and dynamic RAMs will be discussed in detail. Section 1.1 starts with simple SRAM architecture illustrating the general design considerations with circuit implementations elaborated on in Section 1.2. Both NMOS and CMOS circuits are considered. In Section 1.3, more advanced circuit techniques are discussed for either reducing power or improving performance. This brings us to the concept of dynamic memory, and the 4-device cells are briefly discussed in Section 2. Section 3 considers the basic 1-device DRAM, which is by far the most popular approach to MOS dynamic memories. One of the most interesting circuits in DRAM design is the dynamic sense amplifier discussed in Section 3.3. Section 3.4 is devoted to the advances in dynamic RAM architectures.

Before we get into details, a brief review of the notations is in order. Voltage pulses are represented by Greek letters ϕ or ψ with subscripts. For example, ϕ_{CS} is a voltage pulse that enables a memory chip. We also use the mathematical notation $T/C(A_0, A_1, \ldots, A_n)$ to denote the set of all logic product terms derived from signals $A_0, A_1, \ldots, A_n$ and their complements. "T/C" means true and complement. Signals are physically generated by a circuit called the T/C generator (T/C GEN). A mathematical concept needed in stability analysis is the norm of a vector and the distance between two vectors. Let $\mathbf{x} = (x_1, x_2)$ and $\mathbf{y} = (y_1, y_2)$ be two points in the plane. Considered as vectors, the norm of $\mathbf{x}$, denoted by $|\mathbf{x}|$, is defined as $|\mathbf{x}| = (x_1^2 + x_2^2)^{1/2}$, and the distance between $\mathbf{x}$ and $\mathbf{y}$, denoted $d(\mathbf{x}, \mathbf{y})$, is the norm of $(\mathbf{x} - \mathbf{y})$. Thus, $d(\mathbf{x}, \mathbf{y}) = [(x_1 - y_1)^2 + (x_2 - y_2)^2]^{1/2}$. It can be proved that $\mathbf{x} = \mathbf{y}$ if, and only if, $d(\mathbf{x}, \mathbf{y}) = 0$.

Since the threshold voltage of an n-channel enhancement mode device is positive, most MOS circuits employ positive logic. High voltage levels are equated to logic 1, and low voltage levels are equated to logic 0. In this book we also employ positive logic convention. Corresponding logic and voltage levels are used interchangeably throughout.

1. STATIC RAMS

1.1 Basic SRAM Architecture

The design of a RAM starts with its memory element. Each element holding one bit of information is called a (memory) cell. One of the most popular SRAM cells is the so-called 6-device cell, which consists of a cross-coupled latch and two pass devices whose logic representation is shown in Figure 1. The latch holds the information at one of its stable equilibrium states and the pass devices control the data I/O.

From a logic point of view, a memory cell is simply a pair of cross-coupled NORs that can lock on fixed states to store information. In fact many memory functions in logic chips are implemented by interconnecting two NOR gates. In stand-alone memories where density is of paramount importance, however, the NORs are laid out with two inputs shorted as a unit cell and replicated throughout the whole chip to form an array. All control functions, such as address decoding and data sensing, are built around the array. The area ratio between the array and "supporting cir-

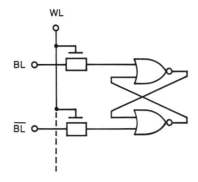

Figure 1. A generic SRAM cell.

cuitry" is an important part of the RAM design. Obviously the bigger the array area, the larger the memory capacity and, hence, the lower the memory cost per bit.

Three kinds of memory cells are shown in Figure 2. In each cell the inverters (T_1, T_3) and (T_2, T_4) form the latch with (T_5, T_6) as the pass devices. The line that controls the pass devices is called the *wordline* (WL), and the lines carrying the data flowing into/out of the cell are called the *bitlines*. Since the data signals are in dif-

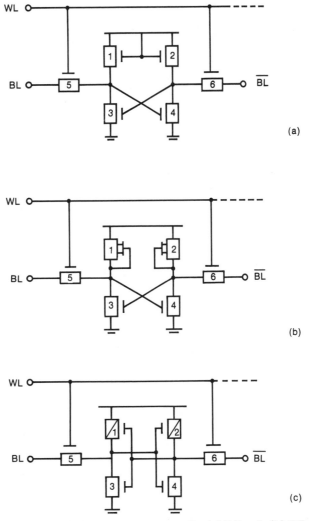

Figure 2. MOS 6-device memory cells. (a) E/E cell, (b) E/D cell, and (c) CMOS cell.

ferential form, the bitlines are designated as BL and $\overline{BL}$. To select a particular cell, drive the WL high to gain access to the cell and then gate the signals into/out of the cell on BL and $\overline{BL}$. The addresses of WL and BL then form a row-column or x-y coordinate system in the array.

The architecture of SRAM is best illustrated by a simple example. Consider the 8-bit SRAM with a 4 × 2 array as shown in Figure 3. Since the array has four rows and two columns, two address bits, A_0 and A_1, are needed to define a row. One bit, A_2, is needed to define a column. The output of each row decoder controls a WL, and the output of each column decoder controls a pair of BLs. To understand the operation of the row address decode better, neglect the effect of $\overline{\phi}_{CS}$ for a moment. When A_0 and A_1 enter the chip, one and only one member of T/C(A_0, A_1) is equal to 00. Since the decoders are basically two input NORs, only the WL whose decoder input equals 00 goes high, turning on the pass devices of all cells on the WL. All other WLs remain low, isolating their cells from the bitlines. This function accomplishes the row selection.

Row address decode allows all cells in the row to access their BLs. The column address decode then specifies the pair of BLs to be connected to the I/O buses by turning on their bit switches through column decoders. Once selected, the cell can be either read or written from I/O pins via the I/O buses.

The write switch and the sense amplifier are connected to the I/O buses. During a write operation (WRITE), the write switch will charge one bitline and discharge the other in accordance with the input data: DI. Since the write switch does not dissipate DC power and there is only one circuit per DI, switch devices can be designed with a large β to ensure fast performance. Indeed, chip performance is seldomly limited by WRITE. During a read operation (READ), the write switch is disabled and the bitlines and the I/O buses are charged/discharged by the memory cell. The rise/fall of the I/O buses will then be detected by a sense amplifier. If the number of cells connected to each bitline and the number of bitline pairs multiplexed to the I/O buses are both small enough that the parasitic capacitances from the cell to the sense amplifier are small, the cell will force the sense amplifier to conform to its state. For SRAMs of practical sizes, however, the loading on the bitlines and I/O buses are much too heavy to be affected by a small cell; READ can be very slow. Furthermore, if the residual charges left on the bitlines by the previous operations are of the opposite polarity, the cell may be overridden by bitlines, resulting in false data. Cell stability then is also a potential problem.

To improve the performance and ensure cell stability, most SRAMs precharge and equalize the bitlines and I/O buses before the cell is accessed. Performance is improved by the fast discharge through the pulldown devices.

The precharge scheme, however, does not come free. "Restoring" the bitlines and I/O buses after each READ results in chip delay. A gating pulse, ϕ_{CS}, is now needed to enable the chip. As shown in Figure 3, chip select ϕ_{CS} is inverted and applied to the precharge devices T_{40}–T_{45}. Prior to chip select, $\overline{\phi}_{CS}$ is high, turning on T_{40}–T_{45} and precharging the bitlines and I/O buses to $(V_H - V_T)$. As soon as ϕ_{CS} goes up, T_{40}–T_{45} turn off, switching control of bitlines to the cells. Since all bitlines have to be restored in each cycle, an additional input controlled by $\overline{\phi}_{CS}$ has to be applied to all NOR decoders to ensure that all wordlines, including the one whose decoder inputs are all 0's, be held low during bitline restore.

In the RAM specification, the time elapsed from chip select to output data valid is called *chip select access time*, t_{ACS}, and the minimum time between any two consecutive operations is called *chip cycle time*, t_{CL}. Since the chip has to be "deselected" in each cycle to allow bitline restore, the chip cycle time is the sum of the access time and restore time. A RAM may have a short access time but long cycle time. In most applications it is the access time that is of primary concern because the chip is selected only once in each machine cycle. In other cases, however, both

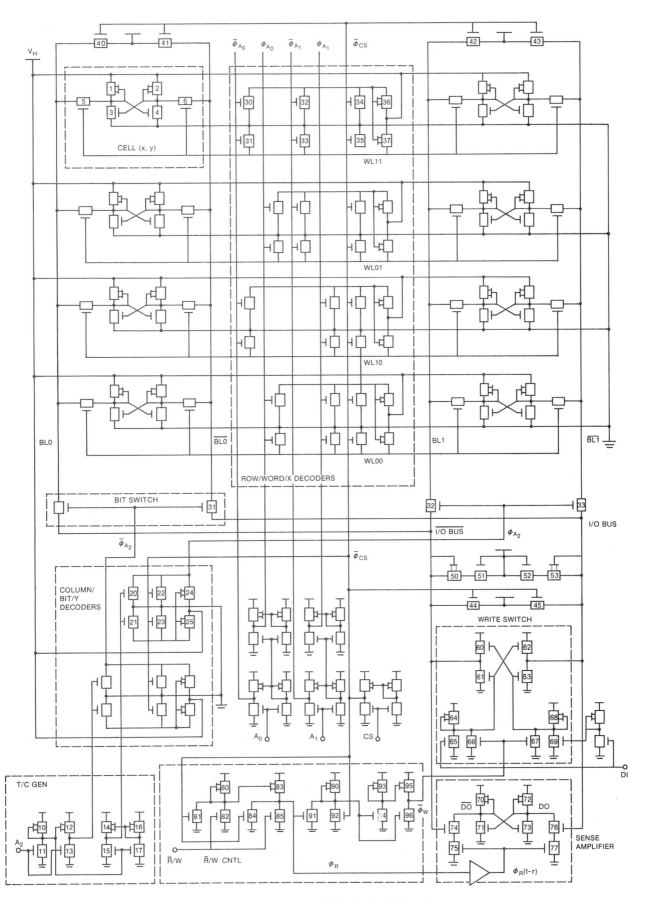

Figure 3. An 8-bit SRAM with E/D cells.

access and cycle times are important. Means of reducing the precharge overhead, such as the address transition detection (ATD) technique, will be discussed in Section 1.3.

The foregoing discussion is best summarized by the timing diagrams shown in Figure 4, where all I/O signals are represented by simplified waveforms. The figure shows the cycle time determines the maximum repetition rate of chip select and the access time as part of the cycle time. Each input pulse is characterized by its minimum/maximum rise/falltimes, its minimum valid time (pulsewidth), and its timing relationship with chip select *CS*. Some of the input pulses must be valid before *CS* goes up. These are called *setup times*. Some of the pulses must be held valid for a certain amount of time to account for internal circuit delays. These are called *hold times*. The *minimum valid time or pulsewidth* is therefore the sum of setup and hold times. All these timing specifications must be followed carefully by *the user* to ensure proper RAM operation.

In addition to timing specifications, the I/O levels in most MOS memories are TTL compatible with MPDL = 0.6 V and LPUL = 2.4 V. In case the MPDL is too close to the threshold voltage of the devices, special OCRs may be used to increase the noise margins (see Chapter 5).

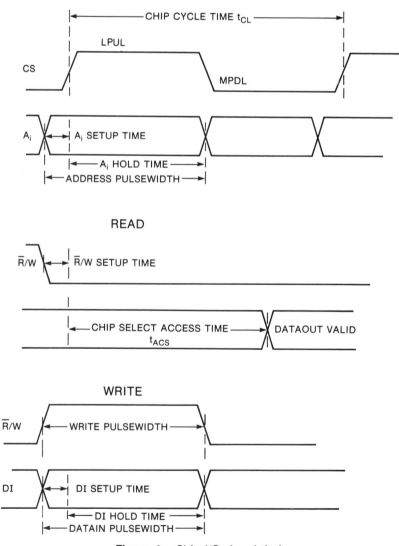

Figure 4. Chip I/O signal timing.

1.2 Circuit Implementations

The circuits in the previous example are the simplest and perhaps the fastest NMOS implementation of SRAMs. Most circuits are static. Static circuits do not make use of temporary charges on a capacitor to maintain voltage levels. Since there are no precharge/discharge requirements, static circuits can operate without clocks. All voltage levels are held indefinitely as long as the inputs do not change. The only timing limitation is the normal circuit delay.

In this section we shall discuss the circuits in each functional block in Figure 3. All inputs, such as CS, A_i, and DI, are assumed to be obtained from appropriate OCRs. We shall also assume that the dataout DO, once becoming valid, stays constant until the next READ cycle. Such an output configuration works fine if the RAM is accessed via dedicated datalines such as that between a cache memory and the CPU of a microprocessor. If the RAM is connected to a common data bus shared by many circuits, DO must be left in a high impedance state (that is, floating) as soon as the chip is deselected to free the bus to other users. The design of tristate OCDs was discussed in Chapter 5. Their timing control will be discussed in more detail in Section 3.

(1) True/Complement Generator: T/C GEN: Both CS and A_i's are needed in true/complement form. The first circuits they encounter are therefore the T/C GENs. In Figure 3 these circuits are implemented with two pushpull E/D inverters. For better timing control, the input of T_{15} and T_{17} comes from T_{10}, instead of T_{12}. Since the T/C GENs drive all decoders, they must be very-high-power drivers.

Simple inverter implementation minimizes circuit delay and hence the access time of the chip. Due to the lack of memory, however, it also imposes a minimum pulsewidth requirement on its input. For example, in a READ the A_i's must be held constant until the sense amplifier latches up.

Since one of the inverters is always on (output low), the DC power dissipation is high. Various dynamic circuits such as those discussed in Chapter 5 have been used to cut down the DC power in many practical designs. At the same time, however, dynamic circuits require more elaborate timing and usually run slower than their static counterparts.

An equivalent T/C circuit in CMOS is shown in Figure 5. It is fully static and does not dissipate DC power. The reader should have little difficulty determining the device sizes for any load capacitance C_L by following the approach suggested in Chapter 4.

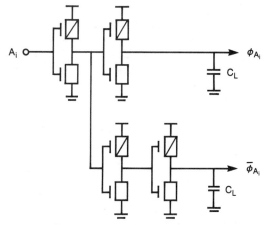

Figure 5. A CMOS T/C GEN.

(2) Decoders: DEC: From a logic viewpoint, a decoder can be implemented either with multiinput NORs or multiinput ANDs. Despite their simple logic structure, decoders are subject to a severe layout restriction: *the pitch constraint*. The pitch of a periodic structure in a specific direction is the minimum period at which the structure repeats itself. For example, the vertical pitch of a group of horizontal, uniformly spaced, identical wires is simply the sum of the line width and the spacing between two adjacent wires. In Figure 3 the height of the row decoders is obviously restricted by the vertical pitch of the cells. Since the layout of the cells is very compact, the pitch constraint can be a severe restriction on decoder size as we will see shortly in CMOS circuits.

Because of the β ratio requirement, decoders in NMOS technology are almost exclusively implemented with multiinput NORs as illustrated in Figure 3. Since the wordlines are connected to the gates of the pass devices, chip density can be greatly improved in silicon gate technology by forming the wordlines with polysilicon. Then no extra contacts are needed. The high resistivity of the polylines, plus the large thin oxide gate capacitances of the pass devices, forms a heavy distributed RC load to each word decoder. This heavy load, together with the limited power available to each decoder, results in considerable chip delay. In fact, in most cases wordline decoding is a major portion of the chip access time. Sometimes M1 or M2 stitches or salicide (Chapter 2) are also used to reduce wordline resistance.

A CMOS counterpart of the word decoder is shown in Figure 6. In this case an AND circuit is used because n-channel devices are considerably faster than p-channel devices. NAND circuits that stack n-channel devices are faster than NOR circuits which stack p-channel devices.

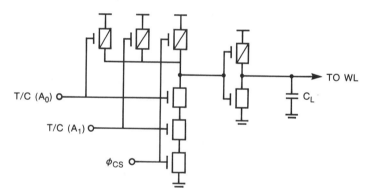

Figure 6.   A CMOS AND decoder.

The 3-input NOR can be easily generalized to accommodate more inputs as shown in Figure 7. As the number of inputs increases, however, so does the total internal parasitic capacitance. To maintain the same speed, the size (that is, W/L) of the pullup has to be increased. This increase, in turn, demands bigger pulldowns. Increasing the pulldown sizes, of course, increases the parasitic capacitances even more. Fortunately, in practice, the iteration eventually ends with a moderate increase

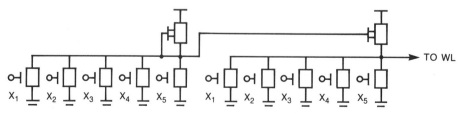

Figure 7.   A 5-input NMOS NOR decoder.

in power. In NMOS technology, 9–12 input NOR decoders of reasonable power are quite common.

Generalization of CMOS AND decoders, however, is much more limited. For the same speed, the pulldown sizes in an n-input AND are roughly *n* times that of a single input inverter. The internal parasitic capacitances, which are proportional to both the number of inputs and the device sizes, grow rapidly with *n*. The circuit designer soon discovers that stacking devices results in a huge device size, diminishing noise margins, and excessive circuit delay.

One way to alleviate this multiinput problem is to divide the inputs into small groups each of which forms a NAND of three or less inputs and then NOR the outputs of the NANDs together as shown in Figure 8(a). Such circuits are larger and slower than a simple NOR configuration but at least these now fit the cell pitch. Another approach is to use the NMOS NOR configuration with a p-channel pullup as shown in Figure 8(b). The gate of the p-channel device is tied to the ground so that it is always on. This circuit is much smaller than the first one, however, as might have already been discovered by the reader, it burns DC power.

The same comments generally apply to bit decoders as well, except that loading to bit decoders are much lighter. Since bit decode is performed simultaneously with word decode, it does not contribute extra delays to the access time.

(3) The Cells: The cell design is closely related to the layout ground rules. There are five "wires" associated with each cell: the V_H bus, the *GD* bus, one wordline, and two bitlines. As has been pointed out before, most wordlines are polysilicon lines. Most bitlines, on the other hand, are diffusion lines which can easily branch into cells to form the pass devices, or metal. The V_H and *GD* buses must be "solid" to

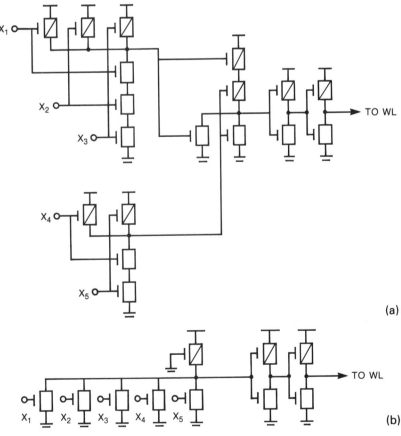

(a)

(b)

Figure 8. (a) A composite CMOS 5-input decoder. (b) A CMOS NOR decoder with grounded-gate p-channel pullup.

avoid coupling noises. They are either formed with metal or with wide conductor lines. In addition to performance, the choice of each wire is also influenced by minimum geometries, cell reliability, and the feasibility of sharing common structures among adjacent cells. For example, metal contacts are best placed on flat surfaces, and contacts to V_H and GD buses should be shared by as many cells as possible. Two practical cell layouts are shown in Figure 9.

Stability and power dissipation dictate a large β ratio for the inverters in E/E or E/D cells. With bitline restore, the channel length of the pullups can be made very long to conserve DC power without sacrificing performance. For CMOS cells the power is not a concern, but n-well or p-well arrangements for complementary devices can still be a challenging task even for a well-experienced designer.

As was mentioned in the previous discussion, bitline restore helps to stabilize the cells. Before we delve into the details of this problem, however, let us introduce some basic concepts of differential equations needed in the next discussion. For more information on the subject of differential equations and stability theory, the reader is referred to the references listed at the end of this chapter.

Consider the system of two-dimensional differential equations:

$$\dot{\mathbf{x}} = \mathbf{f}(\mathbf{x}), \quad \mathbf{x}(0) = \mathbf{x}_0; \qquad \mathbf{x} \in R^2$$

where $\dot{\mathbf{x}} \triangleq d\mathbf{x}/dt$ is the time derivative of $\mathbf{x}$ and $\mathbf{f}: R^2 \rightarrow R^2$ is a piecewise continu-

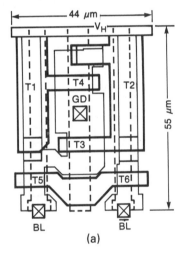

(a)

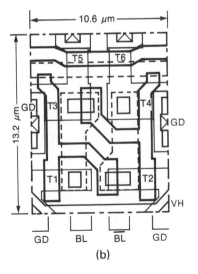

(b)

Figure 9. Examples of SRAM cell layout. The cells are also displayed inside the back cover of this book. (a) An E/D cell (© 1977 *Electronics,* August 1977). (b) A CMOS cell in a p-well (not shown) process [11] (© 1986 IEEE *ISSCC Dig. of Tech. Papers*).

ous function of $\mathbf{x}$. The vector $\mathbf{x}$, which is a function of t, is called the *state* of the system at time t, and $\mathbf{x}_0 = \mathbf{x}(0)$ is therefore called the initial state. Let $\mathbf{x} = (x_1, x_2)$ and then each state $\mathbf{x}(t)$ can be represented by a point in a two-dimensional vector space called the *state space*. Since vectors in a two-dimensional space form a plane, a two-dimensional state space is sometimes also referred to as a *phase plane*.

The behavior of the system is determined by the differential equation and the initial conditions. The *trajectory* of the system from *a particular initial state* is the locus of $\mathbf{x}(t, \mathbf{x}_0)$ plotted in the phase plane with t as the parameter. Since $\mathbf{x}(t, \mathbf{x}_0)$ is a continuous function of t, the trajectory is also a continuous curve in the phase plane.

Trajectories describe the behavior of the system in a graphical manner. Many interesting system characteristics can be observed by plotting its trajectories for different initial conditions. For example, since the solution of a physical system is unique once the initial condition is specified, two trajectories either completely coincide or are mutually exclusive (distinct trajectories never intersect). Another phenomenon vividly illustrated by trajectories is oscillation. Obviously, whenever the trajectory forms a loop, the system oscillates.

A state $\mathbf{x}_e$ at which $\mathbf{f}(\mathbf{x}_e) = \mathbf{0}$ is called an *equilibrium state*. A system may have any number of equilibrium states. The trajectory of the system with $\mathbf{x}_0 = \mathbf{x}_e$ consists of only one point; that is, $\mathbf{x}(t, \mathbf{x}_e) = \mathbf{x}_e$ for all $t \geq 0$. As a matter of fact, in many cases, $\mathbf{x}(t, \mathbf{x}_0)$ stays close to $\mathbf{x}_e$ as long as $\mathbf{x}_0$ is sufficiently close to $\mathbf{x}_e$, and, in some cases, even approaches $\mathbf{x}_e$ as time goes on. These equilibrium states are then said to be *stable*. Formally, an equilibrium state is said to be stable if, given any $\epsilon > 0$, there exists a $\delta > 0$ such that $|\mathbf{x}_0 - \mathbf{x}_e| < \delta$ implies $|\mathbf{x}(t, \mathbf{x}_0) - \mathbf{x}_e| < \epsilon$ for all $t > 0$. The state is said to be *asymptotically stable* if, in addition, $\lim_{t \to \infty} \mathbf{x}(t, \mathbf{x}_0) = \mathbf{x}_e$ for all $\mathbf{x}_0$'s that are sufficiently close to $\mathbf{x}_e$. An equilibrium state that is asymptotically stable behaves like a center of attraction. All trajectories starting sufficiently close to it will be attracted by, and eventually absorbed into, the equilibrium state.

Let $\mathbf{x}_e$ be an asymptotically stable equilibrium state. A *domain of attraction of* $\mathbf{x}_e$, denoted by $\mathcal{R}(\mathbf{x}_e)$, is a neighborhood of $\mathbf{x}_e$ such that if $\mathbf{x}_0 \in \mathcal{R}(\mathbf{x}_e)$, then $\lim_{t \to \infty} \mathbf{x}(t, \mathbf{x}_0) = \mathbf{x}_e$. Thus, all trajectories originating in $\mathcal{R}(\mathbf{x}_e)$ eventually converge to $\mathbf{x}_e$. Let $\mathbf{x}_0 = \mathbf{x}_e$. Any disturbance at $t = 0$ that displaces the system from $\mathbf{x}_e$ to $\mathbf{x}(0+)$ will eventually die out as long as $\mathbf{x}(0+)$ is within $\mathcal{R}(\mathbf{x}_e)$. For this reason, $\mathcal{R}(\mathbf{x}_e)$ is also referred to as a *region of stability of* $\mathbf{x}_e$.

All equilibrium states are not stable, and all stable equilibrium states are not asymptotically stable. Those which are not stable are said to be unstable. The equilibrium states of a cross-coupled latch provide a good example as we shall now see.

Consider the E/D latch shown in Figure 10(a). Since it is a perfectly symmetric circuit, its equilibrium states, which are solutions of the network equation $\mathbf{f}(\mathbf{x}) = \mathbf{0}$, must also be symmetric; that is, if (v_1, v_2) is an equilibrium state, then so is (v_2, v_1). Geometrically this means all equilibrium states must be symmetric with respect to the straight line $\mathcal{L}$ defined by the equation $v_1 = v_2$. Indeed, the latch has three equilibrium states: $\mathbf{x}_1 = (3, 0.046)$, $\mathbf{x}_M = (1.0056, 1.0056)$, and $\mathbf{x}_2 = (0.046, 3)$. State $\mathbf{x}_M$, which is located exactly on $\mathcal{L}$, represents a perfectly balanced condition. The only way the circuit can move into this state is to follow a perfectly balanced path, $\mathcal{L}$ itself. Thus, $\mathcal{L}$ is also a trajectory. Since trajectories do not intersect, $\mathcal{L}$ divides the first quadrant of the plane into two halves: $\mathcal{H}_1 = \{(v_1, v_2), v_1 > v_2\}$ and $\mathcal{H}_2 = (v_1, v_2), v_1 < v_2\}$. No trajectories originating in $\mathcal{H}_1$ can move into $\mathcal{H}_2$, and vice versa. In other words, if $\mathbf{x}_0 = (v_1(0), v_2(0))$ with $v_1(0) > v_2(0)$, then $v_1(t) > v_2(t)$ for all $t > 0$. Now let $\mathbf{x}_0 \in \mathcal{H}_1$. As time t increases, the differential voltage $(v_1 - v_2)$ grows through regenerative action, and the circuit eventually settles down at the equilibrium state $\mathbf{x}_1$. State $\mathbf{x}_1$ is therefore asymptotically stable with a region of stability $\mathcal{R}(\mathbf{x}_1) = \mathcal{H}_1$. Similarly, $\mathbf{x}_2$ is also asymptotically stable with a region of stability $\mathcal{R}(\mathbf{x}_2) = \mathcal{H}_2$. State $\mathbf{x}_M$, however, is unstable because $\mathbf{x}(t, \mathbf{x}_0)$, no matter how close $\mathbf{x}_0$ is to $\mathbf{x}_M$, eventually diverges away from $\mathbf{x}_M$ unless $\mathbf{x}_0$ locates on $\mathcal{L}$ exactly.

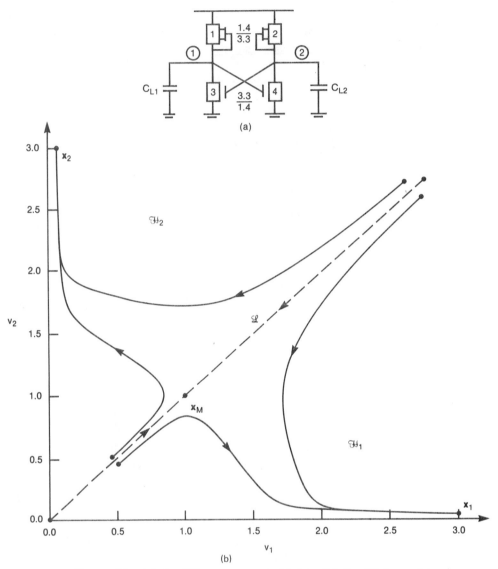

Figure 10. (a) An E/D cross-coupled latch. (b) Equilibrium states and trajectories from symmetrically opposite initial conditions.

The equilibrium states and some of the trajectories of the latch are shown in Figure 10(b).

Before being selected, the latch in the cell settles at a stable equilibrium state. As the wordline goes up and the pass devices turn on, the latch is disturbed by the charge redistribution on the bitlines. The perturbation from its equilibrium state depends not only on the size of the pass devices and the capacitive loading of the bitlines, but also on the risetime of the wordline. Obviously the bigger the pass devices, the heavier the bitline loading and the shorter the risetime of the wordline, the larger the perturbation. Since the bitlines are precharged to the same level, however, the state of the latch immediately after the selection always stays on the same side of $\mathcal{L}$ as the original state if both the cell and its capacitive loads are perfectly symmetric. Thus, a perfect cell always returns to its original state after a READ.

In practice there are always device mismatches among supposedly identical devices. In the worst case analysis, the mismatches between (T_1, T_2) and (T_3, T_4) distort and split $\mathcal{L}$ into two curves $\mathcal{L}'_1$ and $\mathcal{L}'_2$ as shown in Figure 11. Since the actual mismatches can be "in favor" of either x_1 or x_2, the region bounded by $\mathcal{L}'_1$ and $\mathcal{L}'_2$ becomes an uncertain region in the sense that a trajectory originating here may

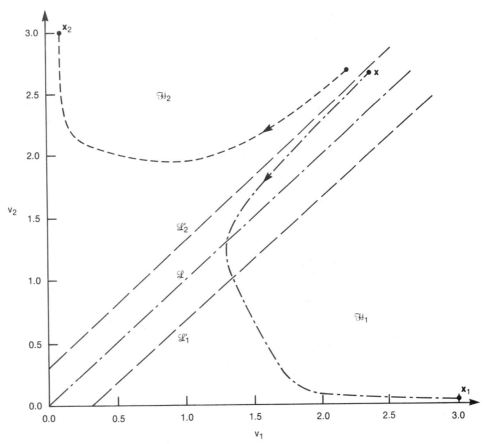

Figure 11. Splitting of $\mathcal{L}$ into $\mathcal{L}_1'$ and $\mathcal{L}_2'$ due to 5% device mismatches. Notice the switching over of the trajectory from $\mathbf{x}_2$ side to $\mathbf{x}_1$ side.

move to either $\mathbf{x}_1$ or $\mathbf{x}_2$. Since a good design ensures that the state of a cell stays out of the uncertain region under all circumstances, mismatches do impose constraints to circuit designs. In particular, to reduce the initial cell disturbance, pass devices cannot be arbitrarily bigger than the pulldowns in the latch. While the 6-device cell in our example is not overly sensitive to the effect of device mismatches, other cell configurations *are* susceptible to stability problems if the device mismatches are large (see Section 2.1).

(4) Read/Write Control: $\overline{R}/W$ CNTL: The $\overline{R}/W$ CNTL decodes the $\overline{R}/W$ command. With ϕ_{CS} high, $\overline{R}/W$ CNTL either activates the sense amplifier by pulling up ϕ_R or releases the write switch by pulling down $\overline{\phi}_W$. The logic equations are given by

$$\phi_R = \phi_{CS}\,\overline{\overline{R}/W}$$

and

$$\phi_W = \phi_{CS}\,\overline{\phi}_R = \overline{\overline{\phi_{CS} + \phi_R}}$$

Since $\overline{R}/W = 0$ corresponds to READ, and $\overline{R}/W = 1$ corresponds to WRITE, the input command is designated as $\overline{R}/W$, which is referred to as "negative READ/positive WRITE."

Since it is not latched, the $\overline{R}/W$ pulse must be held constant until the completion of the READ/WRITE operation. Furthermore, both *DI* and $\overline{R}/W$ must precede

ϕ_{CS} because any change of DI or $\overline{R}/W$ after ϕ_{CS} has already gone up will invalidate the access and cycle time specifications.

(5) Sense Amplifier: In READing, signals on the bitlines are latched up by the sense amplifier. After the address decoding is completed, one of the bit switch pairs opens and differential voltage starts to develop across the I/O buses. The set pulse $\phi_R(t - \tau)$ then goes up, turning on devices T_{75} and T_{77}. With the differential voltage on the gates of T_{74} and T_{76}, the internal regenerative action of the latch amplifies the voltage difference and drives the latch to one of its equilibrium states.

Ideally the set pulse $\phi_R(t - \tau)$ should come up right after, but not prior to, the pass device opening of the cell selected. The delay time τ from $\phi_R(t)$ has to be carefully controlled. On the one hand, τ must be long enough to avoid premature latch setting. On the other hand, an overly long delay affects the access time. In our small RAM, a simple inverter with appropriate capacitive loading suffices to provide the delay. In sophisticated, high-capacity RAM, an extra row of dummy cells which are not connected to the bitlines can be added to "sample" the wordline delay. The wordline of these cells, called *sample wordline,* is monitored by a Schmitt trigger as shown in Figure 12(b) and is selected by ϕ_{CS} all the time. As soon as the voltage on the sample wordline reaches a certain level, the Schmitt trigger snaps, gating the set pulse to the sense amplifier.

The design of the sense amplifier depends on the strength of the signals. If the signals are fast and strong, the simple latch shown in Figure 3 can latch up with little delay. In many practical designs, however, hundreds of cells are attached to each bitline, and the total parasitic capacitance of the bitlines multiplexed to the I/O buses is

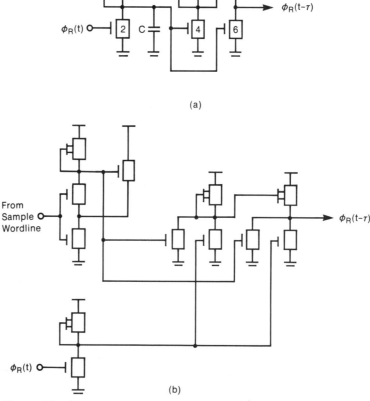

(a)

(b)

Figure 12.   Implementation of $\phi_R(t - \tau)$. (a) Simple inverter delay; (b) sample wordline.

very large. Development of the differential voltage can be very slow. The "preset latch" shown in Figure 13 can be used to amplify initial differential voltage in tens of millivolts quickly to a level that can trigger the main sense amplifier. Operation of such dynamic sense amplifiers is discussed later in this chapter.

Preamplifiers can also be implemented with differential pairs. A single-ended differential pair is shown in Figure 14(a) where (T_1, T_2) and (T_3, T_4) are identical. When the chip is in the standby mode, ϕ_{CS} is low and device T_5 is off, no current flows, and DC power is zero. As soon as ϕ_{CS} goes up, the differential pair starts to conduct current, and the node voltages v_1 and v_2 move toward their DC bias point $V_1 = V_2$.

Now let BL and $\overline{\text{BL}}$ be precharged to V_{BL}. Assuming T_1 and T_3 operate in the saturation region, the bias V_1 is determined by the β ratio between T_1 and T_3:

$$\frac{1}{2}\beta_1(V_1 - V_H - V_{TP})^2 = \frac{1}{2}\beta_3(V_{BL} - V_{TN})^2$$

$$\Rightarrow \qquad V_1 = V_H + V_{TP} - \sqrt{\frac{\beta_3}{\beta_1}}(V_{BL} - V_{TN})$$

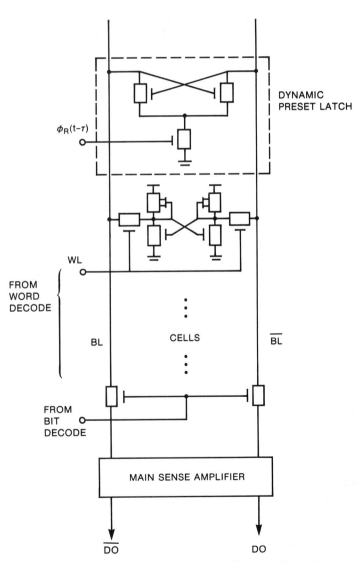

Figure 13. A two-stage sense amplifier configuration.

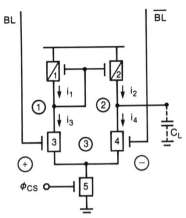

Figure 14(a).   A single-ended differential pair.

where V_{TP} and V_{TN} are the threshold voltages of the p-channel and n-channel devices, respectively, and T_5 is assumed large enough so that V_3 is almost zero. Now, let the bit switches open at $t = 0$ and a small differential voltage across BL and $\overline{\text{BL}}$ start to develop. Define

$$\Delta v_{BL} \triangleq v_{BL} - v_{\overline{BL}}$$

Since

$$v_{BL} + v_{\overline{BL}} = 2V_{BL}$$

we have

$$v_{BL} = V_{BL} + \frac{1}{2}\Delta v_{BL}$$

and

$$v_{\overline{BL}} = V_{BL} - \frac{1}{2}\Delta v_{BL}$$

The currents flowing through T_3 and T_4 are given by

$$i_3 = \frac{1}{2}\beta_3 \left(V_{BL} + \frac{1}{2}\Delta v_{BL} - V_{TN} \right)^2$$

and

$$i_4 = \frac{1}{2}\beta_4 \left(V_{BL} - \frac{1}{2}\Delta v_{BL} - V_{TN} \right)^2$$

Since T_1 and T_2 have the same gate voltage, $i_2 = i_1$. Thus

$$i_2 = i_1 = i_3 = \frac{1}{2}\beta_3 \left(V_{BL} + \frac{1}{2}\Delta v_{BL} - V_{TN} \right)^2$$

Let C_L be the capacitive loading at node ② due to (T_6, T_7) and the parasitics; we then have

$$C_L \frac{d}{dt} v_2 = i_2 - i_4$$

$$= \frac{1}{2}\beta_3\left(V_{BL} + \frac{1}{2}\Delta v_{BL} - V_{TN}\right)^2 - \frac{1}{2}\beta_4\left(V_{BL} - \frac{1}{2}\Delta v_{BL} - V_{TN}\right)^2$$

$$= \beta_3(V_{BL} - V_{TN})\Delta v_{BL}$$

Hence

$$v_2(t) = V_1 + \frac{\beta_3}{C_L}(V_{BL} - V_{TN})\int_0^t \Delta v_{BL}(t')\, dt'$$

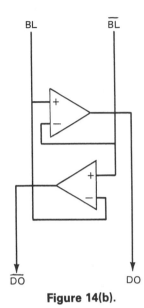

Figure 14(b).

The above equation is valid as long as the devices are in saturation. In case the bitlines are precharged too high, n-channel source followers can be used to bring V_{BL} down to an appropriate level. Obviously the speed of response at the output *DO* depends also on the bias voltages V_{BL} and V_1. To obtain true differential signals DO and $\overline{DO}$, a pair of differential pairs can be used as shown in Figure 14(b). Common centroid layout should be used to ensure good device parameter tracking.

(6) Write Switch: In the WRITE mode, $\overline{\phi}_W$ is low, releasing the write switch to the input data *DI* and $\overline{DI}$. As mentioned earlier, WRITE switch devices are large to ensure performance.

(7) Holdup Devices: The control of I/O buses switches from precharge devices T_{44} and T_{45} to the selected cells during a READ. As $\overline{\phi}_{CS}$ goes down, the turning off of T_{44} and T_{45} couples charges out of the I/O buses, resulting in a voltage droop. After that, if the address decoding cannot be completed in time, the I/O buses will be left floating. The potentials on the buses are subject to noise and will continue to fall due to leakage currents. To stabilize the buses, a pair of small devices in diode configuration (T_{51}, T_{52}) and a pair of small depletion mode devices (T_{50}, T_{53}) for partial isolation are used to hold the I/O buses up. These holdup devices, also known as "bleeding devices," are small enough that they do not interfere with normal operations.

(8) Bitline and I/O Bus Restore: As soon as the chip is deselected with *CS* going down, all bitlines and the I/O buses are charged up to $(V_H - V_T)$. Since the load is heavy, it takes some time for $\overline{\phi}_{CS}$ to complete the operation. If $\phi_R(t - \tau)$ falls before the I/O buses are equalized, the output of the sense amplifier will stay constant regardless of the restore operation; otherwise, the output is indeterminate. If this is the case, a pulldown circuit controlled by $\overline{\phi}_{CS}$ to turn off T_{75} and T_{77} in time must be added to the delay circuit of Figure 12. The reader is invited to modify this circuit so that $\phi_R(t - \tau)$ goes down as soon as $\overline{\phi}_{CS}$ goes up. Interested readers may find the techniques shown in Figure 16 of the next section helpful.

1.3 Advances in SRAM Architecture and Circuits

The architecture and circuit implementation of SRAMs vary widely with application. The example just considered is best suited for "onchip RAM" applications such as the scratch pad memory of a microprocessor. Its static I/O circuits are aimed at fast performance, and the bitline restore scheme allows more cells to be attached to each bitline for high density. The size of such a RAM is limited mainly by its power consumption. Since the RAM is in NMOS technology, the cells, as well as the supporting circuitry, consume DC power. If each NMOS circuit is replaced by its CMOS counterpart as discussed previously, the power will be much smaller, and the chip area is likely to be considerably larger. Large chip areas contain more defects that can lower the yield. The density of a chip, measured by area per bit (the ratio of the total chip area over the number of bits), has a direct impact on any design: the cost per bit. In practice, therefore, the design of a RAM is always a compromise among competing factors such as power, performance, and density. In this section we shall discuss architecture and circuit techniques that alleviate some of these problems. Our discussion, however, will be necessarily incomplete because the design of SRAM is still a dynamic area of research and development. Interested readers are strongly encouraged to consult the current literature.

1.3.1 Power Reduction. One of the major concerns in NMOS SRAM design is power consumption. Since the chip is not selected all the time, a standby powerdown feature is usually available in large standalone SRAMs. A typical powerdown circuit is shown in Figure 15(a). When the chip is in the standby mode with ϕ_{CS} low, the power supply is cut off from the circuit and the pulldown device controlled by $\overline{\phi}_{CS}$ discharges v_O to GD. For circuits with necessarily high outputs in the standby mode, the power control device should be placed between the circuit and GD so v_O will be charged up to V_H by the pullup. These powerdown techniques can be applied to almost all supporting circuits. They typically reduce chip power from several watts to milliwatts in standby. Notice, however, that since power control uses an enhancement mode device, the final uplevel of the circuit in Figure 15(a) is only $(V_H - V_T)$. In high performance SRAMs, low-V_T devices are often used for T_1 and T_2 to reduce the performance impact. A NOR decoder with powerdown feature whose output is driven to V_H by bootstrapping is shown in Figure 15(c). The reader is invited to explain the timing of this circuit in the light of the principle of invert and delay.

Combination of powerdown circuits and pass devices can accomplish considerable savings in both power and area. For example, the 2-input decoder with chip select control shown in Figure 16(a) and (b) performs exactly the same function as the row decoders in Figure 3 with much less area and power consumption. While the pushpull inverters in Figure 16(b) seem unnecessarily complicated with only one input, they are multiinput NORs in practice with other row decoder lines as inputs. Again, since pass devices do limit currents, their impact to chip performance must be carefully assessed.

Another useful observation from the example in Figure 3 is that although the pass devices in both sides of the row decoders have been turned on, only the cells in the side selected by the column decoders are connected to the I/O buses. The decoding scheme shown in Figure 17 divides the memory array into two halves so only the cells in the side selected by the column decoders are affected. This power saving feature, which partitions the memory array into blocks through selective decoding, is very popular in large standalone SRAMs.

All the foregoing schemes have been successfully implemented in practical SRAM designs. Since the only control needed is ϕ_{CS}, switching between the standby and active modes is transparent to the user.

1.3.2 Performance Enhancement. One architecture feature directly affecting the performance of a RAM is its organization. The organization of a RAM is the number of bits simultaneously available at its I/O. A K-bit RAM may be organized as $K \times 1$, $(K/4) \times 4$, or, in general, $(K/M) \times M$, where M is the number of bits available at the I/O simultaneously. The most popular and certainly most flexible architecture is the $K \times 1$ (by 1) organization. Each time the RAM is accessed, one and only one bit is affected. Such a RAM needs the least number of I/O pins, but is

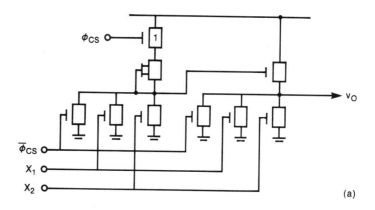

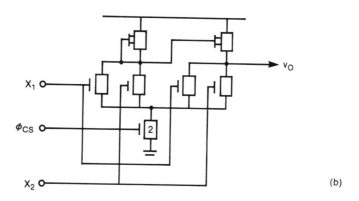

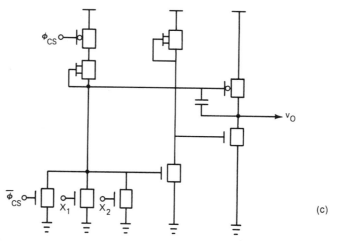

Figure 15. Typical powerdown circuits. (a) Output low in standby; (b) output high in standby; (c) a bootstrap NOR decoder with powerdown feature.

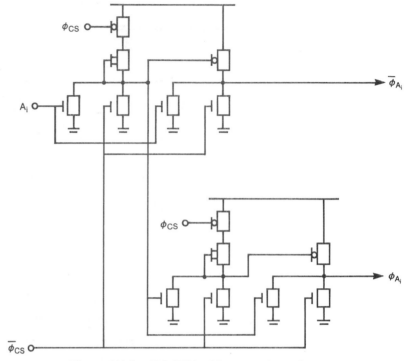

Figure 16(a).   T/C GEN with powerdown feature.

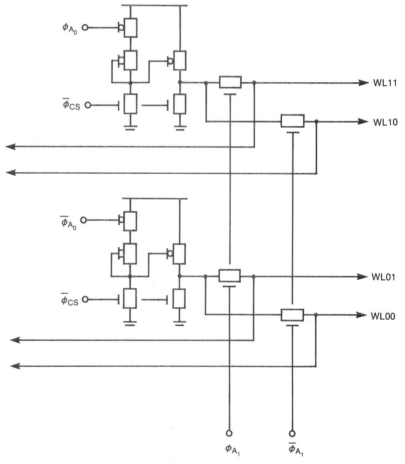

Figure 16(b).   An equivalent 2-input decoder.

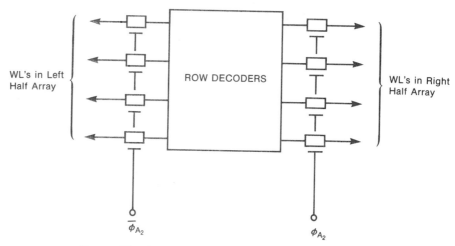

WL's in Left Half Array

ROW DECODERS

WL's in Right Half Array

$\overline{\phi}_{A_2}$

ϕ_{A_2}

Figure 17. Block addressing with pass device decode.

also the most difficult to optimize in circuit design. Since all bitlines must be multiplexed to the single I/O bus, capacitive loading to the cells is extremely heavy. Layout of the chip is complicated, and access time is long.

Since the machine wordlengths are generally in the powers of 2, the ×4, ×8, and ×9 organizations are also available. The ×9 organization represents an 8-bit byte plus a parity bit. A ×M organization divides the whole chip naturally into M "quadrants," all sharing the same decoder lines but having different I/O buses. Performance of the RAM improves proportionally with M. The disadvantages of a wide organization are the large number of I/O pins, less flexibility in some system applications, and higher power.

Another "architectural fix" to improve performance is address access. The standard synchronous operation discussed so far uses the chip select pulse as a clock. A memory cycle coincides with a full cycle of CS. When CS is high, the chip performs READ/WRITE functions, and when CS is low, it does bitline restore. The maximum repetition rate of chip select is inversely proportional to the chip cycle time t_{CL}, which is the sum of the chip access time and bitline restore time. As we also pointed out earlier, a large portion of the access time is allocated to wordline decoding. A typical partition of the chip cycle time is shown in Figure 18(a).

The idea of address access is to perform bitline restore at the same time as wordline decoding so that the chip cycle time becomes identical to the *address access time* t_{AA}. In this scheme, chip select CS stays active as long as the chip is to be selected. An address transition detection circuit associated with each address recognizes the change of address and immediately triggers the bitline restore circuit, which then charges up the bitlines and I/O buses simultaneously with wordline decoding. The partition of the chip cycle time in this scheme is shown in Figure 18(b).

Figure 19 shows a typical address transition detection (ATD) circuit. For an input A_i that has lasted longer than τ', the inputs to the exclusive NOR are out of phase and the output ϕ_i is low. As soon as a transition occurs, however, the inputs to the exclusive NOR circuit are momentarily in phase with each other and ϕ_i goes up. Since all ϕ_i's are ORed together, ϕ_{ATD} goes up. This output pulse will also last for a period of τ'. During this time the bitline restore circuits are triggered. Restore of bitlines and I/O buses immediately begins. The bitline restore circuit has a negative feedback mechanism which shuts itself off as soon as the potential of the bitlines and I/O buses reaches a certain level. All these operations, of course, are performed synchronously with wordline decoding. Detailed implementation of the circuits in Figure 19 is left to the reader as an exercise.

Notice that even though the chip is controlled by a chip select signal, the operation of the memory function is really asynchronous. From a circuit viewpoint each

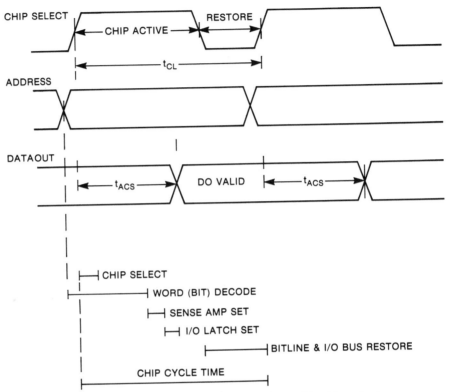

Figure 18(a).   Partition of chip cycle time with chip select access.

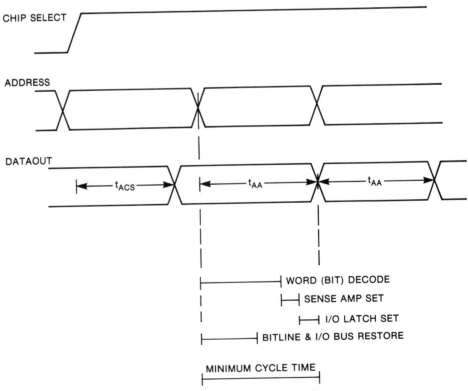

Figure 18(b).   Partition of chip cycle time with address transition access.

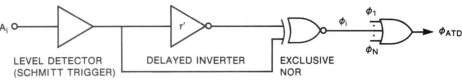

Figure 19. An ATD circuit.

transition at the address also serves the purpose of chip select. There is obviously a limit to the skews among all address inputs and the user now has to take full responsibility for deciding when the output data are valid.

2. STATIC VERSUS DYNAMIC: 4-DEVICE CELLS

2.1 Static 4-Device Cells

The memory cells we have considered so far are called 6-device cells. In Figure 2 the inverters (T_1, T_3) and (T_2, T_4) are cross coupled to form a latch. Since one of the inverters always consumes DC power, in E/D cells the β of the load devices must be small. A device with small β does not imply small area. As a matter of fact the area consumed by the load device is much larger than that of the active device in a typical cell because of its long channel length. Small β, however, does limit the speed of the cell. To improve performance, as we saw in the previous sections, the bitlines are precharged high. During READ the cell discharges the bitline in its low side while leaving the other essentially undisturbed. With bitline precharge, however, device T_1 and T_2 have little effect on performance. Their sole function is to maintain the equilibrium states against leakage in internal nodes of the cell.

One popular approach in large standalone SRAM cells is to replace T_1 and T_2 by polysilicon resistors of very high resistance. As shown in Figure 20, resistors R_1 and R_2 are lightly doped polysilicon bars with resistances in $M\Omega$'s so very small current, and hence very little power, flows through each cell. The area used by the resistors are much smaller than that used by the devices so chip density is also greatly improved. Such cells are called static 4-device cells with poly loads.

Operation of an SRAM with poly loads is essentially the same as 6-device cells. One popular feature that is different from the basic architecture discussed in the previous section in many practical designs is the bitline restore scheme. In Figure 3 the bitlines are precharged by $\overline{\phi}_{CS}$ in standby. As soon as this chip is selected, $\overline{\phi}_{CS}$ goes down, control of bitlines is then switched to the cells selected by the wordline. This is called *dynamic restore* or *dynamic bitline*. In many practical designs, however, to reduce the bitline restore time the bitlines are connected to V_H permanently through devices or resistors. For example, the gates of T_{42} and T_{43} in Figure 20 are now tied to V_H instead of $\overline{\phi}_{CS}$ so that the bitlines are constantly pulled up by V_H. This is called *static restore* or *static bitline*. Right after wordline selection, control of the bitlines is shared by the cell and the bitline pullups. Since the load resistors in the cell are very large, circuit response is essentially controlled by the pulldown and the pass devices of the cell.

Static bitline affects cell stability in READ. Configuration of the cell after wordline has gone up is shown in Figure 21(a) where the resistors are neglected. Since the bitline capacitances C_{BL} and $C_{\overline{BL}}$ are much larger than the node capacitances at ① and ②, respectively, stability of the cell is practically determined by the β ratio $\beta_r = \beta_3/\beta_5 = \beta_4/\beta_6$. In Figure 21(b) the circuit is redrawn with (T_{42}, T_{43}) and $(C_{BL}, C_{\overline{BL}})$ replaced by a constant voltage source $V'_H = V_H - V_{T42}$. The state equation of the circuit is given by

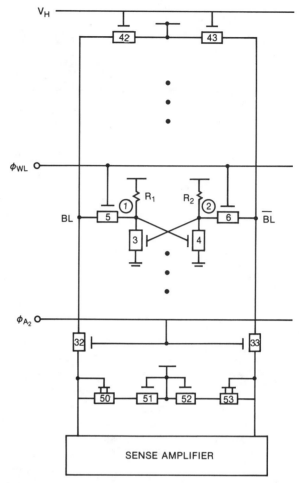

Figure 20. Static 4-device cell with poly loads and static bitlines.

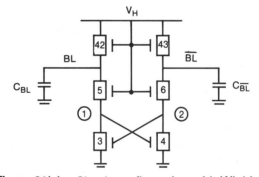

Figure 21(a). Circuit configuration with WL high.

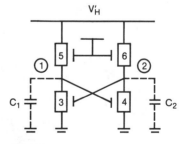

Figure 21(b). Simplified circuit for stability analysis.

$$\frac{d}{dt}\begin{bmatrix} v_1 \\ v_2 \end{bmatrix} = \frac{1}{C}\begin{bmatrix} i_5 - i_3 \\ i_6 - i_4 \end{bmatrix} \triangleq \begin{bmatrix} f_1(v_1, v_2) \\ f_2(v_1, v_2) \end{bmatrix}$$

where $C \triangleq C_1 = C_2$ is the total capacitance at nodes ① and ②. The equilibrium states of the circuit are obtained by solving the equation

$$\mathbf{f}(\mathbf{v}) = \begin{bmatrix} f_1(v_1, v_2) \\ f_2(v_1, v_2) \end{bmatrix} = 0$$

In Section 2 we had a case in which $\beta_r \gg 1$ with three equilibrium states: $\mathbf{x}_1$, $\mathbf{x}_2$, and $\mathbf{x}_M$. Two of them, $\mathbf{x}_1$ and $\mathbf{x}_2$, at which only one side of the latch conducts, are asymptotically stable, and $\mathbf{x}_M$, which locates on line $\mathcal{L}$ defined by $v_1 = v_2$, is unstable. Since device currents are proportional to their sizes, the location of the equilibrium states vary with β_r. As β_r decreases, $\mathbf{x}_1$ and $\mathbf{x}_2$ move closer to each other toward $\mathbf{x}_M$. Eventually both sides of the latch conduct and finally, as $\beta_r \to 1$, $\mathbf{x}_1$ and $\mathbf{x}_2$ merge into $\mathbf{x}_M$. The resulting system now has only one equilibrium state, $\mathbf{x}_M^0$, which is *asymptotically stable*. Any disturbance that turns on all devices in the cell will drive the circuit to $\mathbf{x}_M^0$, and the data stored in the cell will be completely lost.

To find the critical β_r below which the latch has only one equilibrium state, assume all devices are in the saturation region around $\mathbf{x}_M^0$. Let $\mathbf{x}_M^0 = (v_1^0, v_2^0)$. The state equation around $\mathbf{x}_M^0$ is given by:

$$\frac{d}{dt}\begin{bmatrix} v_1 \\ v_2 \end{bmatrix} = \frac{1}{C}\begin{bmatrix} (1/2)\,\beta_5\,(V_H - v_1 - V_T)^2 - (1/2)\,\beta_3\,(v_2 - V_T)^2 \\ (1/2)\,\beta_6\,(V_H - v_2 - V_T)^2 - (1/2)\,\beta_4\,(v_1 - V_T)^2 \end{bmatrix} = \begin{bmatrix} f_1(v_1, v_2) \\ f_2(v_1, v_2) \end{bmatrix}$$

and the Jacobian matrix of $\mathbf{f}$ is equal to

$$\frac{\partial \mathbf{f}}{\partial \mathbf{v}} \triangleq \begin{bmatrix} \partial f_1/\partial v_1 & \partial f_1/\partial v_2 \\ \partial f_2/\partial v_1 & \partial f_2/\partial v_2 \end{bmatrix} = -\frac{1}{C}\begin{bmatrix} \beta_5(V_H - v_1 - V_T) & \beta_3(v_2 - V_T) \\ \beta_4(v_1 - V_T) & \beta_6(V_H - v_2 - V_T) \end{bmatrix}$$

The equilibrium state $\mathbf{x}_M^0$ is asymptotically stable if and only if the matrix $-\partial \mathbf{f}/\partial \mathbf{v}(\mathbf{x}_M^0)$ is positively definite. Since the matrix is symmetric with $v_1^0 = v_2^0$, it is positively definite if and only if all the principal determinants are positive. Thus we obtain

$$\beta_5(V_H - v_1^0 - V_T) > 0$$

and

$$\beta_5^2(V_H - v_1^0 - V_T)^2 - \beta_3^2(v_2^0 - V_T)^2 > 0$$

$$\Rightarrow \qquad v_1^0 < V_H - V_T$$

and

$$\beta_r < \frac{V_H - v_1^0 - V_T}{v_1^0 - V_T}$$

On the other hand, $i_5 = i_6$ at $\mathbf{x}_M^0$ and we have

$$v_1^0 = v_2^0 = \frac{V_H + (\sqrt{\beta_r} - 1)V_T}{1 + \sqrt{\beta_r}}$$

Substituting v_1^0 into the inequality of β_r, we obtain

$$\beta_r < \sqrt{\beta_r} \quad \Rightarrow \quad \beta_r < 1$$

Our simple analysis indicates then that as β_r approaches 1, $\mathbf{x}_1$ and $\mathbf{x}_2$ merge into $\mathbf{x}_M^0$, which becomes asymptotically stable. The reader is now invited to verify the stability of $\mathbf{x}_M^0$ by solving the the eigenvalues of $\partial \mathbf{f}/\partial \mathbf{v}$.

EXERCISE

Prove that the eigenvalues of $\partial \mathbf{f}/\partial \mathbf{v}(\mathbf{x}_M^0)$ are given by

$$\lambda_{1,2} = \beta_5(V_H - v_1^0 - V_T) \pm \beta_3(v_1^0 - V_T)$$

and derive the condition of β_r so that both eigenvalues are negative.

Here, the size of the pass devices becomes critical. For better performance large pass devices are desirable. The larger the pass devices, however, the closer $\mathbf{x}_1$ and $\mathbf{x}_2$ are to $\mathbf{x}_M$ and the less stable the circuit. After WRITE in standby, the node voltages (v_1, v_2) are such that one of the pulldowns is on and the other is off. When the cell is selected, the turnon transient and the change of circuit configuration disturb the cell. One technique for practical design is to limit the size of the pass devices so that the disturbance to the low side of the cell shall not be large enough to turn on the pulldown on the high side. Two interesting cases are shown in Figure 22. For $W(T_5) = 2.8$, device T_3 barely turns on and the cell retains its state, whereas for $W(T_5) = 2.9$, T_3 turns on and the regenerative action of the latch drives the circuit to $\mathbf{x}_M^0 = (1.4, 1.4)$. In practice the exact value of β_r must be determined by extensive circuit simulation with device mismatches.

2.2 Dynamic 4-Device Cells

Since resistors have little effect on performance they can also be eliminated. The result, a dynamic 4-device cell, is shown in Figure 23(a). When storing information one of the devices T_3 or T_4 is on and the other is off. The polarity of the differential voltage, $v_1 - v_2$, represents the data.

To WRITE, the wordline goes high and (v_1, v_2) reaches one of the states $\mathbf{y}_1$ or $\mathbf{y}_2$ determined by the bitlines of the latch shown in Figure 23(b). Like in the static 4-device cells, pass devices and bitline pullups are now an integral part of the cell design.

Let the cell be at $\mathbf{y}_1$ for $t < 0$. At $t = 0$ the wordline falls, cutting off T_5 and T_6. Node voltage v_1 droops due to capacitive coupling and v_2 falls to GD. Thus, right after WRITE the state of the cell is displaced to a new position $\mathbf{x}_1$ as shown in Figure 23(c). Since there is no conducting path from ① to V_H, node voltage v_1 starts to fall because of junction leakage. While the leakage current has many contributors in different regions around the node, a worst case analysis can be made by summing all leakage currents under the worst case voltage conditions and highest junction temperature. Assuming a worst case leakage current I_l and a constant node capacitance C around node ①, the rate of decrease of v_1 is given by

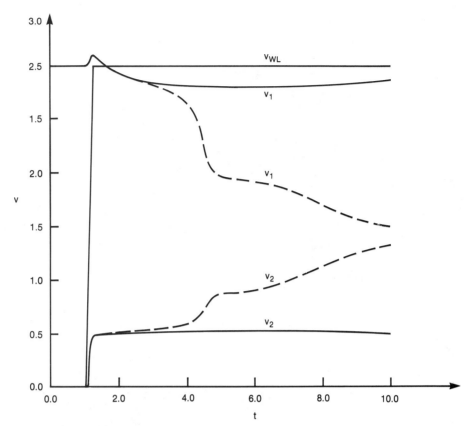

Figure 22. Waveforms of v_1 and v_2 in Figure 21(a) with $W/L(T_5) =$ 2.8/1.4 (solid curves) and $W/L(T_5) = 2.9/1.4$ (dashed curves). $v_H = 2.5$ v.

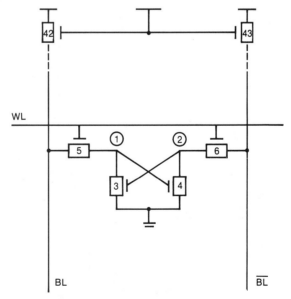

Figure 23(a). A 4-device dynamic cell.

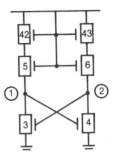

Figure 23(b).   Equivalent circuit with WL high.

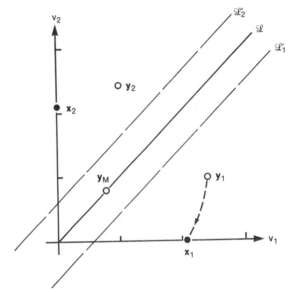

Figure 23(c).   States of (b).

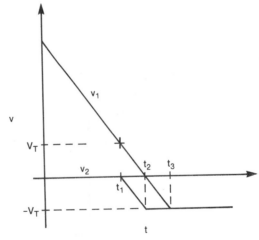

Figure 23(d).   Waveforms of v_1 and v_2 affected by constant leakage.

$$\frac{\Delta v_1}{\Delta t} = \frac{I_l}{C}$$

where Δt is the time elapsed after the wordline has reached ground. As v_1 decreases, v_2 stays at ground as long as $v_1 > V_T$. At $t = t_1$, v_1 falls below V_T. Device T_4 is cut off and v_2 starts to fall. Then at $t = t_2$, v_2 falls to $-V_T$. Assume the wordlines are held to ground. Device T_6 will turn on maintaining v_2 at $-V_T$ for all $t > t_2$. Similarly, v_1 eventually falls to $-V_T$ and stays at $-V_T$ maintained by T_5 after time t_3. Beyond t_3, $v_1 = v_2 = -V_T$ and the information stored in the cell is completely lost. The effect of leakage on v_1 and v_2 is illustrated in Figure 23(d).

This cell has to be refreshed regularly to hold data. The *data retention time*, t_R, defined as the longest time period the cell can endure before it loses its information, is a function of many factors including the power supply tolerance, the junction temperature, device mismatches, and circuit design. In practice t_R is much shorter than t_3. With I_l on the order of low nA's, t_R is of the order of mS's.

A refresh function can be performed by a forced READ either internally or via external circuitry. In each refresh cycle one wordline is selected. The rise of the wordline helps the bitlines, which have been precharged high, to "stretch" the internal nodes of the cell and re-establish the voltage levels. A counter or shift register either onchip or offchip is used to keep track of the address for refresh operation.

3. DYNAMIC RAMS

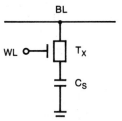

Figure 24. A generic 1-device cell.

The most popular dynamic RAM design is the so-called 1-device cell. A generic 1-device cell, shown in Figure 24, consists of a pass device T_X that controls the cell I/O and a capacitor C_S that stores information. Binary information is represented by the presence/absence of a charge in C_S. The cell is accessed by turning on T_X via the wordline: WL. Charges are transferred into/out of C_S on the bitline: BL. Since C_S is isolated from the rest of the circuit when T_X is off, the amount of charge in C_S decreases constantly because of junction leakage. As a result, the cell must be refreshed regularly to restore its original condition; hence the name dynamic.

The sense amplifier (S/A) shown in Figure 25 is used to detect the charge stored in the cell. On one side of the amplifier is a memory cell (M.C.) that either stores a full amount of charge or is empty, depending on the binary value of the data. On the other side of the amplifier is a dummy cell (D.C.) that stores half the amount of the charge effectively. In the standby mode the BLs on both sides are precharged to the same potential, say, $V_H - V_T$. During READ both WL and D WL go high, allowing the charge in the cells to redistribute along the BLs. After the voltage drop has settled down, since the charge in D.C. is half the amount, $v_2 > v_1$ in case M.C. is empty, and $v_2 < v_1$ in case M.C. is full. The S/A, which detects and

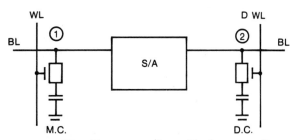

Figure 25. Charge sensing with dummy cells.

amplifies the voltage difference between the BLs, will then latch up in conformity with the data in M.C.

The sense amplifier not only captures the data but also replenishes the cell with fresh charges. Indeed, most DRAMs refresh their cells with forced READ. Cell refresh can be carried out either onchip with internal circuitry or offchip by the user with forced READ.

The simple structure of the 1-device cell leads to extremely high density. Minimum layout ground rules are used wherever possible to maximize the number of cells in the array. The associated large number of I/Os and supporting circuits, however, impose a severe restriction on the power and area of each individual circuit. Most circuits in a DRAM are dynamic in the sense that no DC power is dissipated even in the active mode. Since dynamic circuits must use clocks, performance of a DRAM is greatly influenced by the design of its internal timing. In practice the number of clocks on a chip ranges from as small as 6 to as many as more than 30.

Internal clocks or "phases" are generated sequentially in a driver chain with proper delays. High-power dynamic drivers are used to drive large capacitive loads. These driver chains, called *timing chains*, control the operation of the whole chip like a computer program. The timing chains are interlocked at critical points to avoid race conditions among signals generated from different chains.

In a DRAM, each operation is carried out typically in three stages controlled by clocks: ϕ_1, in which the circuits evaluate their inputs and set their drivers; ϕ_2, in which the drivers drive their outputs according to the input condition; and ϕ_3, in which the circuits are restored back to the standby mode. For example, word decoding starts when the external command $\overline{RAS}$ (row address selection/strobe) goes low. With $\overline{RAS}$ low, all word decoders are preconditioned by the word address T/C generators. After all decoders have settled down, a clock pulse comes on to drive the WLs. The decoders are preconditioned in such a way that only the WL that is selected be driven high, turning on all pass devices connected to the WL. After all subsequent operations such as bit decoding and S/A latching have been completed, all WLs will be pulled down by a restore pulse at the end of the memory cycle upon the rise of $\overline{RAS}$.

This evaluation-drive-restore scheme is used in almost all DRAM designs both in NMOS and CMOS technology. In some cases two operations can be accomplished in the same clock period. In most cases, however, each operation must be carried out in a different clock period.

The external timing of a DRAM is also rather complex. Because of the large number of address inputs (e.g., a 1-megabit RAM needs 20 address bits) almost all practical designs time multiplex row and column addresses on the same pins. The row address selection is carried out first with $\overline{RAS}$ going low. This initiates the word decoding. After the word addresses have been latched in T/C generators or trapped on the word address buses, the same pins are used for column address inputs. Column address selection is signified by $\overline{CAS}$ (column address selection/strobe) going low. $\overline{CAS}$, being an external command, initiates the bit decoding and controls the data I/O operations.

A typical READ cycle is illustrated in Figure 26. Notice that the $\overline{CAS}$ follows the $\overline{RAS}$. The READ cycle time t_{RC} starts with $\overline{RAS}$ going low and ends with $\overline{CAS}$ going high. Two access times can be defined: t_{RAC} for *row access time*, and t_{CAC} for *column access time*. Both $\overline{RAS}$ and $\overline{CAS}$ pulses are characterized by minimum pulsewidths t_{RAS} and t_{CAS}, respectively. Since the READ/WRITE command $R/\overline{W}$ interacts with $\overline{CAS}$, it must be valid before $\overline{CAS}$ goes low. Since the I/O is also controlled by $\overline{CAS}$, the output data DO is valid only when $\overline{CAS}$ stays low. It returns to a high impedance state as soon as $\overline{CAS}$ goes high. The skew between $\overline{CAS}$ and DO data valid (t_{OFF} in Figure 26) is a result of internal circuit delay.

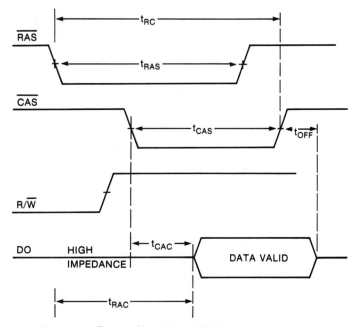

Figure 26. Chip READ timing.

A typical WRITE cycle is shown in Figure 27. For WRITE, $R/\overline{W}$ goes low. The minimum setup time t_{WCS} of $R/\overline{W}$ relative to $\overline{CAS}$ is usually 0. As shown in the figure, input data DI must be valid before $R/\overline{W}$ goes low. It stays valid for a period of time to account for the internal circuit delays. The sum of the DI setup time, t_{DS}, and the DI hold time, t_{DH}, is therefore the minimum pulsewidth for DI. The WRITE cycle time, t_{WC}, is usually shorter than the READ cycle time t_{RC}.

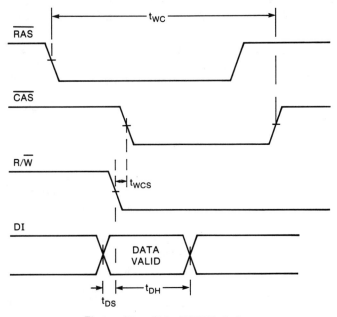

Figure 27. Chip WRITE timing.

3.1 Basic DRAM Architecture

In this subsection we shall study the basic architecture of DRAM with a simple example. Figure 28 displays a small 16-bit DRAM with a 4×4 memory array. Two row address bits, A_0 and A_1, share the same input pins with the column address bits A_2 and A_3, respectively. In our example each address has its own T/C GEN. Sharing T/C GENs between row and column addresses is possible with more elaborate timing.

It is a good practice to lay down the ground rules for naming the timing pulses before the circuit design work begins. In the following discussion we shall use ϕ and ψ for *RAS* and *CAS* timing, respectively. Since *RAS* and *CAS* timings are interrelated, a single numerical subscript system is used to order the pulses in time. Thus, for example, ϕ_2 follows ϕ_1 and ψ_3 follows ϕ_2, and so on. Note that no pulses are named ψ_1 or ψ_2. Complete timing of the RAM is shown in the bottom of Figure 28.

The memory cycle starts when $\overline{RAS}$ goes low. The $\overline{RAS}$ command is inverted and adjusted to the internal MOS level (GD to V_H) by a phase generator PH GEN. The output of PH GEN, ϕ_{RAS}, drives the T/C GENs of the row (word) addresses A_0 and A_1. The T/C GENs, as mentioned earlier, set the conditions for the decoders.

ϕ_{RAS} initiates the RAS timing chain. In addition to driving the T/C GENs it also drives a high-power delayed-driver DR D. The output of the driver, ϕ_1, goes up after a predetermined amount of delay needed for the word decoders to settle down.

ϕ_1 drives the wordline and dummy cells. During ϕ_1, the charge of the selected and dummy cells on opposite sides of the sense amplifiers are redistributed on the bitlines.

Continuing with the timing chain, ϕ_1 also drives another DR D, which generates ϕ_2 as the sense amplifiers are set by a pulse ϕ_2'. After all sense amplifiers have settled down, RAS timing is terminated.

All drivers in the RAS timing chain remain set until $\overline{RAS}$ goes high. As soon as $\overline{RAS}$ moves up, ϕ_R goes up. ϕ_R, a strong *restore* pulse generated by the restore generator R GEN, resets all DR Ds in the chain and pulls all wordlines to the ground. ϕ_R is followed by ϕ_R', which precharges the bitlines and discharges the dummy cells in preparation for the next memory cycle.

CAS timing starts when $\overline{CAS}$ goes low. The first CAS clock, ψ_{CAS}, drives the T/C GENs of the column (bit) decoders. ψ_{CAS} also drives the DR D whose output is ψ_3. The function of ψ_3 is to connect the sense amplifier to the I/O buses by turning on the bit switches on the selected bitlines. To avoid premature connection before the sense amplifier settles, ψ_3 is interlocked with RAS timing. The dominant reset of DR D for ψ_3 is controlled by ϕ_2' through an inverter, so that ψ_3 goes up only after ϕ_2' has gone up.

The $R/\overline{W}$ control takes effect as soon as the sense amplifier is connected to the I/O buses. In READ the data latched in the sense amplifier will be fetched by the I/O latch set by ψ_4. The output DO becomes valid as soon as the I/O latch settles down. In WRITE the sense amplifier will be overwritten by the WRITE latch, which is controlled by DI through the data T/C GEN.

For the user to fetch data, command $\overline{CAS}$ has to stay low for a period of time after DO has become valid. As soon as $\overline{CAS}$ goes high the data valid period ends (with some delay) and the output returns to a high impedance state.

The CAS restore pulse, ψ_R, resets all DR Ds in the CAS timing chain and precharges the I/O buses for the next memory cycle as soon as $\overline{CAS}$ goes high.

A complete memory cycle begins then with $\overline{RAS}$ going low and ends with $\overline{CAS}$ going high. The rise of each clock is controlled by adjusting the delay of the preceding driver. Clock separation must be accurately assessed based on the actual layout of the chip so sufficient, but not excessive, time is allocated to each operation.

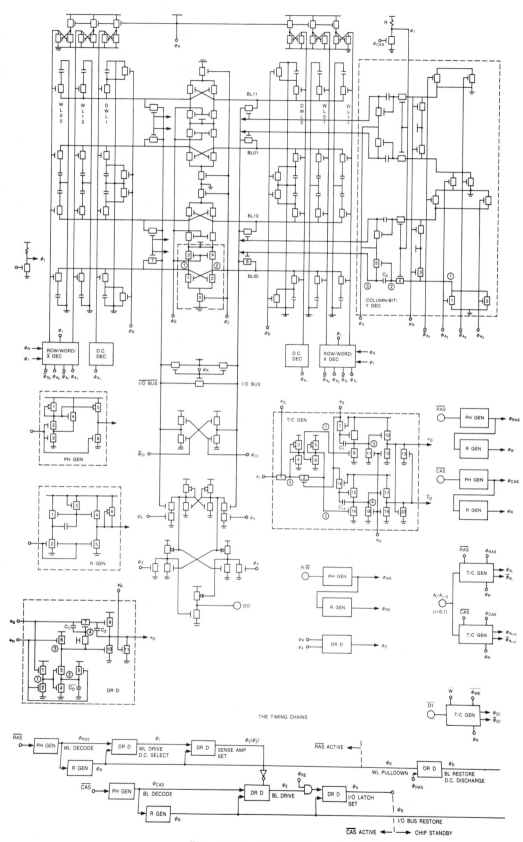

Figure 28. A 16-bit dynamic RAM.

3.2 Circuit Implementations

We are now ready for a more detailed discussion of the operation of each individual circuit. Similar to the previous sections on SRAM, we shall first study the circuits in the example and then investigate some alternatives.

(1) Phase Generator: PH GEN and Restore Generator: R GEN: The PH GEN accepts an external signal, inverting and shifting its level to GD or V_H, and drives it out with high power. Devices T_1–T_4 form a Schmitt trigger and devices T_5–T_6 provide the high-power drive. The R GEN is a standard high-power bootstrap inverter. The output level of R GEN should reach V_H. To maintain the output at V_H in the standby mode, a small holdup device, T_6, is used to compensate for the leakage.

Notice that since the input of R GEN comes from the output of PH GEN, the restore pulses follow the external inputs with some delay. This is one advantage of internal timing as we will see shortly. Notice also that both circuits are static because they are the very circuits that initiate the internal timing chains.

A CMOS implementation of the Schmitt trigger and high-power drivers is shown in Figure 29. Notice that the drivers do not dissipate DC power, the Schmitt trigger may.

(2) Delayed Driver: DR D: The delayed driver DR D is a set/reset driver with a reset dominant control. Two inputs, e_S and e_R, drive the output high (SET) and low (RESET), respectively. The dominant reset, e_R', is connected directly to the output. As long as e_R' is high, the output remains low regardless of other inputs.

In normal conditions e_R' is low, and the inputs e_S and e_R are nonoverlapping pulses. Assume that the driver has first been reset by e_R so that node ① is low and nodes ②, ③, and ④ are at $(V_H - V_T)$; e_R would have gone low before e_S goes high. As soon as e_S moves up, T_1 turns on, which, in turn, turns on T_3 and T_4. Device T_4 and capacitor C_D form a delay circuit that starts to discharge nodes ②, ③, and ④ to ground. The exact amount of delay is controlled by selecting appropriate sizes for T_4 and C_D.

At the same time e_S also charges the bootstrap capacitor C_2 through T_7 and T_{10}. After the delay, nodes ③ and ④ fall below the threshold voltage and cut off T_{10}. The bootstrap action of C_2 and T_9 then drives the output all the way to V_H.

This DR D design is very similar to Example 6 in Chapter 5 except for the delay circuit. Notice that during the delay both T_9 and T_{10} are on, resulting in a DC

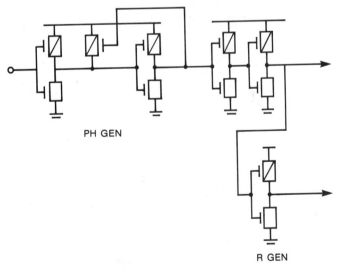

PH GEN

R GEN

Figure 29. A CMOS Schmitt trigger with drivers.

path from V_H to GD. Since these are large output devices, the transient power is significant. As mentioned earlier, excessive delay should be avoided because it results in poor performance and high transient power.

EXERCISE

What is the function of C_1 and T_8 in DR D?

Since enhancement mode NMOS devices belong to a subset of a CMOS technology, this enhancement mode circuit can be used in CMOS DRAM as well. A CMOS set/reset driver that uses a latch to hold its output is shown in Figure 30. The input is ANDed with the dominant reset. Driver delay is controlled by capacitors C_1 and C_2. It is easy to show that the function of this circuit is identical to the previous example.

The acute reader may have noticed that the inputs to DR D never go high simultaneously. Do you think simple inverter drivers could be used instead? The answer is related to the reset of the drivers when the chip exists from the active mode. Detailed explanation is left to the reader as an exercise.

(3) True/Complement Generator: T/C GEN: The T/C GEN in Figure 28 generates the true and complement signals of its input with high power. In active mode, one of its outputs, v_O, is in phase, whereas the other output, $\overline{v}_O$, is out of phase, with input e_I. In the standby mode, both outputs are held low.

Assume that at $t = 0$ the circuit is in standby so that v_O and $\overline{v}_O$ are low with e_I at its downlevel. The circuit samples e_I through the pass device T_1 when e_G is high. The potentials at nodes ① and ② are therefore $v_1 = V_H$ and $v_2 = 0$. Thus, device T_9 is on and T_{16} is off at $t = 0$.

As shown in Figure 28, the T/C GEN is controlled by two slightly overlapped pulses, e_S and e_R. As e_S moves up, capacitor C_7 and C_{14} are precharged to $(V_H - V_T)$ through the paths (T_7, C_7, T_{11}) and (T_{14}, C_{14}, T_{18}), respectively. After the capacitors have been charged up, e_R goes down, turning off T_{11} and T_{18}. Now, since device T_9 is on at this point, node ③ stays low and v_O is not disturbed. On the other side, node ④ starts to rise because T_{16} is off. Capacitor C_{14} and device T_{15} soon bootstrap node ④ to V_H. The rise of node ④ turns on the pullup T_{17} strongly which, in turn, drives $\overline{v}_O$ high.

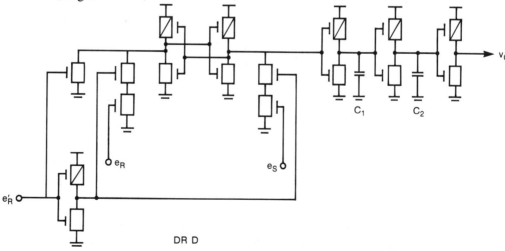

Figure 30. A CMOS reset dominant delayed driver.

Since e_R has to overlap e_S to allow precharge of C_7 and C_{14}, it is important that ϕ_R and ψ_R, being connected to e_R of the T/C GENs, stay high before ϕ_{RAS} and ψ_{CAS} reach their uplevels completely. This explains the connection of PH GEN and R GEN as we mentioned earlier.

At the front end, device T_2 turns on as e_S goes up, locking the input in the latch formed by inverters (T_3, T_4) and (T_5, T_6). In the active mode, e_G should be at its downlevel to prevent v_5 from being changed by e_I. In the output stages since both pullup and pulldown are off after the transient, the output nodes are floating. Devices T_{13} and T_{20} cross couple v_O and $\overline{v}_O$ so that the output node at the downlevel will not creep up by picking up extraneous noise.

A CMOS version of the T/C GEN can be easily implemented by replacing (T_3, T_4) and (T_5, T_6) with CMOS inverters. A popular CMOS T/C GEN used with dynamic CMOS decoders will be described in the following discussion.

(4) Decoders: DEC: Associated with each row and column of the array is a dynamic decoder which, upon selection, drives its corresponding wordline or bitline high. To understand the operation of a dynamic decoder more fully, consider the bitline decoding circuit of BL00 located in the right lower corner of the array. The circuit is in standby before $\overline{CAS}$ goes low. The pulses ϕ_{A_2} and ϕ_{A_3} are low and ψ_R is high, precharging the internal nodes ① and ② to $(V_H - V_T)$. Then as $\overline{CAS}$ and ψ_R go low, device T_3 cuts off and nodes ① and ② are left floating. As we saw in the previous discussion, the falling of ψ_R also releases the T/C GENs driving ϕ_{A_2} and ϕ_{A_3}. If either ϕ_{A_2} or ϕ_{A_3} is high, nodes ① and ② will be discharged to the ground. As ψ_3 moves up, v_3 stays low because device T_5 is turned off. Bitline BL00 is isolated from the I/O buses and remains unselected. On the other hand, if both ϕ_{A_2} and ϕ_{A_3} are low, v_1 and v_2 stay high and device T_5 remains on. In this case v_3 will be boosted to V_H by the bootstrap action of T_5 and C_5 when ψ_3 goes up. The rise of v_3 turns on bit switches T_6 and T_7, connecting BL00 to the I/O buses. The column address $A_2A_3 = 00$ is thereby selected.

There is a minor detail worthy noting in the circuit schematic. Device T_4 is a pass device that isolates node ② during bootstrapping. Due to the slight difference in the threshold voltages of T_3 and T_4, and the out coupling of the charge when ψ_R goes low, node ①, which is precharged to $(V_H - V_{T3})$, may be slightly lower than $(V_H - V_{T4})$. When v_2 starts to move up, charge will "leak out" through device T_4. To maintain high bootstrap efficiency the gate of T_4 is controlled by a level ψ_I, which is slightly lower than V_H when the chip is in active mode, thus ensuring complete cutoff of T_4 during bootstrapping.

Since there are as many decoders as the total number of rows and columns in the array, it is extremely important to minimize the transient power and area consumption of each individual circuit. To satisfy pitch constraints, the pulldown devices should be close to minimal dimensions.

In many practical designs the wordlines are boosted higher than V_H. A decoder with boosted wordline is shown in Figure 40(b) on p. 263.

A CMOS alternative of the 2-input decoder is shown in Figure 31(a). This circuit belongs to a family of dynamic CMOS logic called *domino logic circuits*. Before ψ_3 goes up, the p-channel device T_1 precharges node ① to V_H and the output is low. At the same time, address pulses ϕ_{A_2} and ϕ_{A_3} preset the pulldowns T_2 and T_3. As ψ_3 moves up, device T_1 turns off and T_4 turns on. If both address inputs are high so the pulldowns are on, node ① will discharge to ground and the output v_O goes up. As long as one of the address inputs is low, however, v_1 stays high and v_O remains low. Since the circuit operates dynamically, reasonably good performance can be achieved with relatively small devices. Furthermore, since T_4 is cut off during precharge, the outputs of the T/C GENs do not have to be held low in the standby mode, greatly simplifying the circuit design of T/C GENs.

Charge sharing is a potential problem with domino circuits. For example, let

ϕ_{A_2} and ϕ_{A_3} be low when ψ_3 goes high. Node ① remains high when T_1 turns off. If ϕ_{A_2} goes up a moment later, v_1 dips because of charge sharing between ① and ②. In general, successive turnons of a large number of pulldowns (except the bottom one) may cause a large disturbance on the top node affecting the inverter buffer.

To avoid unpredictable changes of the ϕ_{A_i}'s, all address inputs are latched in their T/C GENs by ψ_{CAS} as shown in Figure 31(a), so the decoder inputs remain stable during the whole CAS cycle.

To improve noise immunity further, a feedback p-channel device, T_5, is tied to node ① to replenish the charge. Devices (T_5, T_6, T_7) form a Schmitt trigger whose trigger points are controlled by adjusting the relative sizes of the devices.

Designs with a large number of inputs can also use the NOR decoder shown in Figure 31(b). This eliminates the charge-sharing problem at the expense of higher transient power. (Why?)

(5) The Cell: Cell design depends heavily on the process. Minimum ground rules are used and common structure sharing, such as contacts to the V_H/GD buses, is exploited carefully to make the density as high as possible. In most cases cells are laid out in pairs or even in quaternary. A typical cell layout is shown in Figure 32. Notice that the "ground" of the capacitor is an AC ground. It can be either the DC ground or V_H. For example, in single polysilicon technology the capacitor is easily formed by extending the source of the pass device as a storage node and then laying a polysilicon layer on its top with thin oxide in between. In this case the "ground" side of the

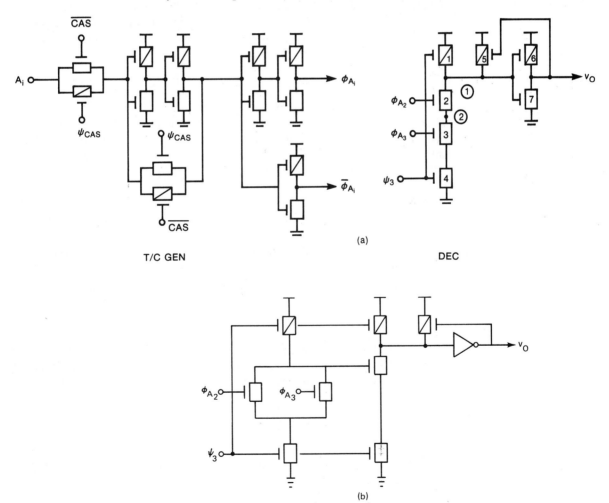

Figure 31. CMOS T/C GEN with dynamic decoders.

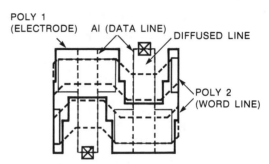

Figure 32. Folded bitline 1-device cell. The cell is also displayed inside the back cover of this book. [8] (© 1980 IEEE *ISSCC Dig of Tech. Papers.,* February 1980)

capacitor may well be tied to V_H. See, however, discussion of Hi-C cells in Chapter 2.

In the following discussions we shall assume that a fully charged cell represents a logic 1 and an empty cell represents a logic 0. Sometimes we also refer to a fully charged cell as a 1-cell and an empty cell as a 0-cell. Such terminology is convenient in discussing the design of cells and sense amplifiers.

Since the cells are subject to leakage, some of the READ cycles must be dedicated to the function refresh. Each forced READ refreshes the cells in an entire row selected by the same wordline. Let n be the number of cells connected to each side of the sense amplifier. It will then take $2n$ cycles to refresh the whole array. The *availability* of the RAM, η_{AV}, is defined as 1 minus the ratio of the total refresh time over the cell retention time; that is,

$$\eta_{AV} = 1 - \frac{2nt_{RC}}{t_R}$$

where t_{RC} is the READ cycle time and t_R is the retention time of the cell. t_R is typically of the order of mS's. In system application an η_{AV} of 95% or higher is desirable.

From the above equation we have

$$t_R = \frac{2nt_{RC}}{1 - \eta_{AV}}$$

Now let I_l be the average leakage current per unit area of the cell, the worst case voltage loss across C_S due to leakage between refreshes is then given by

$$\Delta v_S = \frac{I_l \times t_R}{C'} = \frac{I_l}{C'} \frac{2nt_{RC}}{1 - \eta_{AV}}$$

where C' is the average capacitance per unit area of the cell. Let K be the total number of cells in the array. Since there are $2n$ wordlines, there are $m = K/2n$ bitlines and sense amplifiers in the array. Define the *aspect ratio* of the array by

$$\eta_{AS} = n/m \qquad \text{so that} \qquad n = \sqrt{\frac{\eta_{AS}K}{2}}$$

and let C'_B be the (parasitic) capacitance between each cell and its bitline. The total bitline capacitance is then given by

$$C_B = n \times C'_B = \sqrt{\frac{\eta_{AS}K}{2}} C'_B \tag{1}$$

In a high-density RAM, C_B' is limited by the minimum dimensions of the pass device and hence is fixed by the ground rules. The total bitline capacitance C_B, however, is a circuit design parameter controlled by the aspect ration η_{AS}.

Now let us consider the situation in Figure 33. Let the wordlines go up at $t = 0$ and charges stored in the cells start to redistribute along the bitlines. Consider the left side of the S/A first. Let the potential on the bitline be $v_1(0)$ and the voltage across C_S be $v_S(0)$ at $t = 0$. Since

$$C_S \frac{dv_S}{dt} = -C_B \frac{dv_1}{dt}$$

we have

$$C_S[v_S(t) - v_S(0)] = -C_B[v_1(t) - v_1(0)]$$

Let v_1 be *the final value* of $v_1(t)$ and $v_S(t)$. The above equation then implies

$$v_1 = \frac{C_S}{C_S + C_B} v_S(0) + \frac{C_B}{C_S + C_B} v_1(0)$$

$$= \eta_{TR} v_S(0) + \frac{C_B}{C_S + C_B} v_1(0) \tag{2}$$

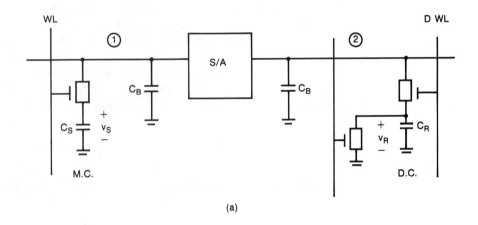

(a)

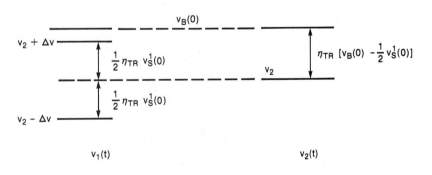

(b)

Figure 33.

where

$$\eta_{TR} \triangleq \frac{C_S}{C_S + C_B}$$

is called the *transfer ratio*. The initial voltage across C_S, $v_S(0)$, is equal to the precharged voltage of C_S minus Δv_S for a 1-cell and is equal to 0 for a 0-cell. We shall denote $v_S(0)$ of 1-cells and 0-cells by $v_S^1(0)$ and $v_S^0(0)$, respectively. In our case, since both the bitlines and C_S are precharged to $(V_H - V_T)$, we have

$$v_S^1(0) = V_H - V_T - \Delta v_S$$

$$= V_H - V_T - \frac{I_l}{C'} \frac{2nt_{RC}}{1 - \eta_{AV}} \tag{3(a)}$$

and

$$v_S^0(0) = 0 \tag{3(b)}$$

Now refer to the right side of the S/A. Two different approaches are commonly used to design the dummy cell. In one, the capacitor C_R is a fraction of C_S (approximately one-half in practice), so that if both are charged up to the same potential, the amount of charge stored in C_R is proportionally a fraction of that in C_S. The challenge of this approach is the construction of a small capacitor about half the size of an already small one. All the parasitic capacitances have to be considered to ensure the right capacitance ratio. In the other approach, the capacitors are identical, but C_R can only be charged up to half the high voltage of C_S by controlling the gate potential of its pass device. This approach alleviates the layout problem at the expense of extra circuits.

Figure 28 illustrates the first approach. The dummy cells are predischarged to ground by the restore pulse ϕ_R' when the chip is in the standby mode. Following the argument used for Equation (2), we obtain v_2, *the final value* of $v_2(t)$:

$$v_2 = \frac{C_B}{C_R + C_B} v_2(0) \tag{4}$$

Subtracting Equation (4) from Equation (2), and noting that $v_1(0) = v_2(0) \triangleq v_B(0)$, yields

$$\Delta v \triangleq v_1 - v_2$$

$$= \eta_{TR} v_S(0) + \left(\frac{1}{C_S + C_B} - \frac{1}{C_R + C_B} \right) C_B v_B(0) \tag{5}$$

There are two possible values for Δv: Δv^1 for a 1-cell with $v_S(0) = v_S^1(0)$, and Δv^0 for a 0-cell with $v_S(0) = v_S^0(0)(= 0)$. For optimal sensing margins, $|\Delta v^1| = |\Delta v^0|$. Substituting Equation (3(a)) and (3(b)) into Equation (5) and letting $\Delta v^1 = -\Delta v^0$, we have

$$\eta_{TR} v_S^1(0) + \left(\frac{1}{C_S + C_B} - \frac{1}{C_R + C_B} \right) C_B v_B(0) = -\left(\frac{1}{C_S + C_B} - \frac{1}{C_R + C_B} \right) C_B v_B(0)$$

Solving for C_R, we obtain

$$C_R = \frac{C_B C_S [2v_B(0) - v_S^1(0)]}{2C_B v_B(0) + C_S v_S^1(0)} \tag{6}$$

Thus, for instance

$$C_R \cong \frac{C_S C_B}{C_S + 2C_B}, \qquad \text{if } v_S^1(0) \cong v_B(0)$$

and

$$\cong \frac{C_S}{2}, \qquad \text{if, furthermore, } C_B \gg C_S$$

Once C_R has been determined, we can evaluate v_2. Substituting Equation (6) into Equation (4) yields

$$v_2 = \frac{1}{2} \eta_{TR} v_S^1(0) + \frac{C_B}{C_S + C_B} v_B(0) \tag{7}$$

Substituting Equation (6) into Equation (5) describes the final differential voltage across the S/A

$$|\Delta v^1| = |\Delta v^0|$$
$$= \frac{1}{2} \eta_{TR} v_S^1(0) \tag{8}$$

and the levels of v_1 are symmetrically located around v_2 as shown in Figure 33.

From Equations (2) and (7) it can be seen that the transfer ratio η_{TR} measures the voltage attenuation from cells to the S/A caused by charge redistribution. Substituting Equation (3(a)) into Equation (8) yields the following design equation:

$$\Delta v = \frac{1}{2} \eta_{TR} \left(V_H - V_T - \frac{I_l}{C'} \frac{2nt_{RC}}{1 - \eta_{AV}} \right) \tag{9}$$

where Δv represents either Δv^1 or $|\Delta v^0|$.

While Equation (9) is a good estimate of the differential voltage available to the sense amplifier, in reality it is further reduced by noises as we will see shortly in the discussion of sense amplifiers.

EXERCISES

1. Derive a relation between η_{AS} and η_{TR} in terms of the basic parameters C_S and C_B'.
2. Substitute the above result into Equation (9) and rewrite Δv as a function of η_{AS}.
3. Substitute the above result into Equation (9) and rewrite Δv as a function of η_{TR}.
4. Refer to Figure 28. Complete the discussion of the RAM operation in terms of $R/\overline{W}$ control, WRITE operation, and the tristate output.

3.3 Dynamic Sense Amplifiers

One of the most difficult tasks in DRAM design is the dynamic sense amplifier. In a high-density DRAM the initial differential voltage across the sense amplifier can be of the order of tens of millivolts. The sense amplifier must detect and amplify this small signal with reasonable speed. On the other hand, since sense amplifiers are located between bitlines, the height of the layout is limited by the bitline pitch. It is these exact circuit requirements together with the extremely tight layout constraints that makes the design of sense amplifiers most interesting and challenging.

The design of sense amplifiers will be discussed in this subsection. We will start with the simple cross-coupled latch of Figure 28 to illustrate the basic design concepts. The reader will see a clear tradeoff between density and performance as the analysis moves along. Since the sensitivity of a sense amplifier is basically limited by the device mismatches, a first-order analysis of mismatches is also carried out in detail. Finally, some elaborate designs such as the bucket brigade and CMOS half V_H sensing are included toward the end.

3.3.1 Basic Operations.   Figure 34(a) is the redrawn sense amplifier used in Figure 28 with lumped bitline capacitances. Before the chip is selected both nodes ① and ② have been precharged to $(V_H - V_T)$ by ϕ_R' through T_3 and T_4. As ϕ_1 goes up, exactly one wordline on each side of the latch rises. Charges stored in the cells immediately redistribute along the bitlines. It takes a short but finite amount of time for the potentials on the bitlines to settle. By the time ϕ_2' goes up to set the latch, $v_1(t)$ and $v_2(t)$ should be very close to their final levels; thus, for all practical purposes assume the bitlines have completely settled when the latch is set. Incomplete

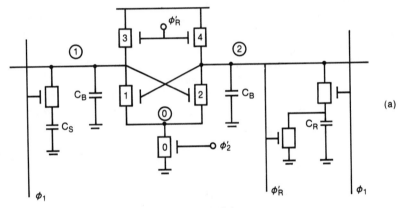

Figure 34(a).

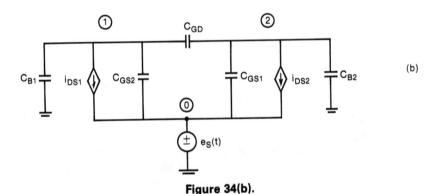

Figure 34(b).

bitline settling and bitline droop will be discussed in the next subsection as part of noise analysis.

Let us assume the latch is set at t = 0. The circuit model for the latch in Figure 34(a) for $t > 0$ is shown in 34(b) where the bitline and cell capacitances are lumped together and the pulldown T_0 and ϕ_2' are replaced by a voltage source $e_s(t)$. Thus $e_s(t)$, called the set pulse, starts to go down at $t = 0$ by definition. Without loss of generality, we shall also assume that $v_1(0) > v_2(0)$ so that $v_d(t) \triangleq v_1(t) - v_2(t) = \Delta v > 0$ *at* $t = 0$ where an estimate of Δv is given by Equation (9).

As soon as e_s falls, T_1 and T_2 turn on. The differential voltage $v_d(0)$ will be further reduced by the charge sharing between the gate capacitances of T_1 and T_2 and the bitlines. It is easy to show that

$$v_d(0+) = f\Delta v$$

where

$$f \triangleq \frac{C_B}{C_{GS} + C_B}$$

with

$$C_{GS} \cong \frac{2}{3}C_o'\text{WL}$$

being the gate capacitance of T_1 and T_2. Notice that in the above equation we have neglected C_S and C_R in the first-order analysis because they are much smaller than C_B.

Initially both T_1 and T_2 operate in the saturation region. Since $v_1 > v_2$, T_2 turns on harder than T_1. As a result, v_2 falls faster than v_1 and the difference $v_d(t)$ increases steadily. As v_1 and v_2 continue to fall, T_2 enters the linear region at some time $t = t_{sat}$ at which $v_d(t_{sat}) \cong V_T$, and, finally, T_1 is cut off at some time $t = t_l$ at which $v_2(t_l) \cong V_T$. From $t = t_l$ on, v_1 falls only slightly due to capacitive coupling and v_2 falls rapidly to ground. The latch is set when v_2 reaches ground and v_d reaches its final value V_D. Since theoretically it takes an infinite amount of time for the latch to settle completely, it is customary to define the latch set time $t_{set} = t_{sat} + t_l$. This is a good definition because $v_2(t)$ goes down to ground very quickly for $t > t_l$.

The latch is controlled by the set pulse $e_s(t)$. Since the final differential voltage, V_D, drives the I/O latch, it must be reasonably large. The determination of the waveform of $e_s(t)$ can be formulated as a time optimal control problem:

Given a voltage V_D^*, find the optimal waveform of $e_s(t)$ that amplifies $v_d(0+)$ to V_D^* in the shortest possible time.

The optimal waveform, $e_s^*(t)$, can be easily calculated by circuit simulation. The final differential voltage V_D is actually slightly larger than V_D^* with $e_s^*(t)$, however, because after v_d has reached V_D^*, v_2 is still falling and the latch still "works" until $v_d(t)$ settles to V_D at $t = \infty$.

Since a detailed derivation of the optimal control goes too far afield, we shall content ourselves with a worked out example. Interested readers are referred to the literature at the end of this chapter.[10] In Figure 35 the waveforms of the internal node voltages of the latch are shown for two cases. In (a) the set pulse $e_s^*(t)$ is relatively slow initially with both devices in saturation. The differential voltage $v_d(t)$ nevertheless develops steadily. After $v_d(t)$ has grown to a reasonable size, $e_s^*(t)$ goes down quickly, discharging node ② to ground and leaving node ① at a high potential.

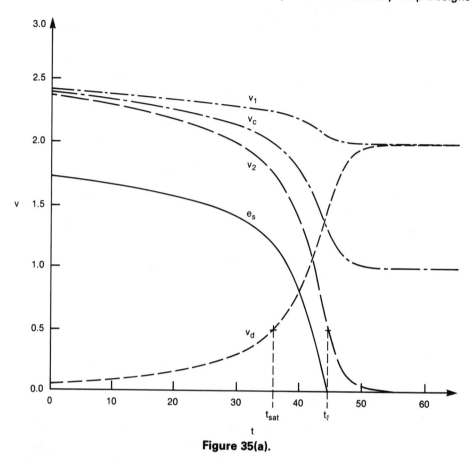

Figure 35(a).

In (b), $e_s^*(t)$ drops suddenly at $t = 0+$ and then falls rapidly with a fairly constant slope. In this case the latch sets much faster but the final differential voltage V_D is also much smaller. The faster the latch sets then, the smaller the final differential voltage V_D. Since V_D sets the I/O latch, a careful tradeoff must be made to optimize overall performance.

Two important facts can be observed in Figure 35. In both cases $v_d(t)$ develops slowly for $t < t_{sat}$, which is therefore the major portion of the total set time. Another fact is that the average $v_c(t) = (v_1 + v_2)/2$ follows $e_s^*(t)$ fairly closely for $t < t_{sat}$, and the lower voltage $v_2(t)$ follows $e_s^*(t)$ for $t_{sat} < t < t_l$. Both of these facts will be used to simplify analysis in the following discussions.

Now refer to Figure 34(b). Applying KCL to nodes ① and ②, we obtain

$$(C_{GS2} + C_{B1} + C_{GD}) \frac{d}{dt} v_1 - C_{GD} \frac{d}{dt} v_2$$

$$= -\frac{1}{2} \beta_1 (v_2 - e_s - V_{T1})^2 + C_{GS2} \frac{d}{dt} e_s \quad (10(a))$$

and

$$-C_{GD} \frac{d}{dt} v_1 + (C_{GS1} + C_{B2} + C_{GD}) \frac{d}{dt} v_2$$

$$= -\frac{1}{2} \beta_2 (v_1 - e_s - V_{T2})^2 + C_{GS1} \frac{d}{dt} e_s \quad (10(b))$$

$$t \leq t_{sat}$$

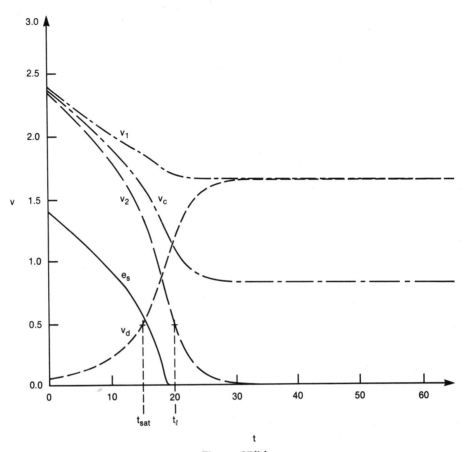

Figure 35(b).

where V_{T1} and V_{T2} are the threshold voltages of T_1 and T_2, respectively. The above equations assume both T_1 and T_2 are in saturation and hence are valid for $t \leq t_{sat}$.

The following discussion assumes the latch is perfectly symmetric so that we can drop all numerical subscripts on the parameters. Let us restate the differential and average voltages:

$$v_d = v_1 - v_2$$
$$v_c = \frac{v_1 + v_2}{2} \tag{11}$$

Subtracting Equation (10(b)) from Equation (10(a)), we obtain

$$(C_{GS} + C_B + 2C_{GD}) \frac{d}{dt} v_d = \frac{1}{2} \beta [(v_1 - e_S - V_T)^2 - (v_2 - e_S - V_T)^2]$$
$$= \beta v_d (v_c - e_S - V_T) \tag{12}$$

Since $v_c(t)$ follows $e_S(t)$, solve Equation (12) under the assumption that

$$K_1 \triangleq v_c - e_S - V_T$$

is a constant. Rewrite Equation (12) as

$$\frac{1}{v_d} dv_d = \frac{\beta K_1}{C_{GS} + C_B + 2C_{GD}} dt$$

and integrate, we obtain

$$\ln \left[\frac{v_d(t)}{v_d(0+)} \right] = \frac{\beta K_1}{C_{GS} + C_B + 2C_{GD}} t$$

Therefore

$$t = \frac{C_{GS} + C_B + 2C_{GD}}{\beta K_1} \ln \left[\frac{v_d(t)}{v_d(0+)} \right] \tag{13}$$

and

$$v_d(t) = v_d(0+) e^{A_1 t}$$

where

$$A_1 \triangleq \frac{\beta K_1}{C_{GS} + C_B + 2C_{GD}}$$

The node voltages are therefore given by

$$v_1(t) = K_1 + V_T + e_S(t) + \frac{v_d(0+) e^{A_1 t}}{2}$$

and

$$v_2(t) = K_1 + V_T + e_S(t) - \frac{v_d(0+) e^{A_1 t}}{2}$$

Since $v_d(t_{sat}) = V_T$ by definition, from Equation (13)

$$t_{sat} = \frac{C_{GS} + C_B + 2C_{GD}}{\beta K_1} \ln \left[\frac{V_T}{v_d(0+)} \right] \tag{14}$$

Thus, t_{sat} is inversely proportional to K_1. As far as the initial differential voltage is concerned, it affects t_{sat} through $\ln [v_d(0+)]$.

For $t > t_{sat}$ the assumption that $v_c(t)$ follows $e_S(t)$ is no longer valid because of the drastic change of the slope of $e_S(t)$. For $t_{sat} < t < t_l$, however, v_2 follows $e_S(t)$ *more or less*. Let us therefore solve the circuit equation under the assumption that $K_2 \triangleq 2(v_2 - e_S - V_T) = $ constant. Applying KCL to nodes ① and ② yields

$$(C_{GS} + C_B + C_{GD}) \frac{d}{dt} v_1 - C_{GD} \frac{d}{dt} v_2$$

$$= -\frac{1}{2} \beta K_2^2 + C_{GS} \frac{d}{dt} e_S$$

$$-C_{GD} \frac{d}{dt} v_1 + (C_{GS} + C_B + C_{GD}) \frac{d}{dt} v_2$$

$$= -\beta \left[v_1 - v_S - V_T - \frac{1}{2}(v_2 - v_S) \right](v_2 - v_S) + C_{GS} \frac{d}{dt} e_S$$

$$= -\beta \left[v_d + \frac{1}{2}K_2 - \frac{1}{2}\left(\frac{1}{2}K_2 + V_T\right) \right]\left(\frac{1}{2}K_2 + V_T\right) + C_{GS} \frac{d}{dt} e_S$$

Subtracting the second from the first, we obtain

$$(C_{GS} + C_B + 2C_{GD}) \frac{d}{dt} v_d = \beta \left[\left(\frac{1}{2}K_2 + V_T\right)v_d - \frac{1}{2}V_T^2 \right]$$

Solving the above equation with the initial condition $v_d(t_{sat}) = V_T$, we obtain

$$t - t_{sat} = \frac{1}{A_2} \ln \left[\frac{(K_2 + 2V_T)v_d(t) - V_T^2}{V_T(K_2 + V_T)} \right]$$

where

$$A_2 \triangleq \frac{\beta(K_2 + 2V_T)}{C_{GS} + C_B + 2C_{GD}}$$

By comparing the waveform of $v_d(t)$ for $t < t_{sat}$ we see that $v_d(t)$ increases much faster after device T_2 has entered the linear region.

At $t = t_l$ the differential voltage $v_d(t_l) = (V_D - V_T)$ where V_D is its final value because for $t > t_l$ node voltage v_1 falls only slightly due to capacitive coupling. Thus we have

$$t_l - t_{sat} = \frac{1}{A_2} \ln \left[\frac{(K_2 + 2V_T)(V_D - V_T) - V_T^2}{V_T(K_2 + V_T)} \right] \tag{15}$$

Another interesting case in sense amplifier design is the so-called "weak conduction condition" in which device T_1 is barely on with negligible current throughout the setting. In this case the gatedrive of T_1: $(v_2 - e_S - V_T) = K_2/2 \cong 0$ for all $t > 0$. We can solve the circuit under the condition $K_2 = $ constant first and then let $K_2 \to 0$. In this case Equation (12) becomes

$$(C_{GS} + C_B + 2C_{GD}) \frac{d}{dt} v_d = \frac{1}{2}\beta[(v_1 - e_S - V_T)^2 - (v_2 - e_S - V_T)^2]$$

$$= \beta v_d \left(\frac{1}{2}v_d + v_2 - e_S - V_T\right)$$

$$= \frac{1}{2}\beta v_d(v_d + K_2)$$

Rewriting the above equation as

$$\frac{1}{v_d(v_d + K_2)} dv_d = \frac{\beta}{2(C_{GS} + C_B + 2C_{GD})} dt$$

and integrating, we obtain

$$\ln \frac{v_d(t)[v_d(0+) + K_2]}{v_d(0+)[v_d(t) + K_2]} = \frac{\beta K_2}{2(C_{GS} + C_B + 2C_{GD})} t$$

Therefore

$$t = \frac{2(C_{GS} + C_B + 2C_{GD})}{\beta K_2} \left\{ \ln \left[1 + \frac{K_2}{v_d(0+)} \right] - \ln \left[1 + \frac{K_2}{v_d(t)} \right] \right\} \qquad (16)$$

and

$$v_d(t) = \frac{v_d(0+)K_2 e^{A_3 t}}{v_d(0+)(1 - e^{A_3 t}) + K_2}$$

where

$$A_3 \triangleq \frac{\beta K_2}{2(C_{GS} + C_B + 2C_{GD})}$$

To solve for the node voltages, refer to Equations (10(a)) and (10(b)). Substituting $v_1 = v_d + v_2$ and noting that $dv_2/dt = de_S/dt$, we have

$$(C_{GS} + C_B + C_{GD}) \frac{d}{dt} v_d + C_B \frac{d}{dt} v_2 = -\frac{1}{2} \beta K_2^2$$

$$\Rightarrow \qquad v_2(t) = \left(1 + \frac{C_{GS} + C_{GD}}{C_B} \right) [v_d(t) - v_d(0+)] - \frac{\beta K_2^2}{2C_B} t + v_2(0+)$$

with

$$v_1(t) = v_2(t) + v_d(t)$$

and

$$e_S(t) = v_2(t) - V_T - \frac{1}{2} K_2$$

Again, since $v_d(t_{sat}) = V_T$ by definition, from Equation (16)

$$t_{sat} = \frac{2(C_{GS} + C_B + 2C_{GD})}{\beta K_2} \left\{ \ln \left[1 + \frac{K_2}{v_d(0+)} \right] - \ln \left[1 + \frac{K_2}{V_T} \right] \right\}$$

For weak conduction condition let $K_2 \rightarrow 0$ in the above and we have [2]

$$t_{sat} = \frac{2(C_{GS} + C_B + 2C_{GD})}{\beta} \left[\frac{1}{v_d(0+)} - \frac{1}{V_T} \right]$$

Thus, t_{sat} is inversely proportional to $v_d(0+)$ with a constant offset.

For $t_{sat} < t < t_l$, take the limit of Equation (15) as $K_2 \rightarrow 0$ and we obtain

$$t_l - t_{sat} = \frac{C_{GS} + C_B + 2C_{GD}}{2\beta V_T} \ln \left(\frac{2V_D - 3V_T}{V_T} \right)$$

The weak conduction solution is the time optimal solution under the condition that the final voltage $V_D = v_1(0+)$ which is obviously the maximum voltage the latch can deliver. Since t_{sat} for such case is very long, however, both devices are allowed to conduct in most practical designs. Equation (14) is then a better estimate for t_{sat}.

3.3.2 Noise Analysis. The latching speed of the sense amplifier is not only limited by the final differential voltage V_D. Since the cells are controlled by dynamic circuits, the raw $v_d(0+)$ that drives the sense amplifier is subjected to noises from many sources. Three internal noise sources can be readily identified: (1) device mismatches, (2) pass device feedthrough, and (3) bitline droop.

(1) Device mismatches: To calculate the effect of device mismatches, it is more convenient to use matrix notations. Rewrite Equation (10(a)) and (10(b)) in matrix form:

$$\mathbf{C}\frac{d}{dt}\begin{bmatrix} v_1 \\ v_2 \end{bmatrix} = -\frac{1}{2}\begin{bmatrix} \beta_1(v_2 - e_s - V_{T1})^2 \\ \beta_2(v_1 - e_s - V_{T2})^2 \end{bmatrix} + \begin{bmatrix} C_{GS2} \\ C_{GS1} \end{bmatrix}\frac{d}{dt}e_s$$

where

$$\mathbf{C} \triangleq \begin{bmatrix} C_{GS2} + C_{B1} + C_{GD} & -C_{GD} \\ -C_{GD} & C_{GS1} + C_{B2} + C_{GD} \end{bmatrix}$$

Multiplying both sides by $\mathbf{C}^{-1}$, we have

$$\frac{d}{dt}\begin{bmatrix} v_1 \\ v_2 \end{bmatrix} = -\frac{1}{\triangle}\left\{\frac{1}{2}\mathbf{C}'\begin{bmatrix} \beta_1(v_2 - e_s - V_{T1})^2 \\ \beta_2(v_1 - e_s - V_{T2})^2 \end{bmatrix} + \mathbf{C}'\begin{bmatrix} C_{GS2} \\ C_{GS1} \end{bmatrix}\frac{d}{dt}e_s\right\} \tag{18}$$

where

$$\mathbf{C}' = \begin{bmatrix} C_{GS1} + C_{B2} + C_{GD} & C_{GD} \\ C_{GD} & C_{GS2} + C_{B1} + C_{GD} \end{bmatrix}$$

and

$$\triangle = (C_{GS1} + C_{B2} + C_{GD})(C_{GS2} + C_{B1} + C_{GD}) - C_{GD}^2$$

is the determinant of $\mathbf{C}$. Let

$$x_1 \triangleq (v_1 - e_s - V_{T2})^2 \quad \text{and} \quad x_2 \triangleq (v_2 - e_s - V_{T1})^2$$

and, for each parameter Z_n, $n = 1, 2$, define

$$Z \triangleq \frac{Z_1 + Z_2}{2} \quad \text{and} \quad \Delta Z \triangleq Z_1 - Z_2$$

so that

$$Z_1 = \frac{Z + \Delta Z}{2} \quad \text{and} \quad Z_2 = \frac{Z - \Delta Z}{2}$$

Multiplying out Equation (18), neglecting the second-order terms, and subtracting the second row from the first, it can be easily derived that

$$\frac{d}{dt}v_d \cong \frac{1}{\triangle}\left\{\frac{1}{2}\beta(C_{GS} + C_B)(x_1 - x_2) - \frac{1}{2}[(C_{GS} + C_B)\Delta\beta\right.$$

$$\left. + \beta(\Delta C_{GS} - \Delta C_B)](x_1 + x_2) - 2(C_{GS}\Delta C_B + C_B\Delta C_{GS})\frac{d}{dt}e_S\right\} \quad (19)$$

Now, the reader is invited to show that

$$x_1 - x_2 = (v_d + \Delta V_T)(v_d + K_2)$$

and

$$x_1 + x_2 = \frac{1}{2}[(v_d + \Delta V_T)^2 + (v_d + K_2)^2].$$

Substituting the above equations into Equation (19) yields

$$\frac{d}{dt}v_d \cong \frac{1}{\triangle}\beta(C_{GS} + C_B)(v_d + K_2)\left\{\left[1 - r_1\left(1 + \frac{1 + 2\Delta V_T/v_d}{1 + K_2/v_d}\right)\right]v_d \quad (20(a))\right.$$

$$+ \Delta V_T \quad (20(b))$$

$$- r_1 K_2 \quad (20(c))$$

$$\left. - \frac{r_2}{v_d + 2K_2}\frac{d}{dt}e_S\right\} \quad (20(d))$$

where

$$r_1 = \frac{(C_{GS} + C_B)\Delta\beta + \beta(\Delta C_{GS} - \Delta C_B)}{2\beta(C_{GS} + C_B)}$$

and

$$r_2 = \frac{C_{GS}\Delta C_B + C_B\Delta C_{GS}}{\beta(C_{GS} + C_B)}$$

are constant coefficients of mismatches determined not only by process but also by circuit design. For example, since $C_S \neq C_R$, $|\Delta C_B|$ is always nonzero.

For correct sensing, the sum of all terms in Equation (20) must be greater than zero. Term (a) represents a percentage degradation of v_d due to device mismatches. This degradation is usually of the order of 10–20%. Term (b) is the mismatch between the threshold voltages of T_1 and T_2 that must be directly subtracted from the raw v_d. The ΔV_T is usually 5–10% of V_T. Terms (c) and (d) are effects *proportional to the latching speed*. Obviously, for small $v_d(0)$, the faster the latch sets, the greater these effects.

For the sense amplifier in Figure 28 then, degradation of v_s should be modified to take the above effects into account. Equation (3(a)) now becomes

$$v_S^1(0) = V_H - V_T - \Delta v_S - \Delta v_S' \quad (3(a'))$$

where $\Delta v_S'$ summarizes the total degradation of v_d in Equation (20).

Degradation of the effective $v_d(0+)$ not only causes possible false latching but also impacts the chip performance as well. In practice the set pulse always starts slow so the noise effects of (c) and (d) are minimized. It then goes down quickly after $v_d(t)$ has become reasonably large to shorten the whole set time. A popular approach to implement such a pulse is to use two pulldown devices as shown in Figure 36. Device T_5 is small and T_6 is large. When ϕ_2 goes up T_5 turns on, pulling node ⓪ down slowly. $\phi_2(t - \tau)$ goes up after some delay. Device T_6 then turns on and sets the latch quickly. In Figure 28 the set pulse(s) were specifically designated as $\phi_2(\phi_2')$ to indicate the possible use of this "dual-slope" composite timing.

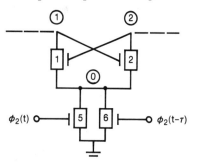

Figure 36.

To alleviate the mismatch effects associated with C_S's and at the same time speed up the latching, pass devices can be used to isolate bitlines when the latch is set. As shown in Figure 37(a) devices T_7 and T_8 are either enhancement mode or depletion mode devices. The gating pulse ϕ_I dips during setting so that the bitlines are cut off from the latch. Generation of ϕ_I is left to the reader as an exercise. Another approach that does not require an additional clock calls for the insertion of a depletion device between the latch and the bitline to limit the current. The bitline capacitance is then partially isolated from the latch as shown in Figure 37(b).

A more sophisticated design that compensates the V_T mismatch between T_1 and T_2 is shown in Figure 38. The latch is set by ϕ_2 at $t = 0$. Before $t = 0$, ϕ_R' has precharged nodes ① and ② to $(V_H - V_T)$ and nodes ③ and ④ to $(v_2(0-) - V_{T1})$ and $(v_1(0-) - V_{T2})$, respectively. Since both are just cut off, T_1 and T_2 would turn on simultaneously at $t = 0$ if $v_1(0-)$ were equal to $v_2(0-)$, regardless of device threshold voltage differences.

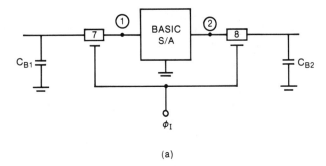

(a)

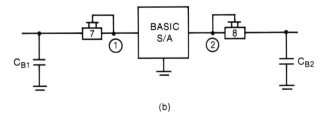

(b)

Figure 37.

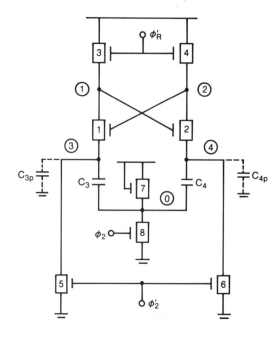

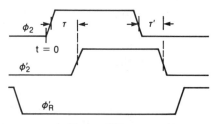

Figure 38.

As ϕ_2 goes up, node ⓪ goes down. Since nodes ③ and ④ are controlled by node ⓪ via capacitor coupling, v_3 and v_4 go down according to the capacitive ratio C_{3p}/C_3 and C_{4p}/C_4, respectively, where the C_{ip}'s are the parasitic capacitances associated with node ⓘ, i = 3,4. The differential voltage $\Delta v = v_1 - v_2$ is also amplified to some degree. Then at $t = \tau$, ϕ_2' goes up, pulls nodes ③ and ④ down toward ground and further amplifies Δv. This is the major amplification and finally pulls v_2 to ground.

The restore starts when ϕ_2 goes down and v_0 moves up. Since ϕ_2' is delayed from ϕ_2 by τ', nodes ③ and ④ are held to ground as v_0 moves up to $(V_H - V_T)$. This ensures that v_3 and v_4 will be charged up to $(v_2 - V_T)$ and $(v_1 - V_T)$, respectively, by ϕ_R' as we have assumed in the previous discussion. Were ϕ_2 and ϕ_2' to go down simultaneously, v_3 and v_4 might be pushed up to some high voltage level by v_0 such that T_1 and T_2 could not be turned on when ϕ_R' goes up. The devices would then lose control of their source voltages, and the latch would not be able to function properly. Timing among the control pulses is important.

Compensation of small voltage differences with capacitors deserves special attention. Noises picked up by the capacitors during clocking as well as insufficient time for small voltage settling may easily offset, if not upset, the would-be benefit.

(2) Pass Device Feedthrough and Incomplete Bitline Restore: MOS devices are also good capacitors. The gate capacitances C_{gd} and C_{gs} store charges when the device is on. Charges are coupled into the source/drain when the device is turned on and out of the source/drain when the device is turned off. Such a phenomenon is particularly

troublesome when the source or drain is left floating, such as in the storage node in a 1-device cell. Refer to Figure 24 again. When WL goes low, charges are coupled out of C_S. Since C_S is very small, v_S will go negative if the cell is originally empty. If v_S is too low, subthreshold current becomes significant and the cell will discharge the bitline through T_X. If there are many 0-cells on one side of the sense amplifier, successive reading of these cells can create enough leaky devices to prevent the bitline from a complete restore. The potential on the bitline is lower than that on the other side after restore. If the next cell to be read is a 1-cell on the same side, this potential difference will offset the differential voltage developed by the cell. Another problem of charge coupling occurs when the bit switch goes off. The charge out-coupling on the bitline may increase the subthreshold currents in the T_X's of the 1-cells nearby. Again, v_d can be severely reduced.

Feedthrough problems can be reduced by slowing down the gate pulse to allow sufficient time for the charge to redistribute between the source and drain.

(3) Bitline Droop and Storage Charge Degradation: While successive reading of 0-cells can cause unbalance between the bitlines, successive reading of a 1-cell may also degrade v_d. Consider Figure 34 again, where, as we always assume, $v_1(0) > v_2(0)$. During the latch set, v_1 droops due to capacitive coupling and certain conduction of T_1. Since v_1 is written back to the cell after the latch has been set, the amount of charge stored in the cell is reduced. If the cell is read successively without being refreshed by a WRITE command, the charge stored in the cell decreases monotonically and the differential voltage available to the sense amplifier degrades. This is called the *bitline droop* problem.

To evaluate the storage charge degradation, let us assume that both the cell and the bitline are charged to $(V_H - V_T)$ at the beginning of the first READ. The total charge stored in C_S and C_B is given by

$$Q_0 = C_B(V_H - V_T) + C_S(V_H - V_T)$$

$$\Rightarrow \qquad C_B(V_H - V_T) = (1 - \eta_{TR})Q_0$$

Let ΔQ be the amount of charge lost through the sense amplifier during each READ; the charge stored in the cell at the end of the first READ is then given by

$$\frac{Q_0 - \Delta Q}{C_S + C_B} C_S = \eta_{TR}(Q_0 - \Delta Q)$$

Thus, at the beginning of the second READ, the total charge in C_S and C_B becomes

$$Q_1 = C_B(V_H - V_T) + \eta_{TR}(Q_0 - \Delta Q)$$
$$= (1 - \eta_{TR})Q_0 + \eta_{TR}(Q_0 - \Delta Q)$$
$$= Q_0 - \eta_{TR}\Delta Q$$

It is left to the reader as an exercise to show that at the beginning of the nth READ the total charge is given by

$$Q_{n-1} = Q_0 - (\eta_{TR} + \eta_{TR}^2 + \eta_{TR}^3 + \cdots + \eta_{TR}^{n-1})\Delta Q$$

Taking the limit as $n \to \infty$ we have

$$Q = \lim_{n \to \infty} Q_n = Q_0 - \frac{\eta_{TR}}{1 - \eta_{TR}}\Delta Q$$

Hence after a large number of successive READs the voltage across the storage capacitor C_S is reduced by

$$\Delta v_s'' = \frac{\eta_{TR}}{1 - \eta_{TR}} \frac{\Delta Q}{C_S}$$

and Equation (3(a')) should be further modified to be

$$v_s^1(0) = V_H - V_T - \Delta v_S - \Delta v_S' - \Delta v_S'' \qquad (3(a''))$$

Bitline droop can be solved by two circuit techniques: (1) capacitor boost and (2) active pullup.

1. *Capacitor Boost*. The bitline droop can be alleviated by boosting v_1 and v_2 with coupling capacitors as shown in Figure 39. Since the coupling capacitors C_1 and C_2 cannot be too large because of layout constraint, bitlines are isolated from the latch by $\phi_{1.4}$ during boosting. Also, since small devices have larger percentage mismatches for fixed tolerances, it is important to lay out the capacitors as symmetrically as possible to ensure that the capacitor mismatch does not contribute significantly to the overall device mismatch budget.

2. *Active Pullup*. The active pullup not only eliminates the bitline droop but can also allow a full V_H charge-up for the 1-cell. As shown in Figure 40(a), after the latch has been set by ϕ_2', the pass devices are turned back on by $\phi_{1.5}$. At this time, in the low side, v_2 and v_4 are at ground, and in the high side v_1 and v_3 are at some high level. Device T_1 is off and T_2 is on. Now $\phi_{2.4}$ goes up. Since T_1 is off, v_5 is disturbed only slightly and stays high. Node voltage v_6, however, is discharged to ground through the

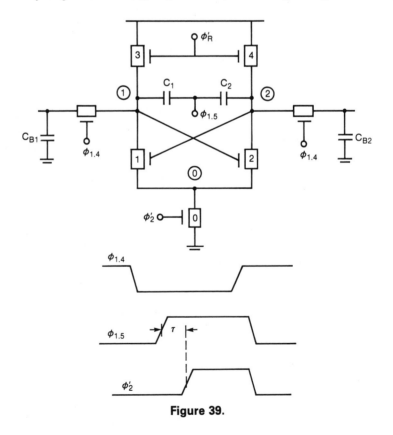

Figure 39.

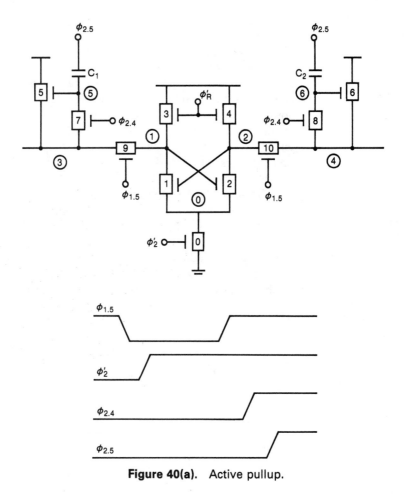

Figure 40(a). Active pullup.

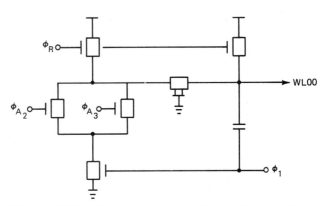

Figure 40(b). Address decode with boosted wordline.

path (T_8, T_{10}, T_2, T_0). Thus, when $\phi_{2.5}$ goes up, v_5 will be boosted high, turning T_5 on strongly which, in turn, charges bitline ③ to V_H. Device T_6, on the other hand, remains off, and bitline ④ stays at ground. The active pullup enhances the uplevel of the latch while leaving the downlevel undisturbed.

In case the wordlines are also boosted higher than V_H, a full V_H level can be written into the 1-cell, increasing the raw differential voltage to the sense amplifier. A dynamic decoder with boosted wordline is shown in Figure 40(b).

3.3.3 Variations in Design

(1) Charge Transfer Sense Amplifier: The *charge transfer sense amplifier,* also called the *bucket brigade sense amplifier,* is especially suited for detecting small signals. As shown in Figure 41(a), two transfer devices, T_5 and T_6, are inserted between the bitlines and the sense amplifier. Unlike the isolation devices in Figure 37, however, these devices are biased by constant voltage and are used as charge transfer devices similar to those found in MOS bucket brigade. Charges can be transferred by the bitlines to the sense amplifier inputs, nodes ① and ② without attenuation regardless of the bitline capacitances and any geometric mismatch between the transfer devices.

To understand charge transfer better, first, let the wordlines go up at $t = 0$. Prior to $t = 0$ the node voltages v_1 and v_2 have been precharged to some high level $v_1(0) = v_2(0)$. The voltages on the bitlines, v_3 and v_4, are charged to $(V_H' - V_{T5})$ and $(V_H' - V_{T6})$, respectively, where V_H' is a constant bias. At $t = 0$ charges in the cells start to redistribute along the bitlines. Let us consider the left side of the sense amplifier first. Assume C_S is a 0-cell so $v_S(0-) = 0$. Since $C_S \ll C_{B1}$, v_S will quickly reach v_3. Thus $v_S(0+) = v_3(0+)$. By charge conservation

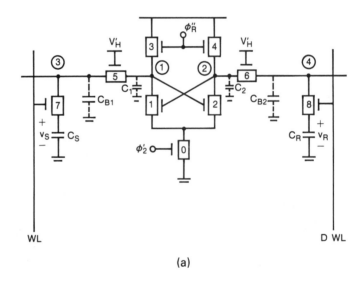

(a)

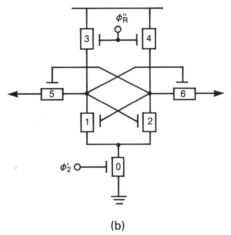

(b)

Figure 41. Charger transfer sense amplifiers.

$$C_{B1} v_3(0-) = (C_S + C_{B1}) v_3(0+)$$

$$\Rightarrow \qquad v_3(0+) = \frac{C_{B1}}{C_S + C_{B1}} (V_H' - V_{T5})$$

Since $v_3(0+) < (V_H' - V_{T5})$, T_5 turns on and current starts to flow from ① to ③. Now if V_H' biases T_5 in the saturation region all the time, node ③ *will be charged back to* $(V_H' - V_{T5})$ and node ① falls. Let v_1 and v_3 be the final values of $v_1(t)$ and $v_3(t)$, respectively. We then have

$$C_1 v_1(0) + (C_S + C_{B1}) v_3(0+) = C_1 v_1 + (C_S + C_{B1}) v_3$$

Substituting the values of $v_3(0+)$ and v_3 into the above equation, it can be shown that

$$\Delta v_1 = v_1 - v_1(0) = \frac{C_S}{C_1} (V_H' - V_{T5})$$

Thus the voltage drop of node ① is independent of C_{B1}. This is to be expected because v_3 is charged back to $v_3(0-)$ after the charge redistribution, so the net charge transfer to C_{B1} is zero.

Similarly, if C_S is a 1-cell and $v_S(0-) \neq 0$, the voltage drop of ① is given by

$$\Delta v_1 = v_1 - v_1(0) = \frac{C_S}{C_1} [V_H' - V_{T5} - v_S(0-)]$$

On the right side of the sense amplifier, the voltage drop of ② can be obtained in the same fashion

$$\Delta v_2 = v_2 - v_2(0) = \frac{C_R}{C_1} [V_H' - V_{T6} - v_R(0-)]$$

The differential voltage across the sense amplifier is thus equal to

$$\Delta v = v_1 - v_2 = \Delta v_2 - \Delta v_1$$

If $C_R = C_S$ and $v_R(0-) = v_S(0-)/2$ the differential voltage becomes

$$\Delta v = \frac{C_S}{C_1} [v_R(0-) + V_{T5} - V_{T6}]$$

Since C_S and C_1 are of comparable magnitudes, $|\Delta v|$ is much larger than that obtained by charge redistribution. The major drawback of the charge transfer amplifier, however, is its slow performance. Consider the left bitline for $t > 0$. Neglecting v_{DS} of T_7, the circuit equations are given by

$$C_1 \frac{d}{dt} v_1 = -i = -\frac{1}{2} \beta_5 (V_H' - v_3 - V_{T5})^2 \qquad (21)$$

$$(C_{B1} + C_S) \frac{d}{dt} v_3 = i = \frac{1}{2} \beta_5 (V_H' - v_3 - V_{T5})^2 \qquad (22)$$

Integrating Equation (22) we obtain

$$\frac{1}{V'_H - v_3(t) - V_{T5}} = \frac{1}{2}\frac{\beta_5}{C_{B1} + C_S}t + \frac{1}{V'_H - v_3(0+) - V_{T5}}$$

Substituting the above into Equation (21) and integrating, the reader is invited to show

$$\Delta v_1(t) = v_1(t) - v_1(0)$$

$$= \frac{C_S}{C_1}(V'_H - V_{T5})\left(\frac{1}{t/\tau + 1} - 1\right)$$

where

$$\tau \triangleq \frac{2(C_S + C_{B1})}{\eta_{TR}\beta_5(V'_H - V_{T5})}$$

Thus the transfer ratio, η_{TR}, still affects the performance even though there is a perfect charge transfer. Since τ is inversely proportional to β_5, the transfer devices should be made as large as practical.

Because of the long "time constant," in practice the latch is set even before the differential signal has completely settled. The device mismatches, then, do affect the differential signal available to the cross-coupled pair. The exact timing depends on both transfer ratio and device tracking.

Since the charge transfer sense amplifier is particularly suited for sensing small signals, it is important that the threshold voltage mismatches between critical devices are small. If this is not so, the cross-coupled charge transfer sense amplifier shown in Figure 41(b) can be used. The regenerative action inside the latch not only compensates the mismatches, but also improves the performance.

(2) CMOS Half V_H Sense Amplifier: In the design of Figure 28 the bitlines are precharged to $(V_H - V_T)$. After the sense amplifier has been set, there is degradation of the 1-level signal written back to the cell. To make up for the charge loss active pullups can be used as discussed in the previous subsection. In CMOS technology active pullups can be easily implemented with a pair of p-channel devices shown in Figure 42. To exploit the full strength of the p-channel devices, however, bitlines

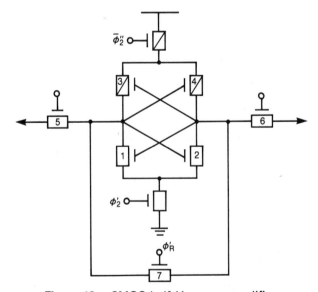

Figure 42.   CMOS half V_H sense amplifier.

are precharged to approximately half way between V_H and GD. This is called half V_H sensing.

As shown in the same figure, ϕ_2' turns on the n-channel devices T_1 and T_2, which amplify the initial differential voltage $v_d(0)$. After $v_d(t)$ has grown to a reasonably large magnitude such as $v_d(t) = V_{TN}$, ϕ_2'' goes down, turning on T_3 and T_4. The regenerative action between the p-channel devices reinforces the same between the n-channel devices and pulls the high voltage node to V_H quickly. After sensing, restore is accomplished by turning on T_7 with ϕ_R'. Voltages v_1 and v_2 will then be equalized to approximately $V_H/2$ level.

Thus the sense amplifier is first set by n-channel devices. The p-channel devices restore the uplevel. Both devices are not turned on simultaneously because of process considerations. In single-well CMOS processes, only one type of device can be optimized. For example, in many n-well CMOS processes, the n-channel devices are normal surface devices, but the p-channel devices are so-called buried channel devices. As a result, tracking between the n-channel devices is much better than that between the p-channel devices. Turning on both types of devices simultaneously requires more initial differential voltage to overcome WC threshold voltage mismatches. It also causes turnon noises generated by the device itself. Since the latch is extremely sensitive to noise in the initial setting, turning on more devices may adversely affect the final result.

The drawback of CMOS half V_H sensing is the smaller transfer ratio. Since the bitline capacitance is inversely proportional to the square root of the voltage between bitline and substrate, precharging the bitlines to $V_H/2$ results in larger bitline capacitance and hence a smaller transfer ratio. Half V_H sensing, however, generates less noise during the set of sense amplifiers because half as much charge is now being dumped to the ground. The transient power is therefore smaller and the ground is more stable.

3.3.4 Bitline Configurations. The bitlines in Figure 28 that stretch out evenly from both sides of the sense amplifier are in an *open-bitline configuration*. As sense amplifier design becomes more complicated, the layout of the circuit becomes more stringent. To alleviate layout constraints, most designs fold the bitlines in the middle and attach the sense amplifier to the end as shown in Figure 43. The sense amplifier can now be laid out in two bitline pitches instead of one. Better yet, if another folded bitline is attached to the other side, the same sense amplifier can be shared by the two with pass gate selection. In Figure 43 pass devices (T_1, T_2) and (T_3, T_4) select the bitlines to be attached to the sense amplifier during sensing. After the latch has been set both devices are turned on to allow the signal to pass to the I/O buses to the right of the array.

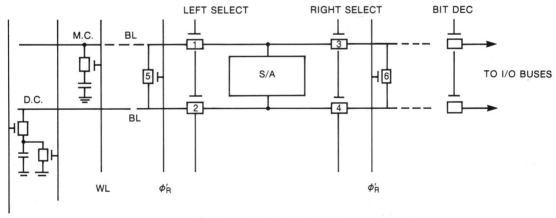

Figure 43. Folded bitline configuration.

Sharing the sense amplifier with *folded bitlines* not only alleviates layout constraints but also enhances noise immunity. Since the bitlines are next to each other, common mode noises are readily rejected by the sense amplifier. Another advantage is the equalization of the bitline potentials. The pass devices T_5 and T_6 equalize the bitlines quickly to $V_H/2$ level, resulting in a short bitline restore time.

3.4 Advances in DRAM Architecture and Circuits

3.4.1 Chip Modes. One of the major timing problems in systems application is the performance gap between processor and memory. Because of its complex internal timing, a DRAM runs far behind its logic cousins in the same technology. To match the memory data rate to the MIPS (mega-instructions per second) of a processor, various memory addressing schemes are used. Bank addressing sends the same address to a bank of memory chips simultaneously and selects the chip through selection lines. Memory interleaving sends addresses to and fetches data from different chips at the same time. All these techniques increase the bandwidth of a memory system (the reciprocal of its data rate). Complex addressing, however, requires tight tolerances in signal skews. In many cases, the raw speed of a DRAM, even if slow, can still not be fully utilized.

To make full use of the memory speed, addressing schemes are best implemented on chip. Since the address decoding is a major portion of access time, many chips allow a READ followed by an immediate WRITE at the same address. This option, called READ/MODIFY/WRITE, can be easily implemented with minimal modification of the basic architecture and is available in most products. Organization can also improve the data rate. The memory array is partitioned into quadrants so that multiple cells can be accessed from the same address. The data are temporarily held in a shift register so they can be either written into or read out in sequence. This is called *serial mode operation.* The most popular serial modes are *nibble* (4 bits) *mode* and *byte* (8 bits) *mode.*

An extremely important observation of the internal operations of a DRAM is the fact that in each cycle all cells in the same row controlled by the same wordline are affected equally as far as RAS timing is concerned. In bit organized memory only one cell is selected by the column decoders in CAS timing. Cells that share the same row address are said to be in the same *page.* By slight modification of CAS timing and I/O circuits, cells in a whole page can be accessed without changing the row address. The following are some popular chip modes that make use of this principle.

(1) Page Mode: In page mode, the row address is gated in by $\overline{RAS}$ staying active (low) throughout the operation. Column addresses are gated in by $\overline{CAS}$ so the cells are accessed without row decoding. A typical page mode timing is shown in Figure 44. Notice that $\overline{CAS}$ must toggle to initiate the precharge of internal nodes of the column decoders.

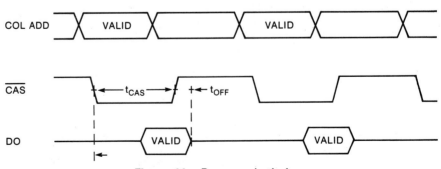

Figure 44. Page mode timing.

(2) Ripple Mode: In page mode, the column decoders are dynamic circuits that operate on the principle of evaluation-drive-restore sequence as discussed early in Section 3. In the ripple mode, column decoders are static circuits. New column addresses are gated in when $\overline{CAS}$ is high. The addresses are latched by the falling edge of $\overline{CAS}$. The column address access time, t_{CAA}, is measured either from the rising edge of $\overline{CAS}$ if the valid column address precedes it or from column address valid if it follows the rising edge of $\overline{CAS}$. A typical ripple mode timing is shown in Figure 45.

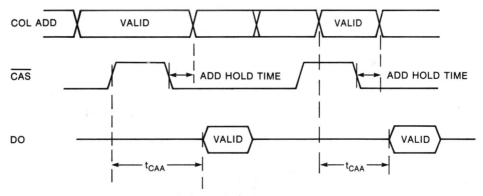

Figure 45. Ripple mode timing.

(3) Static Column Mode: In the ripple mode, $\overline{CAS}$ command toggles to gate in the column addresses. In static column mode both $\overline{RAS}$ and $\overline{CAS}$ stay active (low) for the whole page. Address detection circuits are used to detect and trigger I/O operations at the new column address. An idea borrowed from static RAMs address transition detection, the static column mode becomes a standard feature in CMOS DRAMs. Typical timing is shown in Figure 46.

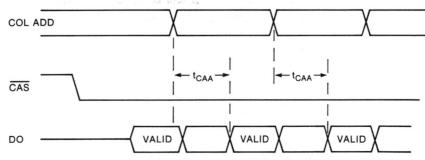

Figure 46. Static column mode timing.

3.4.2 Refresh Schemes. The cells in a DRAM must be refreshed regularly to hold data. Since a READ operation sets the sense amplifiers to replenish charges back to the cells, a refresh can be accomplished by a forced READ. The simplest refresh scheme is the so-called *RAS-only refresh.* The row address for refresh is gated in by $\overline{RAS}$ command. Since all cells in the same row are refreshed simultaneously, only RAS timing is involved inside the chip during a refresh. The $\overline{CAS}$ command is held inactive (high) throughout, so the output at *DO* is in the high impedance state.

Sometimes it is desirable to maintain the valid data at *DO* during a refresh cycle. This requirement, called *hidden refresh* (hidden in the sense that the refresh is carried out during a continuous data stream), is accomplished by system commands as shown in Figure 47. The $\overline{CAS}$ command stays active (low) after a normal READ; $\overline{RAS}$ then toggles to gate in the refresh addresses. Output at *DO* will stay constant during the refresh cycles. The reader is invited to modify the I/O timing and circuit of Figure 28 to implement such a feature as an exercise.

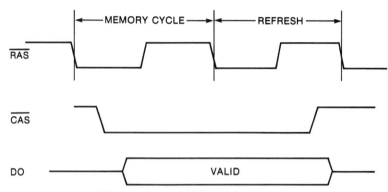

Figure 47. A hidden refresh cycle.

A more sophisticated scheme simplifying the refresh operation on the system level is the *CAS before RAS refresh*. In this scheme a built-in counter is used to keep track of the refresh address. A one-shot circuit detects the incoming order of $\overline{RAS}$ and $\overline{CAS}$ commands. In normal operations in which $\overline{RAS}$ precedes $\overline{CAS}$, this circuit remains inactive. As soon as the order of $\overline{RAS}$ and $\overline{CAS}$ is reversed, however, a pulse is generated by the circuit connecting the refresh address in the counter to the row address decoders. Normal address input pins will be ignored. A subsequent READ then refreshes the cells in the whole row. After the refresh has been completed at the rise of $\overline{RAS}$, the counter advances by one count, waiting for another occasion.

CAS before RAS refresh eliminates the external refresh address counter. It is becoming a popular feature among recent products.

3.4.3 Redundancy.   Because of their extremely small sizes, DRAM cells are very sensitive to chip defects. As the density of DRAM increases, the good chip yield could be very low. In many high-density RAMs, redundant cells forming extra rows or columns are added to the array. Each spare row or column has its own decoder. The address inputs to these decoders are controlled by polysilicon fuses so a particular address pattern can be programmed into the decoders by selectively blowing the fuses either with a large current or a laser beam.

In "standby," spare decoders are held inactive by clamping their outputs via fuses. Once a bad cell is found during wafer probing, a spare row or column is switched in by blowing the clamping fuses and the fuses associated with the inputs of the decoder. The replaced decoder is then either permanently disconnected from the rest of the chip by blowing fuses or simply disabled by the spare decoder each time its address is detected. In the former, a bigger area is needed for the fuses on each decoder whereas in the latter, a small increase of the access time may incur.

Redundancy increases the chip area, and in some cases, the access time as well. It does, however, enhance yield significantly, particularly during the early manufacturing stage of the product.

4.   CONCLUSION

The MOS memory industry is growing at an astounding pace. RAM densities quadruple every three or four years, and performances improve constantly each time a new generation of products is introduced. New technologies, both in process and in circuits, are invented in almost every new product, and new architecture, such as the address transition detection and CAS before RAS refresh, is constantly suggested by innovative chip designers to reduce system overhead.

The development of MOS RAM is, however, not without its challenges. A serious problem looming ahead for all semiconductor memories is soft error. A soft error is an intermittent error caused by nonrepeatable noises. One prominent cause is the α-particle emitted from the residual radioactive materials in the chip package. A high-energy α-particle that penetrates into the silicon surface can travel a long distance before it is stopped by its surroundings. As the particle slows down, energy is released and a large number of electron-hole pairs are generated in the silicon substrate. The electrons (resp. holes) will be collected by the diffusion of n-channel devices (resp. p-channel) and the n-side (resp. p-side) of the pn junctions. Critical nodes nearby, such as the internal nodes of a cell or the sensing nodes of a sense amplifier, may thus be discharged/charged to such an extent that errors occur.

The exact nature of the α-particle problem depends on many factors, including the energy of the particle, the incident angle and total length of the particle's trajectory, the collection efficiency of the critical nodes (proportional to the area and time period the nodes remain floating), and the effectiveness of common mode noise rejection of symmetric structures such as the 6-device static cell and the folded bitlines of a DRAM. In CMOS technology placing the array and critical circuits in protective diffusion wells also enhances α-particle immunity.

As the density increases with lower power supply, however, all semiconductor memories will be susceptible to soft error problems. The soft error rate (SER), measured by the percentage error per thousand power-on hours (% KPOH) has become a standard specification of MOS DRAMs since the 64K DRAM was introduced. Systems that use high-density DRAMs must use ECC (error correction code) circuits to check the data word before they are processed by the CPU. CMOS SRAMs, which have enjoyed α-particle immunity thus far because of their more stable cell structure and well protection, may have to face the problem even sooner. Most SRAMs are used in high-performance applications where the ECC is simply not feasible. The requirement of a zero SER could be a serious challenge to SRAM chip designers.

Innovation is the key.

REFERENCES

1. Stein, K. U., A. Sihling, and E. Doering. "Storage Array and Sense/Refresh Circuits for Single-Transistor Memory Cells." *IEEE J. of Solid State Circuits,* Vol. SC-7, October 1972, pp. 336–340.

2. Lynch, W. T., and H. J. Boll. "Optimization of the Latching Pulse for Dynamic Flip-Flop Sensors." *IEEE J. of Solid State Circuits,* Vol. SC-9, April 1974, pp. 49–55.

3. Heller, L. G., D. P. Spampinato, and Y. L. Yao. "High Sensitivity Charge Transfer Sense Amplifier." *IEEE J. of Solid State Circuits,* Vol. SC-11, October 1976, pp. 596–601.

4. McKenny, V. G. "A 5 V-Only 4-K Static RAM." *ISSCC Dig. of Tech. Papers,* February 1977, pp. 16–17.

5. Steward, R. G. "High Density CMOS ROM Arrays." *ISSCC Dig. of Tech. Papers,* February 1977, pp. 20–21.

6. Heller, L. G. "Cross-Coupled Charge Transfer Sense Amplifier." *ISSCC Dig. of Tech. Papers,* Feburary 1979, pp. 20–21.

7. Gray, K. S. "Cross-Coupled Charge-Transfer Sense Amplifier and Latch Sense Scheme for High-Density FET Memories." *IBM J. Res. & Develop.,* Vol. 24, May 1980, pp. 283–290.

8. Itoh, K., R. Hori, H. Masuda, and Y. Kamigaki. "A Single 5V 64K Dynamic RAM," *ISSCC Dig. of Tech. Papers,* February 1980, pp. 228–229.

9. Eaton, S. S., D. Wooton, W. Slemmer, and J. Brady. "Circuit Advances Propel 64-K RAM Across the 100-ns Barrier." *Electronics,* March 1982, pp. 132–136.

10. Wang. N. N. "On the Design of MOS Dynamic Sense Amplifiers." *IEEE Trans. on Circuits and Systems,* Vol. CAS-29, no. 7, July 1982, pp. 467–477.

11. Lu, N. C., and H. H. Chao. "Half-V_{DD} Bit-Line Sensing Scheme in CMOS DRAM's." *IEEE J. of Solid State Circuits,* Vol. SC-19, no. 4, August 1984, pp. 451–454.

12. Okazaki, N., F. Miyaji, K. Kobayashi, Y. Harada, J-I Aoyama, and T. Shimada. "A 30 ns 256K Full CMOS SRAM," *ISSCC Dig. of Tech. Papers,* February 1986, pp. 204–205.

MOS LOGIC CHIP DESIGNS

INTRODUCTION

The design of a logic chip is greatly influenced by design methodology. Because of the ever-increasing demand for higher density and better performance, logic chips are getting more complex. To keep track of the large number of logic functions, various design aids and checking tools have been developed. Formal methodologies, for both logic and circuit designs, have become a necessity. Design automation (DA), which is the ultimate goal of all design methodologies, has revolutionized the IC industry from the overall system design down to the placement of individual circuits. While the early microprocessors were mostly custom designed with all circuits laid out and wired together by hand, almost all of today's logic chips are designed either partially or fully with an automatic design system. A modern chip designer, in addition to being able to design and test physical circuits, must also be a skillful tool user, and, occasionally, an innovative programmer as well.

Formal methodologies, of course, are not without drawbacks. The exact design rules that facilitate automation impose severe restrictions on circuit design. Since an automatic design system must be used for many different applications, the final chip design may not be optimal for all products. An automatic design system once set up, however, can be used to implement as many designs as the computer resources permit. Furthermore, the *turnaround time* (TAT) from logic entry to hardware is much shorter with a design system than with a full custom system. Since it is very expensive to make *engineering changes* (EC) once the hardware build is underway, the concept of *single pass design*—a design that works the first time it is built—is of paramount importance to system designers. The full custom approach falls far behind in this regard. While it is almost a routine with automatic design systems, a fully custom designed chip rarely works on the first design phase. As a matter of fact, many custom chip designers have finished their last "patch-up" only to discover that the technology used has already passed into oblivion.

Heavy use of automatic design systems does not mean computers will eventually replace *all* human designers. People are still more innovative than computers. Most design systems are "interactive": they accept exceptions at a human designer's command. The human designer still has total control of all design activities, but only within the framework laid down by the formal methodology. Indeed, part of a chip designer's responsibility is to communicate with the design system to ensure all design activities are monitored properly by formal procedures. Human errors and oversights will be minimized in this way.

In this chapter we shall discuss logic chip designs within the framework of a design system. The methodology used here is a compromise between full custom and full automation. The particular approach, the *masterimage,* also known as the *standard cell approach,* is the natural outcome of an effort to meet high performance and density demands (best handled by custom design) and short turnaround time (best handled by an automated design). It is fully automatic in that a complete design can be put together by the system alone. It also has elements of custom design in that chip designers have the option to add their own circuits to the system. Such designs must be done in complete compliance with the rules of the whole system and be used only in a controlled environment at the designer's discretion. In this way, the chip designer can concentrate on more critical custom circuits while adopting standard circuits from the library. Thus, both chip density and performance can be improved without sacrificing turnaround time and design integrity.

1. MASTERIMAGE DESIGN SYSTEM

1.1 Design Systems

A masterimage design system consists of a circuit library, a physical design (PD) system, and a logic design (LD) system. A chip designer uses the LD system to implement his logic design, interacts with the PD system if necessary, and chooses circuits from the library.

The circuit library contains the physical implementations of all commonly used logic functions such as NANDs and NORs. These are the building blocks for all applications. Each library entry is called *a book* or *a macro,* although the latter label is generally reserved for more complex books such as a general-purpose register (GPR) macro, an arithmetic/logic unit (ALU) macro, and so on. The books are predesigned and laid out in accordance with general chip rules so they can be placed and wired together in a grid system on the chip called the *chip image.* Each book is fully characterized in terms of power, area, and performance so a chip designer can calculate the chip size and performance before the hardware is actually built.

The placement and wiring of the books is carried out by the PD system. In addition to this function, the PD system also outlines the chip floor plan, optimizes the layout to the criteria set forth by the chip designer, checks the physical layout against the circuit schematic (physical/circuit schematic check), and checks the interconnections among books against the logic description (logical/physical or L/P check). Although the PD system is fully automated, it does accept commands from the chip designer according to the needs of a particular application.

The LD system, on the other hand, interacts with the chip designer at all times. It enables the chip designer to define, enter, simulate, and check her logic implementation in the design system. Most LD systems also generate test patterns or vectors automatically from the logic design. All logic functions are implemented with books in the library, simulated with the delay information from the books, and finally transformed into physical layout by the PD system.

1.2 Design Procedures

In a masterimage design system a typical design proceeds according to the following flowchart.

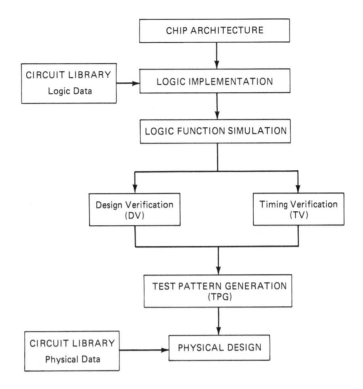

The architecture of a machine is best described by a software model written in some high-level language (HLL). This model allows the chip designer to simulate all machine functions with the efficiency of an HLL. Since the hardware implementation will be based on this model, it is the main blueprint of the machine. In fact, for all practical purposes, the model *is* the machine.

Design automation starts with the logic design. Once the machine architecture is described in some design language, detailed logic implementation in terms of logic gates and elementary functions can be carried out automatically by the LD system. The chip designer uses the DV and TV tools only to audit and, if necessary, to modify the results. As soon as the logic implementation is complete, the design is taken over by the PD system. Chip layouts are generated automatically by the PD system with occasional human interventions such as the placement and wiring of some critical macros. The results are again audited by the system and analyzed by the chip designer with L/P check and TV tools. A chip designer can analyze machine performance in much more detail with the physical layout data. Timing problems can be resolved either by replacement and/or rewiring of critical macros or modifying the logic design, or both. It is in this iterative process that human interventions become critical.

Figure 1 illustrates the relationship between the degree of DA and TAT in arbitrary scales. If the design does not need special attention in PD, the only system the chip designer deals with is the LD system. As soon as the logic design is completed, the PD system handles the physical design automatically. The design work is fully automated, and the TAT is minimal. There are occasions, however, when the PD must be customized. Special requests, such as the preplacement of clock drivers at some particular locations on the chip can be coded into the PD system by the designer. The system will preplace such circuits and arrange others around them in subsequent automatic placement and wire (APW) operations. Finally, if some special circuits aren't available in the library, the chip designer can design them herself

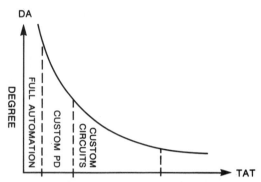

Figure 1. Design automation versus turnaround time.

and later enter them. Such custom circuits, for example, may need special process steps for some critical elements not used in standard books.

2. CHIP IMAGE

To facilitate automatic placement and wiring of circuits, each wiring plane, such as M1, M2, polysilicon, and diffusion, is overlaid with a grid system consisting of all possible circuit contact points and wiring channels in that plane. A *wiring channel,* or *track,* is a space in the plane that accommodates exactly one wire. The collection of all these grid systems, together with the layouts of the power and ground buses, is called the chip image. A chip image is described by chip rules. The rules specify every possible location of contact points between wires in different planes (the via holes) and every possible channel for wiring in each wiring plane.

The construction of the chip image is determined by many factors. The chip image for a technology with three layers of metal is definitely different from one that has only one layer of metal. Another reason for the wide variety of chip images is application. The image of a chip that consists of all random logic gates may very well be different from one that contains large macros such as onchip RAM and ROM. The design system should be flexible enough to accommodate many different configurations while still maintaining a common data base that can be shared by all users.

In this section we shall describe a small image for a simple double-metal, single-polysilicon technology. M1 will be used for horizontal (row, or x-direction) wirings and M2 for vertical (column, or y-direction) wirings. Polysilicon wires will be used mainly for short runs in the vertical direction to connect metal wires to the book I/Os as we will see in more detail next in the section on the circuit library.

As shown in Figure 2, M1 V_H/GD buses run horizontally across the chip forming the book shelves. Books will be placed between the V_H and GD buses in rows. Two shelves are grouped together so books placed back to back can be wired together using polysilicon or diffusion underpassing the V_H buses. The book shelves are separated by wiring bays of horizontal M1 wiring channels. The width of the wiring bay in terms of number of wiring channels is determined by the size of the chip and the density of the book I/Os. In some systems the width of the wiring bay can be adjusted to accommodate the exact number of M1 wires needed to interconnect the books in the whole row. In others, the width of the wiring bay is fixed based on some statistical average. Local overflow of M1 wires will be directed to other wiring bays through M2. One popular empirical formula used in many practical designs is the so-called *Rent's rule,*

$$T = AN^p$$

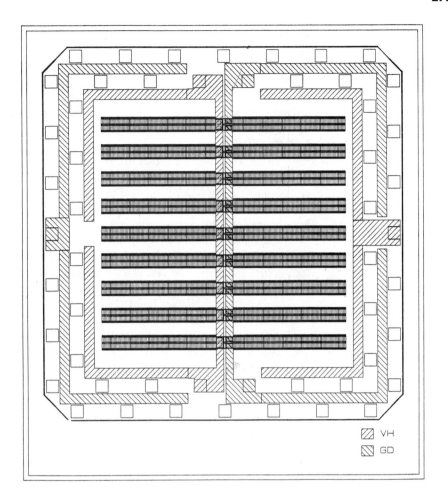

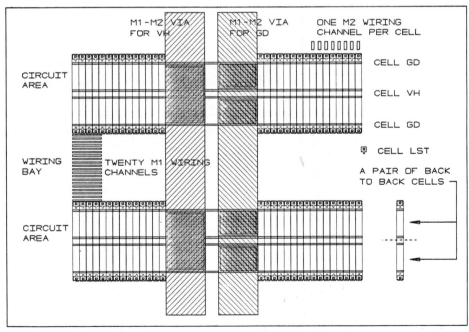

Figure 2. A chip image.

where T is the total number of I/O terminals and N is the number of books in a group. In the equation both A and p are empirical constants, with A varying from 2 to 4 and $p = 0.6$.

Rent's rule is a good estimate for many practical designs. It is also used to estimate the average interconnection wire length in a group of circuits. The constants are determined by substituting statistical data into the equation. In practice the number of wiring channels in the wiring bay varies from 20 to 30 for chip sizes between 5×5 mm^2 and 10×10 mm^2.

M1 V_H/GD buses are tied to the wide M2 buses in the middle of the chip. The M2 wiring channels run in vertical (y-) direction in columns. Each column may consist of one or more M2 wiring channels in parallel. In our case each column contains exactly one M2 wiring channel. This yields the finest granularity for circuit layout but also requires the largest memory storage in the chip data base.

The intersection of a row and a column is called a *cell*. The location of a cell is therefore specified by its column number from the left, and its row number from the top, across the chip. The cell is the smallest area of a circuit. A book may occupy several cells. In our case, since each cell allows only one M2 wire to pass through providing only one contact point for the I/O, the smallest book, an inverter, takes two cells.

I/O contacts to the books are located on the border of the cells facing the wiring bay. These contact points, also referred to as LSTs (logic service terminals), are in either diffusion or polysilicon, with contact holes that can be readily connected to metal or polysilicon lines.

A book that occupies a single row of N cells is called an $1 \times N$ book. A book that occupies a double row of N cells is called a $2 \times N$ book.

Circuits that occupy several rows and columns can be grouped together to form a logic macro. These circuits as a whole can be "prewired" before APW and placed onto the image as a single book. Rules must be generated for the macro to indicate the prewiring and, consequently, the blockage of wiring channels for subsequent APW operations.

Generally speaking, circuits in a book should be laid out in such a way that they stay in cells when placed onto the image with their I/O contact points aligned with the cell LSTs. This ensures the integrity of the V_H/GD buses for all books in the same row. In many cases, however, it is advantageous to spread the layout into the wiring bays. Such is the case, for example, for an onchip RAM. Books that block the V_H/GD buses or the whole M1 wiring bay must be preplaced to the side of the chip so they do not disrupt the power supply or communications of other circuits.

The width of V_H/GD buses is determined by current density. Electromigration of metal lines limits the maximum current density to the order of 10^5A/cm^2. In NMOS technology the V_H drop and GD shift due to DC currents must be evaluated carefully to ensure adequate noise margins. In CMOS technology the DC current density is negligible, but the AC current density is proportional to the operating frequency. Let τ be the pulsewidth of current $i(t)$ and the effective current I_{eff} is given by

$$I_{eff} = \frac{\int_0^\tau i(t)\,dt}{\tau}$$

Let T be the machine cycle time and the equivalent current for electromigration calculation is then given by

$$I_{eq} = \sqrt{\frac{1}{T}\int_0^\tau I_{eff}^2\,dt} = \sqrt{\frac{\tau}{T}}I_{eff}$$

which must obey the technology ground rules.

Another important factor that affects power bus configurations is simultaneous switching. The switching of many high-power books on the same power bus all at once can produce a large voltage spike that may set a quiet latch on the same bus to the wrong state. For large chips, more M2 power bus lines or even M3 power bus meshes may be needed to reduce the noise.

Offchip receivers and drivers are a special class of circuit. The peripheral of the chip is divided into compartments called *I/O cells*. Each I/O cell contains an I/O pad and its surrounding area for a driver/receiver pair. In some applications both the driver and the receiver are used, and the I/O pad is therefore called a *bidi* (bidirectional I/O). In other cases only the driver or the receiver is used and the I/O pad is referred to as a *unidi* (unidirectional I/O). Still, some of the I/O pads are used for power supply and ground connections. In these cases the driver/receiver pair is either disabled or removed. To alleviate the simultaneous switching problem, it is a good idea to place these power I/Os among the signal I/Os to help reduce coupling noises.

3. CIRCUIT LIBRARY

All basic logic functions are implemented and documented according to the chip image and the design rules. Each book is described in at least two distinct ways. To the user, the book is described in terms of its logic function, power consumption, performance, and area so that she can simulate the design and estimate the cost before the hardware is built. Inside the design system, however, the book is described by its physical layout data such as the location of I/Os, the usage, and hence the blockage, of wiring channels in each wiring plane, and general information about size and shape. These data are used by the APW program to wire the book with others. The collection of both rules and information constitutes a complete library.

Most libraries are open-ended in the sense that new books are constantly added to the library as special needs arise. In this section we shall discuss some of the most commonly used books. Before we start, however, defining some terms is necessary. In discussing logic circuits sometimes it is more convenient to use logic values of 0's and 1's for voltage levels. In these cases a node labeled by a number will be prefixed with a P (pin). For example, suppose node $\boxed{10}$ is at its high (low) voltage level $V_H(GD)$. Instead of saying $v_{10} = V_H(0)$, we say P10 = 1(0). As mentioned before, positive logic convention is adopted throughout this book.

3.1 Static Logic Circuits

3.1.1 Multiinput NOR: OIs. NOR gates, also called OIs (OR Invert), are the most popular logic elements in NMOS technology. As shown in Figure 3(a), the output y of the circuit is related to the inputs $x_1, x_2, \ldots, x_m$ by

$$y = \overline{x_1 + x_2 + \cdots + x_m}$$
$$= \bar{x}_1 \bar{x}_2 \cdots \bar{x}_m$$

so that $y = 1$ if and only if all x_i's are 0. In NMOS circuits the value of m varies from 1 to 12. In CMOS technology since an m-input NOR requires stacking of m p-channel devices which are slow, $m \leq 4$ in most practical applications. An m-input (m-way) NMOS NOR is shown in Figure 3(b) and a 4-input CMOS NOR is shown in Figure 3(c).

3.1.2 Multiinput NAND: AIs. NAND gates, also called AIs (AND Invert), are the most popular logic elements in CMOS technology. As shown in Figure

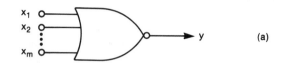

(a)

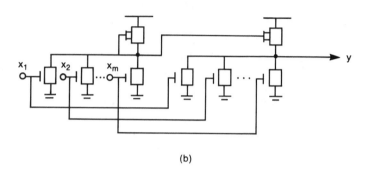

(b)

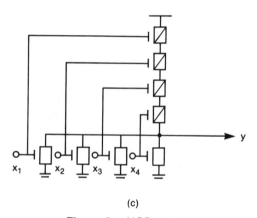

(c)

Figure 3. NOR gates.

4(a), the output y of the circuit is related to the inputs $x_1, x_2, \ldots, x_m$ by

$$y = \overline{x_1 x_2 \cdots x_m}$$
$$= \overline{x_1} + \overline{x_2} + \cdots + \overline{x_m}$$

so that $y = 0$ if and only if all x_i's are 1. In CMOS circuits the value of m varies from 1 to 6. In NMOS technology since an m-input NAND requires stacking of m large pulldown devices demanded by β ratio requirement, $m \leq 3$ in most practical applications. An m-input (m-way) CMOS NAND is shown in Figure 4(b), and a 3-input NMOS NAND with pushpull driver is shown in Figure 4(c).

3.1.3 Two-Level Logic Functional Block: OAIs and AOIs. More complex two-level logic functions can be easily constructed by combining the basic NORs and NANDs in MOS technology. For example, the output of the circuits shown in Figure 5 is given by

$$y = \overline{x_1 x_2 + x_3 x_4}$$

and the output of the circuits in Figure 6 is given by

$$y = \overline{(x_1 + x_2)(x_3 + x_4)}$$

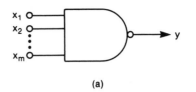

(a)

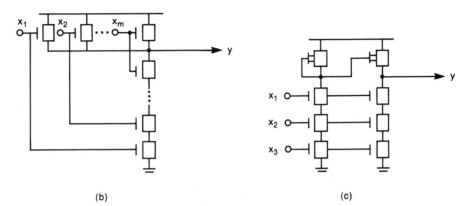

(b)   (c)

Figure 4.   NAND gates.

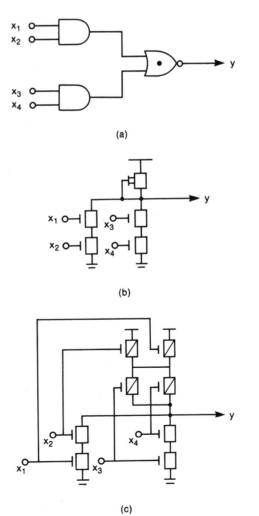

(a)

(b)

(c)

Figure 5.   AOI gates.

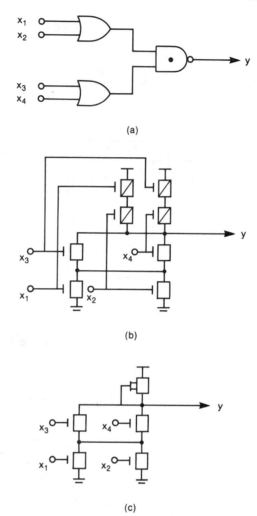

Figure 6. OAI gates.

The former is called an AOI (AND-OR-Invert) and the latter is called an OAI (OR-AND-Invert). The mark inside the output gate indicates that its inputs are dotted.

Logic gates with common inputs can be merged without increasing the height or delay. For example,

$$y = \overline{x_1 x_2 + x_1 x_3 + x_4}$$
$$= \overline{x_1(x_2 + x_3) + x_4}$$

Even though the first expression has two levels of logic and the second has three, both implementations have the same delay.

Complex logic blocks are slower than simple gates. In most applications extra drivers (buffers) are attached to the output of the circuit to improve speed. The effect of internal loading is particularly severe in NMOS circuits with junction capacitances of large pulldown devices. As a result, NMOS complex logic blocks are mostly used inside large macros with controlled loading.

Complex logic blocks, however, are very popular in CMOS designs. Since the DC current is zero, relative sizes of pullups and pulldowns are limited only by noise margins (see Chapter 3). With output buffers, stacking devices up to four levels high is quite common. CMOS logic designs should be partitioned or grouped in a way that takes full advantage of such circuit implementations.

The layout of CMOS functional blocks is particularly efficient in the chip image discussed in Section 2. Since the n-channel and p-channel devices are always in pairs with dual topology (i.e., whenever the n-channel devices are connected in parallel, the p-channel devices are connected in series, and vice versa), layout can be done by first placing devices in a row and then adjusting their relative positions to maximize the sharing of common diffusions. For example, Figure 7 shows two layouts of $y = \overline{x_1(x_2 + x_3) + x_4}$ of the previous example. In (b) the diffusion is broken to isolate node ① from ⑩. In (c) device of x_4 is placed to the left of device of x_1. In this case there is no diffusion breakage and the area is minimized. While there are formal algorithms [12] that optimize the layout in this fashion, a little practice is all that is needed for layout personnel to master the scheme.

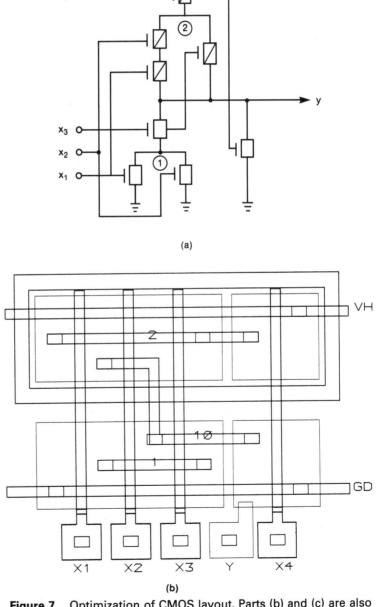

Figure 7. Optimization of CMOS layout. Parts (b) and (c) are also displayed inside the back cover of this book.

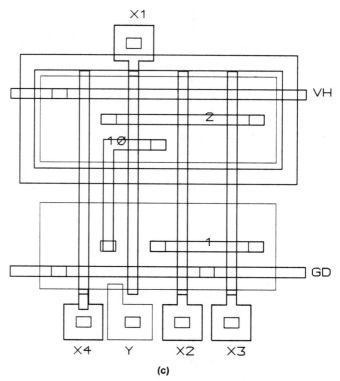

(c)

Figure 7. *(cont.)*

3.1.4 Exclusive OR: XORs and XNORs. Exclusive OR and NOR are very important logic functions. Let $x_1, x_2, \ldots, x_n$ be inputs and the output of an XOR circuit is given by

$$y = (\cdots(((x_1 \oplus x_2) \oplus x_3) \oplus x_4) \cdots \oplus x_n)$$

where

$$x_1 \oplus x_2 = x_1\bar{x}_2 + \bar{x}_1 x_2 = \overline{\overline{x_1 x_2}(x_1 + x_2)}$$

The output of an XNOR is the complement of an XOR. The logic symbols for XOR and XNOR are shown in Figure 8.

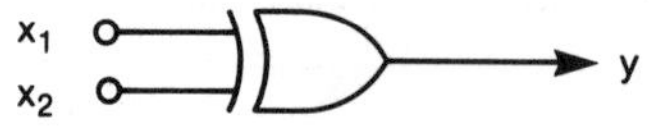

Figure 8(a). A 2-input XOR gate.

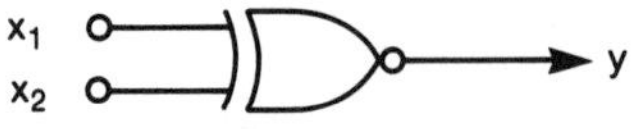

Figure 8(b). A 2-input XNOR gate.

EXERCISE

Prove that the XOR function is commutative and associative; that is,

$$y = (\cdots(((x_1 \oplus x_2) \oplus x_3) \oplus x_4) \cdots \oplus x_n)$$
$$= x_1 \oplus x_2 \oplus \cdots \oplus x_n$$

independent of the order and grouping of the variables.

XOR can be constructed by connecting simple books. In practice, however, because of their heavy usage and the requirement of both true and complement inputs, most libraries offer 2- and 3-input XOR and/or XNOR as individual books. Two 2-input XOR books are shown in Figure 9. Both designs are complex. Two simpler implementations with pass gates are shown in Figure 10. Although the circuit in Figure 10(a) uses fewer devices, there is a "hidden" β ratio requirement left

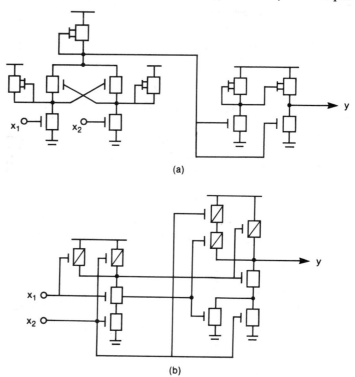

Figure 9.   2-input XOR circuits.

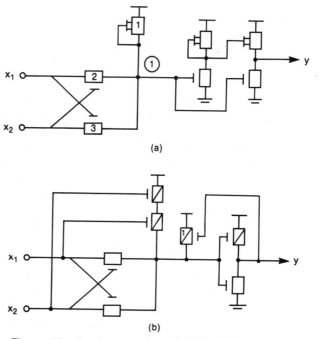

Figure 10.   Implementation of XOR with pass gates.

by the preceding driver. To see the problem, let $x_1 = 1$ and $x_2 = 0$ so that device T_3 is on. Since there is a DC current path from T_1, T_3 and the pulldown of the driver driving x_2, the pulldown of the driver must be large enough to sink the current while still maintaining a reasonable downlevel of x_2 and v_1. Thus, a special driver with a large β ratio is needed. In Figure 10(b) the feedback device T_1 is a small p-channel device that pulls the input of the inverter to V_H when the output is low so that no DC power is consumed.

Performance of circuits with pass gates is sensitive to the proximity of their drivers. As a rule, pass gates are only used inside a macro without being directly controlled by macro I/Os.

3.1.5 Parity Generators. An important function of XOR books is the parity generation. Let $\{x_0, x_1, \ldots, x_{n-1}\}$ be a group of logic variables. The odd (even) parity of the group is equal to 1 if there are an odd (even) number of 1's in the set. Otherwise it is zero. It is easy to see that the odd parity of $\{x_0, x_1, \ldots, x_{n-1}\}$ is exactly equal to

$$x_0 \oplus x_1 \oplus x_2 \cdots \oplus x_{n-1}$$

Then the even parity equals its complement. An 8-bit odd parity generator with an XOR tree is shown in Figure 11.

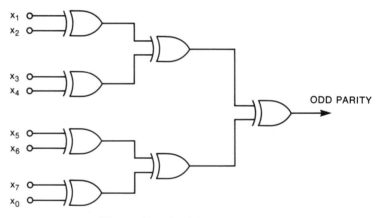

Figure 11. An 8-input parity tree.

3.1.6 Decoders. As shown in Figure 12 an n-bit decoder with inputs x_0, x_1, $\ldots$, x_{n-1} generates 2^n lines each representing a product $y_0 y_1 \ldots y_{n-1}$, where $y_i = x_i$ or $\bar{x}_i$. Decoders are both device and I/O intensive. Most libraries offer a 2-bit decoder as a basic building block. For more bit decoding a cascade of 2-bit decoders with simple logic gates can be used. An example of a 3-bit decoder built with a 2-bit decoder and simple gates is shown in Figure 13. The dynamic decoders discussed in Chapter 6 are also used for large decoders.

3.1.7 Multiplexers. A multiplexer consists of a decoder and a selection network. In Figure 14 the inputs s_0, s_1, $\ldots$, s_{n-1} are selection bits, and D_0, D_1, $\ldots$, D_{2^n-1} are data lines. The selection bits select the particular data line through the decoder and pass it to the output y. If performance is not a big concern, the AND gates can be dotted together forming a wide AOI followed by an inverting buffer or simply implemented with pass gates as shown in Figure 15. The former is popular with NMOS circuits for noise margin, and the latter is popular with CMOS circuits for performance.

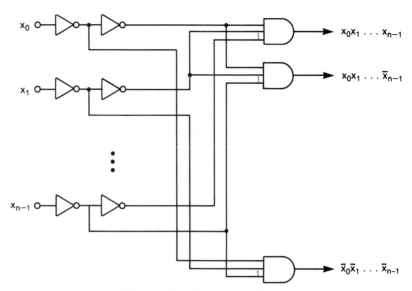

Figure 12. An n-input decoder.

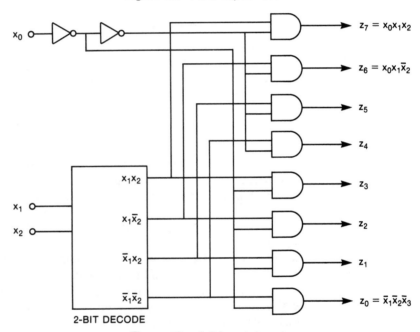

Figure 13. A 3-input decoder.

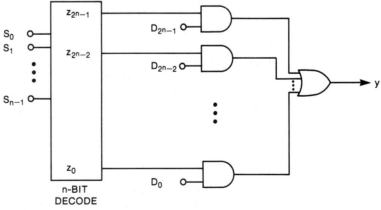

Figure 14. A multiplexer.

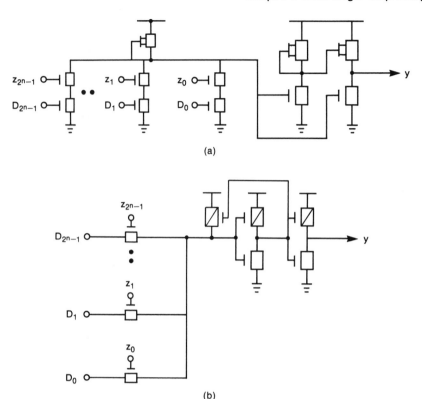

(a)

(b)

Figure 15. Multiplexer circuits.

3.1.8 Rotators and Barrel Shifters. A rotator rotates the bits on a bus. Let $DI_0, DI_1, \ldots, DI_{n-1}$ be the input lines and $DO_0, DO_1, \ldots, DO_{n-1}$ be the output lines and the signals on the I/O lines of an m-bit left rotator are related by

$$DO_{k \oplus m} = DI_k \qquad k = 0, 1, 2, \ldots, n - 1$$

where $m < n$ and $k \oplus m = k + m \bmod n$. In a left shifter $DO_{k+m} = DI_k$ for $k = 0, 1, 2, \ldots (n - 1 - m)$, and DO_l is padded with 0 or 1 for $l \leq m - 1$.

A rotator (shifter) that rotates (shifts) n bits by up to m positions is called an $n \times m$ rotator (shifter). A 4×4 left rotator is shown in Figure 16. Since there are four possible values for m, two control bits are needed. The 2-bit decoder selects the group of pass devices in the cross-bar switch array, which then connects the input lines to the output lines according to the above equation with $m = S_0 S_1$.

A right rotator (shifter) rotates (shifts) the bits to the right. Unlike the left shifters, however, the high-order positions at the output vacated after a right shift are usually filled by the highest-order bit of the input. This is because in computer arithmetic right shifts are used to accomplish division. The sign bit of the dividend must be extended to the leftmost position after each shift.

In Figure 16 the I/O to the rotator, DI_k/DO_k, are individual lines. It is called a *bit rotator*. In practice DI_k/DO_k may be "multilined" bundles. For example, in a *nibble rotator*, each DI_k/DO_k consists of 4 bits. On a 16-bit bus, a 16×4 bit rotator followed by a 4×4 nibble rotator forms a 16×16 bit rotator with just two switching arrays.

3.1.9 Latches. An indispensable member in any logic family is the register. A register is a high-speed memory for temporary data storage. The basic element inside a register is the latch.

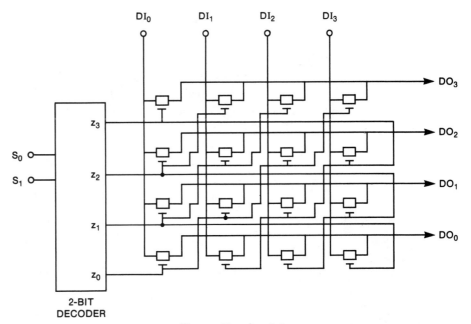

Figure 16. A rotator.

A latch consists of two cross-coupled inverters. As shown in Figure 17, the CMOS latch has three equilibrium states: x_1, at which v_1 is greater than v_2; x_2, at which v_2 is greater than v_1; and x_M, at which $v_1 = v_2$. States x_1 and x_2 are stable in the sense that the latch can stay in either state indefinitely. State x_M, however, is unstable. Any small disturbance can upset the delicate balance and drive the latch to x_1 or x_2. By identifying x_1 and x_2 as two possible values of a logic variable, the latch can be used as a memory device.

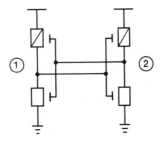

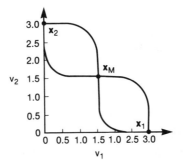

Figure 17. Equilibrium states of a cross-coupled latch.

EXERCISES

1. Consider the E/D latch shown in Figure 18. Plot the transfer characteristics of the latch as a function of the β ratio: $\beta_r = \beta_3/\beta_1$. As β_r decreases, what happens to the equilibrium states? Explain the difference between the CMOS latch and the NMOS latch.

2. Consider an SRAM with 6-device CMOS cells and a *static pullup*. Discuss the stability of the cell as the size of the pass devices vary. Does the number of equilibrium states change as the pass devices increase in size? If so, why?

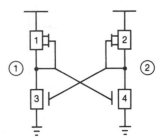

Figure 18.

To avoid timing problems, most latches are controlled by clocks. A typical S-R (set-reset) latch is shown in Figure 19(a). Clock pulse C gates the inputs S and R into the latch at $t = -T$. If $S = 0$ and $R = 1$, the latch moves to $\mathbf{x}_1$ during $-T < t < 0$. At $t = 0$, clock C goes down, leaving P10 = 0 and P20 = 1. The latch has been reset. On the other hand, if $S = 1$ and $R = 0$, the latch will be in state $\mathbf{x}_2$ with P10 = 1 and P20 = 0. It has been set. In case $S = R = 0$, both input devices are off and the outputs remain constant.

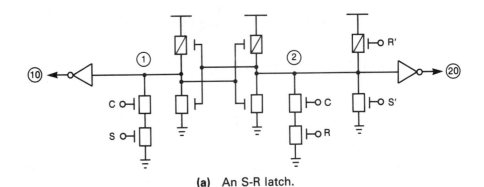

(a) An S-R latch.

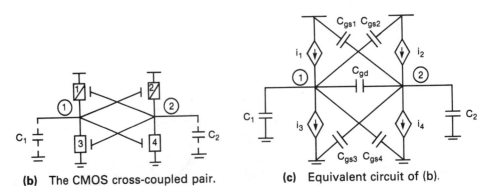

(b) The CMOS cross-coupled pair.   **(c) Equivalent circuit of (b).**

Figure 19. Analysis of the metastability problem.

The case $S = R = 1$, however, causes a problem. With both inputs high, v_1 and v_2 are held low when C is high. As soon as C goes down, v_1 and v_2 bounce up. Since a latch is a symmetric circuit $v_1 = v_2$, the circuit moves into the unstable equilibrium state $\mathbf{x}_M$. While it will eventually be driven into either $\mathbf{x}_1$ or $\mathbf{x}_2$ by noise, it may take a long time for the latch to get out of the unstable region. This problem, called the *metastability problem*, is one of the main causes of system malfunction in asynchronous applications.

To analyze the metastability problem, refer to Figures 19(b) and (c) where all four devices are assumed to be in the saturation region. To find $\mathbf{x}_M$, notice that $i_1 = i_3$ in equilibrium:

$$\frac{1}{2}\beta_1(v_2 - V_H - V_{TP})^2 = \frac{1}{2}\beta_3(v_2 - V_{TN})^2$$

$$\Rightarrow \qquad v_2 = \frac{\sqrt{\beta_1}\,(V_H + V_{TP}) + \sqrt{\beta_3}\,V_{TN}}{\sqrt{\beta_1} + \sqrt{\beta_3}} \tag{1}$$

$$= v_1$$

The state equation of the latch is given by

$$\begin{bmatrix} C_1 + C_{gs2} + C_{gs4} + C_{gd} & -C_{gd} \\ -C_{gd} & C_2 + C_{gs1} + C_{gs3} + C_{gd} \end{bmatrix} \frac{d}{dt} \begin{bmatrix} v_1 \\ v_2 \end{bmatrix} = \begin{bmatrix} i_1 - i_3 \\ i_2 - i_4 \end{bmatrix}$$

By letting $v_d = v_1 - v_2$ and premultiplying both sides by $[1 \quad -1]$, we have

$$(C_1 + C_{gs1} + C_{gs3} + 2C_{gd})\frac{dv_d}{dt} = i_1 - i_3 - i_2 + i_4$$

where it is assumed the devices are perfectly matched. Substituting Equation (1) into the above, we obtain

$$\frac{d}{dt}v_d = \frac{\sqrt{\beta_1\beta_3}}{C_1 + C_{gs1} + C_{gs3} + 2C_{gd}}(V_H + V_{TP} - V_{TN})v_d$$

$$\triangleq av_d$$

Solving for v_d,

$$v_d(t) = v_d(0)e^{at}$$

Thus $v_d(t)$ grows exponentially at $t = 0$. The larger the value of a, the faster the latch gets out of the unstable region. Since

$$C_{gsk} \cong \frac{2}{3}C'_o\,WL\,(\mathrm{T}_k)$$

$$= \frac{2}{3}\frac{L_k^2}{\mu_k}\beta_k, \qquad k = 1, 3$$

where L_k is the channel length of device T_k and $C_{gd} \cong 0$; we have

$$a = \frac{\sqrt{\beta_r}}{\dfrac{2}{3}\dfrac{L_1^2}{\mu_1} + \dfrac{2}{3}\dfrac{L_3^2}{\mu_3}\beta_r + \dfrac{C_1}{\beta_1}}$$

where $\beta_r = \beta_3/\beta_1$. Differentiating the equation with respect to β_r, and then setting $da/d\beta_r = 0$, we obtain the optimal β_r:

$$\beta_r^* = \frac{\dfrac{2}{3}\dfrac{L_1^2}{\mu_1} + \dfrac{C_1}{\beta_1}}{\dfrac{2}{3}\dfrac{L_3^2}{\mu_3}} \tag{2}$$

Thus the speed of the latch is optimized if β_r is chosen according to Equation (2). In case C_1 is small and $L_1 = L_3$, the optimal $\beta_r \cong \mu_3/\mu_1$. In practice, however, the value of a is a weak concave function of β_r around β_r^*. The optimization can be easily overwhelmed by process variations. Any value of β_r close to β_r^* produces much the same result.

To eliminate the case of $S = R = 1$, a built-in inverter can be used to generate $R = \bar{S}$ as shown in Figure 20. This is called a *polarity-hold* or *P-H latch*. The single input is now designated by D for data. In the NMOS latch both D and $\bar{D}$ are used for S and R, respectively. In the CMOS latch, only one input, D or $\bar{D}$, is needed for one side of the latch. The inverter is used as a buffer for speed.

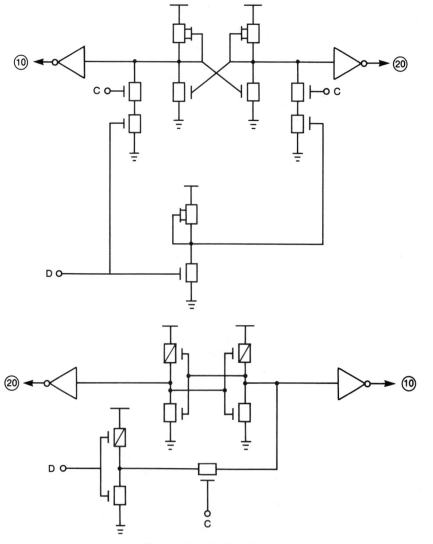

Figure 20. P-H latches.

Another approach that avoids the metastability problem is to use single input with complementary clocks C and $\overline{C}$ as shown in Figure 21. In fact two such circuits have already been used in the T/C GEN circuits in Figure 28 and Figure 31 of Chapter 6. An alternate CMOS implementation that combines the feedback switch with one of the inverters is shown as $L2$ in Figure 23(d).

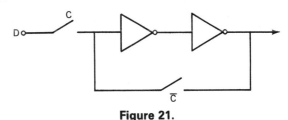

Figure 21.

Figure 22 shows a typical specification of latch timing. To latch data, inputs S, R, or D and the clock C must have a minimum overlap of t_{SET} to account for the internal delay. After the valid data has been latched, the inputs should not change before C goes down. This specification, t_{HOLD}, is usually 0 although it can be slightly negative. The minimum pulsewidth of C, designated t_{PW}, is equal to t_{SET}. In most cases t_{PW} is specified a little larger than t_{SET} to cover uncertainties of the rise/falltime of the pulses.

This discussion has assumed the clocks are positive: the latch responds to the input when the clock stays high. Since negative clocking can be easily obtained by controlling p-channel devices with positive clocks, we shall assume that all clocks are positive. Furthermore, as indicated by S' and R' in Figure 19(a), most latches are equipped with direct set and/or reset lines so they can be initialized to a known state. From now on it will be assumed all latches have direct set/reset lines, although their existence may not be explicitly shown in circuit schematics.

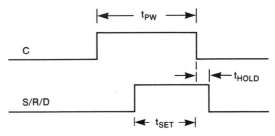

Figure 22. Latch timings.

3.1.10 Shift Register Latch: SRLs. Data can be transmitted either in parallel or in serial in a logic system. In parallel transmission all data bits are transmitted simultaneously. This is the fastest method but also very I/O and hardware intensive. In serial transmission the data bits are transferred one at a time, usually controlled by a clock. Serial transmission is slow but requires minimum hardware. Both types generally exist in a complex system. Shift register chains are then used to convert the parallel data into serial, and vice versa.

The basic element of a shift register chain is the *shift register latch* (SRL). As shown in Figure 23(a), an SRL consists of two latches that can be either *S-R* or *P-H* type, controlled by two clocks. The first latch, L1, is controlled by clock C. It receives data from outside. The second latch, L2, is controlled by clock B. It receives data from L1 and drives the load. The output of the latch can be taken either from L1 or from L2. If the output is taken solely from L2, the SRL is called a *master* (L1)/*slave* (L2) latch.

The relationship between the clocks and data is shown in Figure 23(c). Both t_{SET}, the setup time of L1 and t_D, the delay time of L2 must be minimized because they have to be added to the total delay of all logic paths between SRLs. In addition to t_{SET} and t_D the clock separation t_S is also specified. Since nothing happens during t_S, the "dead time" between C and B should be as short as possible. On the other hand, however, any overlap between C and B may result in *data flush through* to the next SRL.

Sometimes $B = \overline{C}$ is used to control L2. Ideally $B = \overline{C}$ goes down as soon as C goes up, latching the state of L1 into L2. In reality, however, $\overline{C}$ may fall behind C with some delay. With C being high, new data will enter L1, and, if $\overline{C}$ is still high,

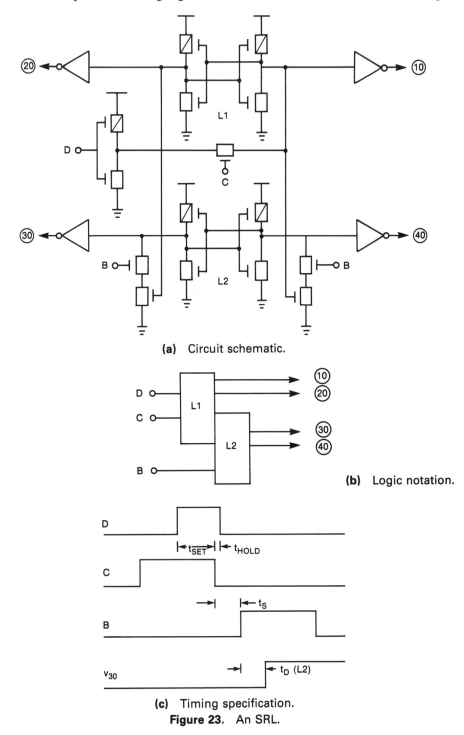

(a) Circuit schematic.

(b) Logic notation.

(c) Timing specification.

Figure 23. An SRL.

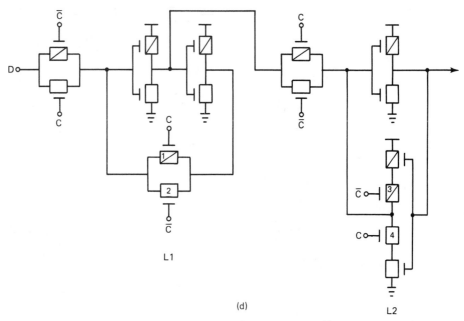

(d) A CMOS SRL with clock $B = \overline{C}$.

Figure 23. (*cont.*)

may "update" L2. This problem, called *clock feedthrough,* can be avoided by balancing the loads and sharpening the rise/fall time of the clocks. The following section describes a clock generation scheme that minimizes the clock skews.

EXERCISE

Consider the CMOS SRL shown in Figure 23(d). Comment on the relative sizes among the inverters. Comment on the sizes of (T_1, T_2) and (T_3, T_4) from t_{SET} and t_D viewpoints. Do you suggest switching the circuits between L1 and L2? Why?

3.1.11 Clock Generators and Drivers. Latches can be used to generate multiphase clocks. Figure 24 illustrates a simple example of a 4-phase clock generated from a 2-phase clock. Two external signals are needed: the clock $C0(\overline{C0})$ and the clock set pulse $S(\overline{S})$. Prior to $t = 0$, the clock set $S = 1$. Nodes ① and ③ are high, and the latches are stable. Now at $t = 0$, the clock set S goes low at the rising edge of $\overline{C0}$, allowing latch II to toggle the state of latch I. Now, node ① goes low while node ② goes high. The state of latch I stays constant until $t = 2T$ at which $\overline{C0}$ rises again. On the other hand, clock $C0$ goes up at $t = T$, allowing latch I to toggle the state of latch II. Then, v_3 goes down and v_4 goes up. The state of latch II stays constant until $t = 3T$ when $C0$ rises again. Thus, latch states toggle at the rising edge of either $C0$ or $\overline{C0}$. The four phases ϕ_1, ϕ_2, ϕ_3, and ϕ_4 can be generated by ANDing the clock with the node voltages of the latches as shown in Figure 24(c).

Any two consecutive phases can be used to drive the SRLs. Figure 24(d) shows a clock distribution network that generates the C and B clocks from ϕ_2 and ϕ_3, respectively. These pulses are then repowered by high-power driver trees. Two identical trees, one for ϕ_2 and one for ϕ_3, are rooted at the clock generator. At the end of each tree branch is a cross-coupled clock driver. The ϕ_2 and ϕ_3 inputs to the driver come from corresponding tree branches. The outputs of the driver are the B and C clocks driving the same SRLs. Since the driver outputs are cross-coupled, clock B cannot go up before clock C goes down, and vice versa, ensuring nonoverlapping clocks at SRL inputs.

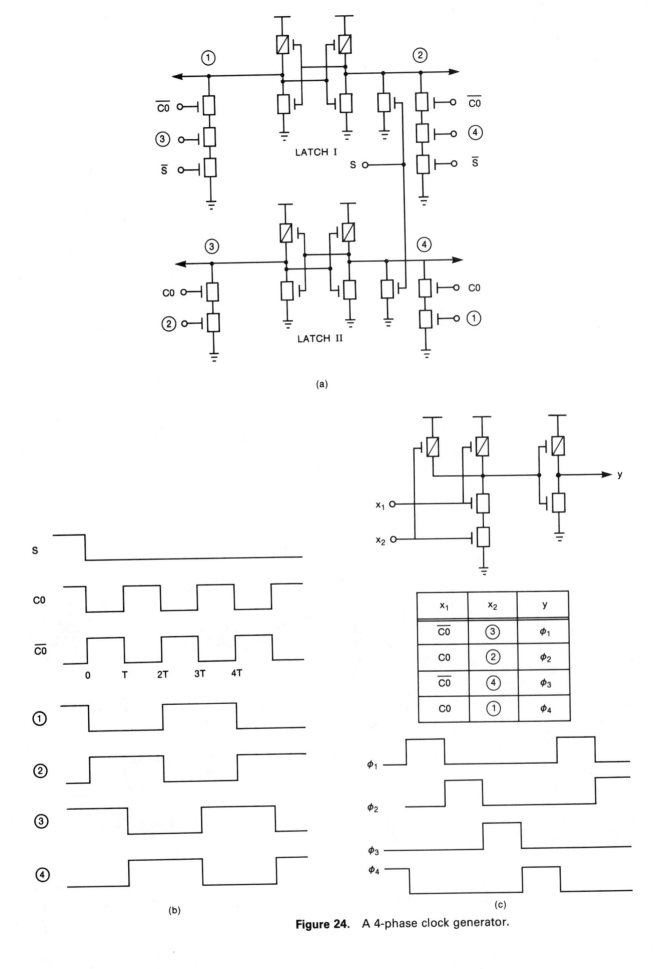

Figure 24. A 4-phase clock generator.

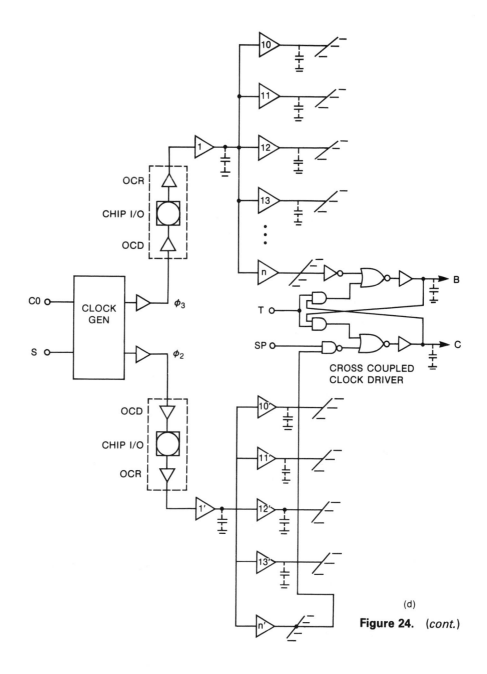

(d)
Figure 24. (*cont.*)

To facilitate chip testing (see Section 4), two control lines are applied to the drivers: line *T* decouples the outputs and line *SP* suppresses the *C* clock. Raw clock signals are also brought to I/O pads right after their generation so they can be observed directly from outside. Other type of multiphase clocks can be generated in the same fashion.

3.1.12 Flip-flops. Flip-flops are SRLs with special built-in logic. The following is a list of commonly used flip-flops:

1. *T-type flip-flop.* The toggle flip-flop, also referred to as the *T*-type, changes state every time input *T* goes high. A logic diagram of a *T*-type flip-flop is given in Figure 25(a).
2. *R-S flip-flop.* This is a master/slave *SRL* with an *S-R* L1.
3. *D-type flip-flop.* The delay or data flip-flop, commonly referred to as the *D*-type, is a master/slave SRL with a *P-H* L1.

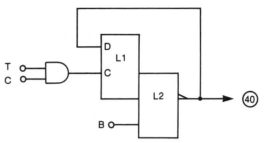

Figure 25(a). A *T*-type flip-flop.

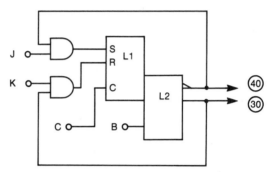

Figure 25(b). A *J–K* flip-flop.

4.   *J-K flip-flop.* A *J-K* flip-flop is a refined *R-S* flip-flop that allows all four combinations of inputs. As shown in Figure 25(b), the output L2 = 1, when $J = 1$ and $K = 0$, and L2 = 0, when $J = 0$ and $K = 1$. The output does not change when $J = K = 0$, whereas it toggles when $J = K = 1$.

In some cases a built-in inverter generates $B = \overline{C}$ so only one clock line is needed. The 2-clock design, however, is more flexible.

3.1.13 Counters.   Counters are used in all kinds of applications, and there are many different kinds. An asynchronous counter updates its count any time an event occurs, whereas a synchronous counter advances with clocks. Depending on the application, the code represented by the counter output may be binary such as a modulo 16 ripple counter or a decimal such as a BCD counter, or just some kind of special codes like the Grey code counter. A counter counts either in ascending or descending order. The former is called *an upcounter* and the latter, *a downcounter*. One that counts in either direction at the command of the user is called *an updown counter.*

Counters can be implemented with SRLs. Figure 26(a) displays a 4-bit *asynchronous ripple counter* implemented with *T*-type flip-flops. A 1 is added to the binary count $y_3y_2y_1y_0$ each time the input receives a 1.

The counter resets to $y_3y_2y_1y_0 = 0000$ after it has received $2^4 = 16$, 1s. It is therefore called a modulo 16, or mod 16, counter. Figure 26(b) displays a 4-bit *synchronous binary counter*. It advances in sync with the clocks.

In Figure 27(a) is an *n*-stage *ring counter,* which, after being initialized to $y_{n-1}y_{n-2} \cdots y_1y_0 = 00 \cdots 01$, propagates the 1 down the SRL chain. It returns to the initial state after *n* shifts. By feeding the complement of the output at the last stage to the input of the first stage, the *twisted-ring,* or *Johnson counter* shown in Figure 27(b), is created. Outputs of this counter change one bit at a time. It counts up to 2*n*.

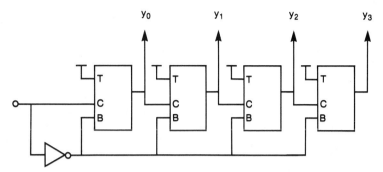

Figure 26(a).   A 4-bit ripple counter.

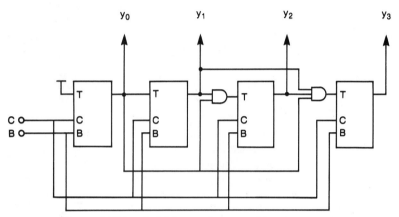

Figure 26(b).   A 4-bit synchronous binary counter.

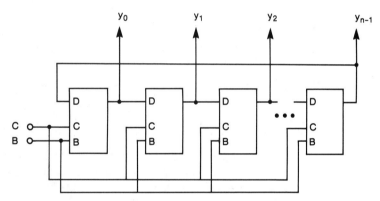

Figure 27(a).   An *n*-stage ring counter.

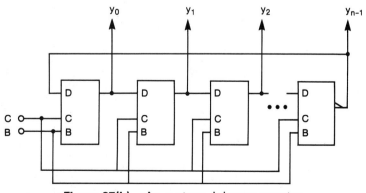

Figure 27(b).   An *n*-stage Johnson counter.

EXERCISE

Let $n = 4$ for the above counter. Start with $y_3y_2y_1y_0 = 0000$. List the outputs of all stages for eight clock cycles.

Some libraries offer 4-bit counters as standard books. The counters can be extended in 4-bit increments by cascading the books. An AND gate is used to detect the maximum count in each book and passes the clock through to the higher-order books. More sophisticated counters can be constructed with SRLs and random logic.

3.1.14 Edge-Triggered Latches. The latches considered earlier are *level sensitive*. The state of the latch is continuously affected by the input as long as the clock stays high. An edge-triggered latch receives data when the clock is in transition. Instead of following the input continuously, however, its state stays constant until the next transition of the clock. In other words, the latch is triggered by the edge of the clock.

An SRL can be triggered either by the clock edge or by clock level. *The latch is level sensitive with respect to clock C but is edge sensitive with respect to clock B.* After the transition following the rise of B, the L2 output stays constant regardless of the inputs to L1 until clock B rises again.

Since the L2 latch is set by L1, which has an internal delay t_{SET}, an SRL used as an edge-triggered latch has a long setup time. As shown in Figure 28, the input to L2 must be latched by L1 before it can be shifted into L2.

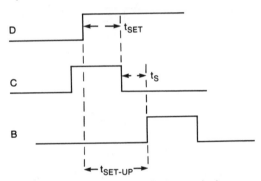

Figure 28. SRL edge trigger timing.

An edge-triggered latch that has a short setup time is shown in Figure 29. It consists of a dynamic sense amplifier followed by a static latch. A single clock ϕ is needed. When ϕ is low, nodes ① and ② are restored and the pass devices T_3 and T_4 are off, so the static latch is not affected. As ϕ goes up at $t = 0$, devices T_1 and T_2 turn on, gating the inputs S and R into the dynamic latch. The latch is then set by ϕ and $\overline{\phi}$ and its output gated into the static latch through the pass gates T_3 and T_4. Devices T_3 and T_4, controlled by clock ϕ', which is a one-shot generated by the rising edge of ϕ, turn off after the static latch has been settled. The outputs of the static latch stay constant afterwards until clock ϕ rises again.

This edge-triggered latch has wide application as synchronizers between two asynchronous systems. Its dynamic front-end minimizes the metastability problem. Its short setup time makes it a good phase comparator in digital phase locked loop applications.

Dynamic latches are less sensitive to metastability problems than are their static counterparts. In Chapter 6 we have shown that the final differential voltage output of a dynamic latch is inversely affected by the speed of the set pulse $e_s(t)$. The faster $e_s(t)$ falls to GD, the less amplification during the setting. In the worst case in which $e_s(t)$ falls to GD instantaneously, the output voltages are practically unchanged after $e_S(t)$ has reached GD. A dynamic latch in which $e_S(t)$ falls to GD instantaneously, of course, becomes a static latch. This is why in applications that are sensitive to metastability problems such as the comparators in an analog-digital converter, dynamic latches incur a much lower soft error rate than static latches.

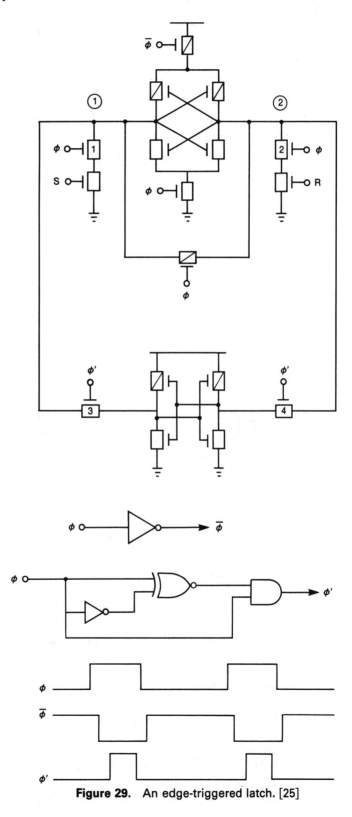

Figure 29. An edge-triggered latch. [25]

3.2 Dynamic Logic Circuits

3.2.1 Dynamic Shifter Registers: DSRs. Dynamic shift registers, or DSRs, are mainly used as digital delay lines or sequential memories. A simple DSR is shown in Figure 30 where circuits I and II are identical. Let input v_I be held on C_1 as ϕ_1 goes up, turning on T_2 and T_4. Voltage v_2, being the output of the inverter (T_2, T_3), will be equal to $\overline{v_I}$. Since T_4 is also on, $v_3 = v_2 = \overline{v_I}$. Clock ϕ_1 subsequently goes down and v_3 is temporarily held at node ③ by C_3. The same action occurs in circuit II in the ϕ_2 period. At the end of ϕ_2, $v_5 = v_4 = \overline{v_3} = v_I$ and the input has been shifted to the right by one bit.

The DSR in Figure 30 is called a *2-phase, ratio-type* design. The holding capacitor C_1 can be just the gate and parasitic capacitances associated with node ①. It needs no extra area. Devices T_2 and T_3, however, must satisfy the β ratio requirement to ensure a reasonable down level for v_2.

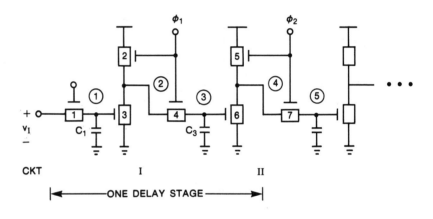

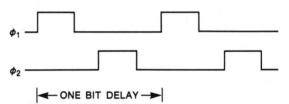

Figure 30. A 2-phase ratio-type DSR.

To improve density, both V_H and GD can be pulsed by clocks as shown in Figure 31. When ϕ_1 goes high node ② is unconditionally precharged to $(V_H - V_T)$ by T_2 and $v_1 = v_I$. Device T_3, with its source tied to ϕ_1, remains off regardless of v_1. The clock ϕ_1 subsequently goes down and v_1 is held on C_1. If v_1 is high, node ② will be discharged to GD. If v_1 is low, node ② stays high. Thus $v_2 = \overline{v_I}$ at the end of ϕ_1.

The same action taken by circuit II during the ϕ_2 period leaves $v_4 = \overline{v_3}$ at the end of ϕ_2. Since $\overline{v_3} = \overline{v_2} = v_I$, the input has been shifted right by one bit.

Since T_2 and T_3 never turn on at the same time, both can be minimum geometry devices. This is therefore a *2-phase, ratioless-type* design. The density improvement, however, is partially offset by the relatively large size of C_2. To determine the size of C_2, assume $v_I = 0$. At the end of ϕ_1 the charge stored in C_2 is given by $C_2(V_H - V_T)$. As soon as ϕ_2 goes up, nodes ② and ③ are shorted by T_4 and the charges are redistributed. If $v_3 = 0$ prior to ϕ_2, the final voltage of v_3 is determined by charge conservation: $C_2(V_H - V_T) = (C_2 + C_3)v_3 \Rightarrow v_3 = (V_H - V_T)/(1 + C_3/$

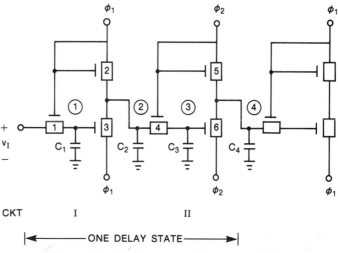

Figure 31. A 2-phase ratioless-type DSR.

C_2). Hence, in practice, C_2 is made much larger than C_3 to ensure a reasonable up level for v_3.

The charge sharing problem can be eliminated by introducing another inverter stage with an additional clock as shown in Figure 32. The reader is invited to show as an exercise, that three phases are now needed to shift one bit of data.

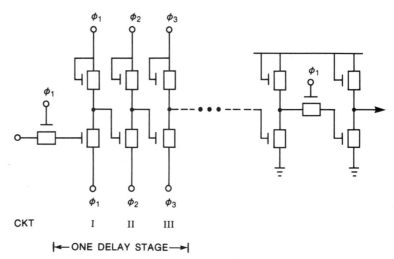

Figure 32. A 3-phase DSR.

A 4-phase DSR with wide application in dynamic logic is shown in Figure 33. Since each delay stage consists of four identical circuits, we shall discuss the operation of one circuit only.

Starting with ϕ_1, as ϕ_1 goes up node ① is precharged to $(V_H - V_T)$. Circuit I is in the *precharge* mode. As ϕ_2 goes up, ϕ_1 goes down. Node ① will stay high or discharge to GD depending on v_I. If v_I is high, v_1 falls to GD. If v_I is low, node ① will share charge with node ①'. If properly designed, the resulting v_1 should be high enough to turn on T_6. Circuit I is now in the *sample or evaluation* mode. At the same time, node ② of circuit II is precharged. Thus, when ϕ_3 goes high, v_1 will be sampled through T_6 by circuit II. Voltage v_1, however, will be held constant throughout ϕ_3 and ϕ_4 until ϕ_1 rises again. Circuit I is thus in a *hold* mode. To accomplish one bit shift the same action must be repeated four times until $v_4 = v_I$ in the following ϕ_1 cycle. The precharge, sample, and hold modes of each circuit during different clock periods are shown in Figure 33(b).

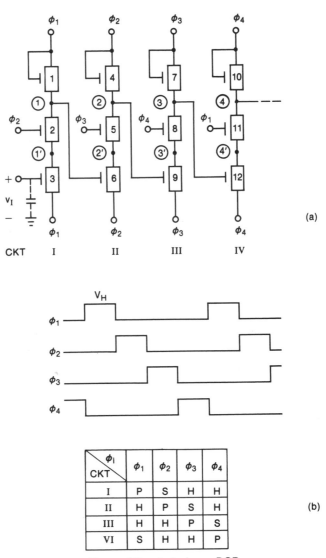

Figure 33. A 4-phase DSR.

Notice that when a circuit is in the sample mode, its previous circuit must be in the hold mode. In other words, a circuit can be used to drive others only when it is in the hold mode. This rule must be observed when employing 4-phase clocks in dynamic logic designs.

3.2.2 Dynamic Logic. One advantage of MOS devices is the capability to hold charges. By incorporating random logic into the inputs of the delay stages, a DSR can be easily modified into a dynamic logic circuit. Compared with its static counterpart, a dynamic logic circuit generally offers higher density, lower power consumption, and less troublesome timing. Since multiphase clocks are involved, dynamic logic circuits are generally slower than the fastest implementation of static logic circuits. By the same token, however, the clocks also ensure the completion of all logic operations within specified time periods. For massive logic functions such as the AND/OR arrays of a PLA or the address decoding of a large memory, dynamic logic often offers both higher density and better performance.

Dynamic logic is particularly useful in implementing synchronous sequential machines. A sequential machine is a logic system that possesses memory. Its output at time *t* depends not only on the input at the moment but also on all inputs taken in the past. A sequential machine consists of a combinational logic network and a memory as shown in Figure 34 in which all variables are vectors of various dimensions. The memory summarizes the effects of all past inputs. Its output, **Q**, defines the *(present) state* of the machine. The output of the *machine*, **Y**, is determined by the input **X** as well as the state through the combinational logic network. At the same time, feedback variables **D** are also generated and fed into the memory to become the next state.

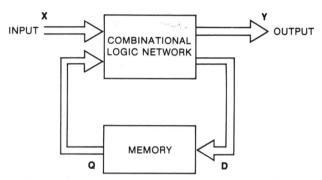

Figure 34. Configuration of a sequential machine.

To ensure that **X** is stable when taken by the combinational network, clocks are used to synchronize **X** and **Q**. Such devices are called *synchronous* machines. Figure 35 shows the general configuration of a synchronous sequential machine where the clocks introduce a delay T and eliminate the race condition between **X** and **Q**. The states of the machine, **Q**, are the outputs of the delay elements. Both the inputs to the delay elements,

$$\mathbf{D} = (D_1, D_2, \ldots D_k)$$

$$= \mathbf{g}(\mathbf{Q}, \mathbf{X})$$

and the machine output

$$\mathbf{Y} = \mathbf{f}(\mathbf{Q}, \mathbf{X})$$

are defined by the combinational logic network.

We shall illustrate the use of dynamic logic with a detailed design of a small synchronous sequential machine: a 4-bit counter. The input X to the counter is a bit stream. The counter counts from 0, advances by one if $X = 1$, and holds the present count if $X = 0$. After four 1's have been recorded, the counter resets and repeats the process. For convenience, allow **Y** = **Q** so that no extra logic is needed for the output.

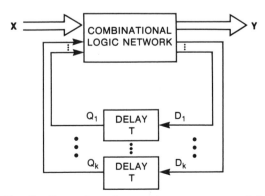

Figure 35. Configuration of a synchronous sequential machine.

The design procedure of sequential machines is well established. The essential steps are shown in the flow chart in Figure 36. We shall follow these steps as we proceed with the design.

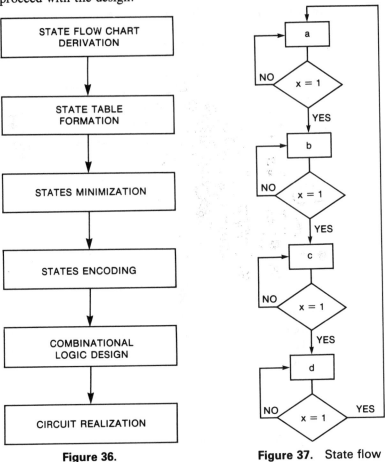

Figure 36.

Figure 37. State flow chart.

1. *State flow chart and flow table.* The enumeration of states is usually the most difficult step. In our example, however, because the counter resets after receiving four 1s, there are obviously four states. Designating the states by a, b, c, and d, the operation of the counter is depicted by the flow chart shown in Figure 37. The output and next state functions **f** and **g** are then derived from the flow chart and listed in a flow table in Figure 38.

2. *State minimization.* After the flow table has been constructed, the number of states should be minimized by merging equivalent ones. *Two states are equivalent, if, responding to the same input, they produce the same output and their next states are either identical or equivalent.* In our example, since all rows are distinct, it is easy to see that the number of states is already minimal.

3. *State encoding.* To construct logic the states must be encoded by 0's and 1's. Each delay element provides one bit. For m states, a minimum of $k = [\log_2 m]$ delay elements are needed where $[z]$ is the smallest positive integer greater than z. In our case two delay elements yield four states. The state assignments are shown in Figure 39.

In principle, any encoding scheme that assigns different codes to different states will be satisfactory. In practice, however, some assignments

Q	x		x	
	0	1	0	1
a	a	b	a	a
b	b	c	b	b
c	c	d	c	c
d	d	a	d	d
		D		Y

Figure 38. State flow table.

	Q_1	Q_2
a	0	0
b	0	1
c	1	1
d	1	0

Figure 39. State encoding.

are more desirable than others. The following rules usually lead to simpler logic implementations:

a. Try to assign codes to the states that correspond to the output so output logic can be reduced. This rule is trivially satisfied by our example.

b. Try to assign adjacent codes to adjacent states in the flow chart so state transition affects only one delay element. For example, since state c follows state b, which is coded as 01, state c is coded as 11 instead of 10. This is particularly critical if X is an asynchronous signal.

c. Try to assign adjacent codes to states with the same next state.

d. Try to assign adjacent codes to states that are next states of the same state. Rules (c) and (d) are partially satisfied in our assignment.

4. *Design of combinational logic*. Substituting the state codes into Figure 39 creates the transition table in Figure 40. We now proceed to implement logic for the delay elements:

				X		
	Q		0		1	
	Q_1	Q_2	D_1	D_2	D_1	D_2
a	0	0	0	0	0	1
b	0	1	0	1	1	1
c	1	1	1	1	1	0
d	1	0	1	0	0	0
				D		

Figure 40. State transition table.

a. *Input logic of D_1*. Extracting the columns of D_1 from Figure 39 produces the subtable D_1 in Figure 41(a). To use NOR logic, we work with the 0's. From the table,

$$\overline{D}_1 = \overline{Q}_1\overline{X} + \overline{Q}_2X \quad \Rightarrow \quad D_1 = (Q_1 + X)(Q_2 + \overline{X})$$

which is shown in Figure 41(b).

Q_1	Q_2	X	
		0	1
0	0	0	0
0	1	0	1
1	1	1	1
1	0	1	0
		D_1	

(a)

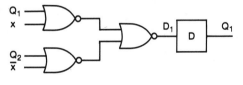

(b)   **Figure 41.**

b. *Input logic of D_2*. Similarly, the input logic of D_2 is given in Figure 42(a) and (b).

		x	
Q_1	Q_2	0	1
0	0	0	1
0	1	1	1
1	1	1	0
1	0	0	0
		D_2	

(a)

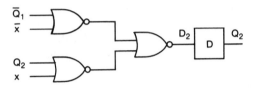

(b)   **Figure 42.**

5. *Circuit realization.* Logic can be implemented with different dynamic circuits featuring different density, performance, and timing. In our simple example, either 2-phase ratio-type or ratioless designs are adequate.

   a. *2-phase ratio-type design.* In a 2-phase DSR each bit delay is accomplished by one complete cycle of ϕ_1 and ϕ_2. The delay element D then can be split into two elements, each of which contributes $\frac{1}{2}$-bit delay as shown in Figure 43. These $\frac{1}{2}$-bit delay elements can be moved around throughout the logic. They can be placed anywhere in the network as long as there is at least one output node between any two delay elements controlled by alternate phases. Each clock phase then applies to all nodes preceding it. Applying this rule to Figure 41 produces the logic diagram in Figure 44. Using the two-phase ratio-type DSR, we obtain the dynamic logic circuit shown in Figure 45. Notice that since only the true input of X is available, an extra inverter is used to generate $\overline{X}$.

   Implementation of Figure 42 for D_2 proceeds in the same fashion. Try this implementation as an exercise.

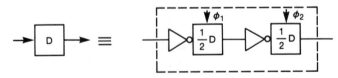

Figure 43.

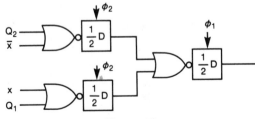

Figure 44.

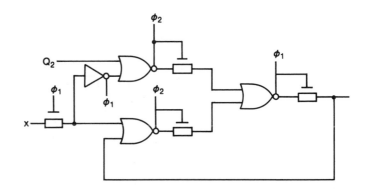

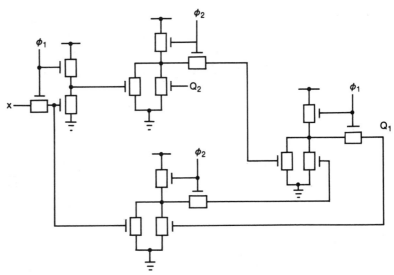

Figure 45.

b. *2-phase ratioless-type design.* Converting a logic diagram into a 2-phase ratioless design follows similar procedures. Figure 43 is also applicable. Normally there is only one level of logic per clock phase because of the noise margin degradation caused by charge sharing. Thus, in moving the $\frac{1}{2}$-bit delay elements throughout the logic, each node must be associated with a $\frac{1}{2}$-bit delay element. Consecutive nodes must therefore be driven by alternate clocks. A complete circuit realization of Figure 44 is shown in Figure 46 with a minimum delay gate for $\bar{X}$. The reader is invited to carry out the rest of the design as an exercise.

While the most complicated logic function in our example is a 2-input NOR, each gate can be complex in practice. The number of clocks can also be a design parameter. For more complex logic, 4-phase clocks should be used.

3.2.3 Domino Logic. In dynamic logic no memory elements such as latches are needed. The clocks themselves provide the delays, and the device capacitances provide the store function of a memory element. Since latches occupy large areas and consume power, elimination of latches improves density and power consump-

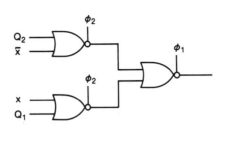

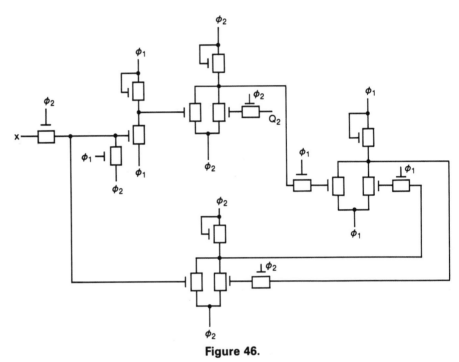

Figure 46.

tion. The disadvantages of dynamic logic, however, include high transient power disturbances and less noise margins in some applications.

Domino logic, being a simplified version of dynamic logic, is best used with latches. Figure 47 shows a typical application. Both the inputs and outputs are latched in registers. The random logic is grouped in circuit blocks. Each block implements a complex logic function in cascode form. A typical block generating the function $x = a + b(c + d)$ is shown in Figure 48(a). Two nonoverlapping clocks are used. The precharge clock, ϕ_p, precharges node ① so that v_{10} is initially low. The evaluation clock, ϕ_e, turns on the bottom device. Node ① will discharge if the logic generates a conducting path from node ① to *GD*. Then v_{10} will go to V_H; otherwise, v_{10} stays low. The outputs of the blocks are selectively pulled up in sequence by the logic like a domino chain; hence the name domino logic.

Since node ① is precharged high, it is important that all inputs that come from other blocks switch only from low to high and all external inputs stay constant during ϕ_e. A block input that switches from high to low may hold a "temporary current path," discharging node ①. Once node ① has been discharged, v_{10} goes up and can no longer be switched low during the rest of ϕ_e.

While the one-way switch in domino logic produces much less noise than the general dynamic logic, it imposes a severe restriction to logic design. The complement of a variable can no longer be generated by a simple inverter. For example,

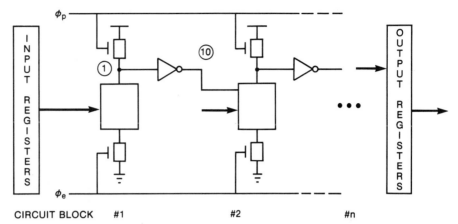

CIRCUIT BLOCK   #1   #2   #n

Figure 47. Domino logic with registers.

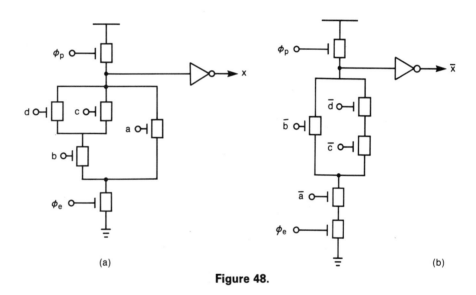

(a)   (b)

Figure 48.

$\overline{x} = \overline{a + b(c + d)} = \overline{a}(\overline{b} + \overline{c}\,\overline{d})$ must be generated by another block as shown in Figure 48(b). In NMOS technology this leads to very poor density compared with static logic circuits.

Domino logic is used mostly in CMOS technology. Notice that the DeMorgan theorem not only changes all inputs to their complements but also transforms the original graph into its dual form. Thus the topological relationship between x and $\overline{x}$ is the same as that between the n-channel and p-channel devices in CMOS functional blocks. The density is therefore comparable to ordinary CMOS static logic. Furthermore, if p-channel devices are used for precharging, a single clock works for precharge and evaluation.

3.2.4 Cascode Logic. The generation of both true and complement forms of logic variables leads to the use of differential cascode logic. A particular approach, called *differential cascode voltage switch (DCVS)*, uses the concept of bipolar cascode current-source trees to implement the cascoding logic of the pull-downs. An example of the function $x = a + b(c + d)$ is shown in Figure 49(a). Here devices T_1 and T_2 are p-channel pullups that predischarge the outputs $P10 = P20 = 0$ when the clock ϕ is low.

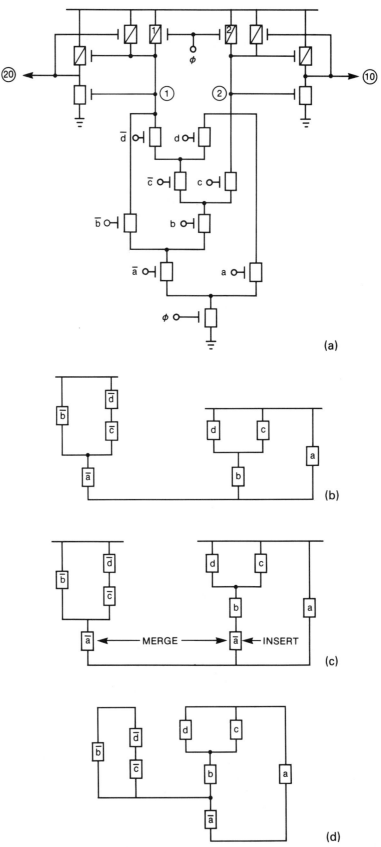

Figure 49. Domino logic versus DCVS. **(a)** A DCVS tree.
(b-f) Transformation from CMOS static logic into DCVS.

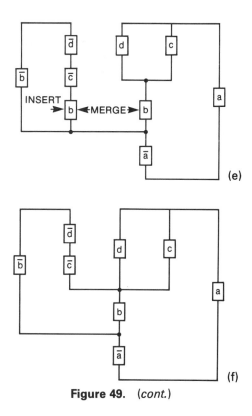

Figure 49. (*cont.*)

Both true and complement forms of the output are available from the cascode logic tree. The topology of the tree allows one and only one conducting path from either node ① or node ② to *GD* for any input. In some cases the outputs are also cross coupled through p-channel pullups.

The height of a tree is defined to be the number of cascoding stages in the tree. For better performance, the tree heights seldom exceed eight stages. This limitation, together with the complexity of wiring differential signal lines, makes it very difficult to manage the complex logic design by hand. Optimization algorithms, which partition the logic to fit the trees with a minimum number of devices and/or delays, have been developed; and automatic PD systems, which use a *masterslice* (gate array) approach, have been implemented.

In a masterslice approach trees of maximum height are laid out in a device array with predefined wiring channels in between. The "raw" chip is partially completed up to the device level. The chip designer first applies the optimization algorithm to partition the logic. She then invokes the PD system to personalize the chip by wiring the devices in M1 and M2 levels according to the partition. Since all trees are of maximum height, not all devices in all trees are used. In practice 70% usage can be achieved with logic optimization.

It is instructive to make a topological comparison between the ordinary domino logic and DCVS. Consider the circuits of x and $\bar{x}$ in Figure 48 which are redrawn in Figure 49(b). In this latter figure the device gates and clock controls are omitted for simplicity. We will change these circuits into the DCVS form. Start with input a and $\bar{a}$. Since $a + \bar{a}z = a + z$ for any z, a device controlled by $\bar{a}$ can be inserted in series with $b(c + d)$ without changing the logic as shown in Figure 49(c). With $\bar{a}$ in both sides, however, it is now possible to merge the two devices and obtain the equivalent circuit as shown in Figure 49(d). Next work on b. By inserting a device controlled by b to the left side and merging it with the one on the right, we obtain the equivalent circuit in Figure 49(f). Same operations on c and d will eventually lead to an equivalent circuit identical to that in Figure 49(a).

Both domino and DCVS logic circuits suffer from charge sharing problems. As the pulldowns turn on, charges stored on the top nodes redistribute and the precharge potential falls, which can cause a misfire of the output inverter. To avoid excessive charge loss, extra capacitances are added to the top nodes and Schmitt triggers are used as output buffers. All these techniques alleviate the problem at the expense of performance, however.

3.3 Large Macros

3.3.1 Adders. An n-bit adder performs binary addition

$$\mathbf{S} = \mathbf{A} + \mathbf{B}$$

where the *addend* $\mathbf{A} = A_{n-1}A_{n-2} \ldots A_1 A_0$ and the *augend* $\mathbf{B} = B_{n-1}B_{n-2} \ldots B_1 B_0$ are *n*-bit numbers. The *sum* $\mathbf{S} = C_n S_{n-1} S_{n-2} \ldots S_1 S_0$ is an $(n + 1)$-bit number. The addition proceeds in the usual pencil and paper manner as shown in Table 1.

Table 1

Addend	$A_{n-1}A_{n-2}$	$\cdots$	$A_1 A_0$
Augend	$B_{n-1}B_{n-2}$	$\cdots$	$B_1 B_0$
Carry	$C_n C_{n-1} C_{n-2}$	$\cdots$	C_1
Sum	$C_n S_{n-1} S_{n-2}$	$\cdots$	$S_1 S_0$

At the *i*th bit position, a sum bit

$$S_i = A_i \oplus B_i \oplus C_i \tag{3}$$

is formed, and a carryout

$$C_{i+1} = A_i B_i + C_i (A_i + B_i) \tag{4}$$

is added to the next-higher-order bit. Equation (3) shows that $S_i = 1$ if there are an odd number of 1's among A_i, B_i, and C_i. Equation (4) shows that $C_{i+1} = 1$ if either it is generated by the condition $A_i = B_i = 1$ or simply being C_i passing through. In this case, either $A_i = 1$ or $B_i = 1$.

A circuit that performs by Equation (4) is called a *carry generator*. A circuit that performs by both Equation (3) and (4) is called a *full adder* (*FA*). An *n*-bit adder can be constructed simply by cascading *n* FAs. Since carries propagate through the FA chain, such an adder is called a *ripple carry adder*.

An FA can be implemented in many ways. A popular style is shown in Figure 50. The carryout, C_{i+1}, is generated first. Its complement, $\overline{C}_{i+1}$ is then used to form S_i according to

$$S_i = \overline{C}_{i+1}(A_i + B_i + C_i) + A_i B_i C_i \tag{5}$$

The reader is invited to show that Equation (5) is equivalent to Equation (3) under the condition of Equation (4). While the topology of n-channel devices is fairly obvious in Figure 50(b), the topology of the p-channel devices may need some explanation. Consider S_i first. Since the topology of the p-channel devices is the same as that of the n-channel devices, the "equation" for the p-channel topology can be drawn from Equation (5) by replacing all AND gates by OR gates, or vice versa. Thus we have

$$(\overline{C}_{i+1} + A_i B_i C_i)(A_i + B_i + C_i) = \overline{C}_{i+1}(A_i + B_i + C_i) + A_i B_i C_i$$

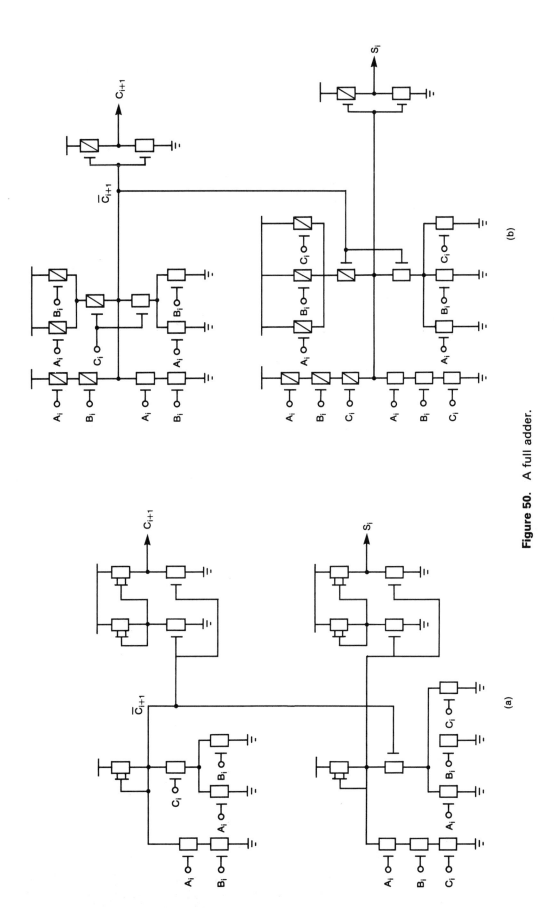

Figure 50. A full adder.

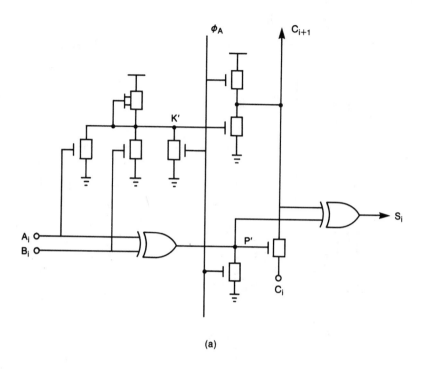

(a)

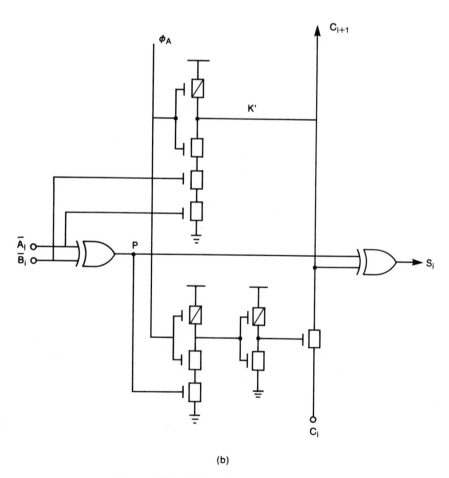

(b)

Figure 51. A Manchester carry chain.

which is identical to Equation (5). The reader should have no difficulty verifying the p-channel device connections for C_{i+1} in the same fashion.

EXERCISE

Prove that the expressions of S_i and C_{i+1} are self-dual; that is,

$$\overline{S}_i = \overline{A}_i \oplus \overline{B}_i \oplus \overline{C}_i$$

and

$$\overline{C}_{i+1} = \overline{A}_i\overline{B}_i + \overline{C}_i(\overline{A}_i + \overline{B}_i)$$

What does this mean to the topology of a CMOS circuit?

The speed of a ripple carry adder is basically limited by the worst case propagation delay from C_1 to C_n. Different schemes have been invented to speed up the propagation of carries. The following are some examples of "fast-carry" designs.

(1) Manchester Carry Chain: In the previous circuits the carry C_{i+1} is generated according to Equation (4). For technologies where the pulldown is much faster than the pullup, node C_{i+1} can be precharged high by a clock ϕ_A when the adder is not in use. As soon as the adder is "selected," ϕ_A goes down and node C_{i+1} is under the control of A_i and B_i according to Table 2:

Table 2

A_i	B_i	C_{i+1}	Comments	
0	0	0	C_i	annihilated
0	1	C_i	C_i	propagated through
1	0	C_i	C_i	propagated through
1	1	1	C_{i+1}	generated

Let us define the logic functions

$$G_i = A_iB_i$$

for carry *generation,*

$$K_i = \overline{A}_i\overline{B}_i$$

for carry *kill,* and

$$P_i = A_i \oplus B_i$$

for carry *propagation.* Since C_{i+1} has been precharged high, only K_i and P_i functions need to be implemented in the adder. Such a scheme, called the *Manchester carry chain,* is very popular with NMOS ripple carry adders. Two such designs are shown in Figure 51. Since the adder is now in a controlled environment, the XORs can be constructed with the pass gates of Figure 10.

(2) Carry Bypass: The idea of carry bypass is to allow a low-order carry to propagate through a group of adder stages without waiting for the generation of local carries. A 4-bit carry bypass configuration is shown in Figure 52. The OR gates associated with A_i and B_i gate the low-order carry bit C_{in}. If the outputs of all OR gates are 1, C_{in} will be propagated to C_{out} with approximately three gate delays. Both 4-bit and 8-bit carry bypass circuits can be implemented as standard books. The improvement in speed is proportional to the total number of bits in the adder.

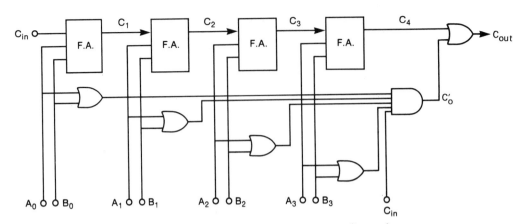

Figure 52. 4-bit carry bypass configuration.

(3) Carry Look-Ahead: The carry bypass is a special case in the family of fast-carry schemes called *carry look-ahead* (CLA). In Figure 52 the carryin C_{in} looks ahead over four adder stages in the sense that the output $\dot{C_o}$ is directly related to inputs A and B and does not depend on any intermediate carries. It is also possible to express the internal carries C_1, C_2, and C_3 as direct functions of A_i, B_i, and C_{in}. The result is then a fast 4-bit carry-look-ahead adder.

As shown in Equation (4), the carry C_{i+1} in each stage is given by

$$C_{i+1} = G_i + P_i C_i \qquad (6)$$

where

$$G_i = A_i B_i$$

and

$$P_i = A_i \oplus B_i, \qquad i = 0, 1, 2, 3$$

Substituting G_i and P_i into C_{i+1} yields

$$C_1 = G_0 + P_0 C_{in}$$
$$C_2 = G_1 + P_1 G_0 + P_1 P_0 C_{in}$$
$$C_3 = G_2 + P_2 G_1 + P_2 P_1 G_0 + P_2 P_1 P_0 C_{in}$$
$$C_4 = G_3 + P_3 G_2 + P_3 P_2 G_1 + P_3 P_2 P_1 G_0 + P_3 P_2 P_1 P_0 C_{in}$$

Thus, in a carry-look-ahead adder all carries are generated by two-level logic as functions of P_i, G_i, and C_{in}.

EXERCISE

Show that the sum bits

$$S_i = P_i \oplus C_{i+1}, \qquad \text{for } i = 0, 1, 2, 3$$

As in the case of carry bypass, the carry-look-ahead scheme does not offer much advantage over the ripple carry adders for small adders. For large adders the power of carry-look-ahead scheme becomes obvious by observing that C_4 can be written as

$$C_4 = GG + GP\,C_{in}$$

where

$$GG = G_3 + P_3 G_2 + P_3 P_2 G_1 + P_3 P_2 P_1 G_0$$

and

$$GP = P_3 P_2 P_1 P_0$$

Thus, we obtain an equation exactly in the same form as Equation (6) except that in this case *GG* represents *group generation* and *GP* represents *group propagation*. Thus, for each group of four 4-bit CLA adders, another level of CLA circuits can be used with *GG* and *GP* as inputs. An example of a 16-bit CLA adder is shown in Figure 53.

Most libraries offer a 4-bit CLA adder "slice" and 1 to 4-bit CLA generators as standard books. Since an n-bit CLA requires n-bit NOR gates, a 4-bit CLA adder slice is a good compromise between density and performance.

(4) Carry Select: Like carry-bypass adders, a carry-select adder is composed of a chain of adder stages. Each stage consists of two identical (usually ripple-carry) FAs. As shown in Figure 54, one FA has a carryin equal to 1 whereas the other has a carryin equal to 0. The final sum **S** and group carryout C_{out} are then selected by

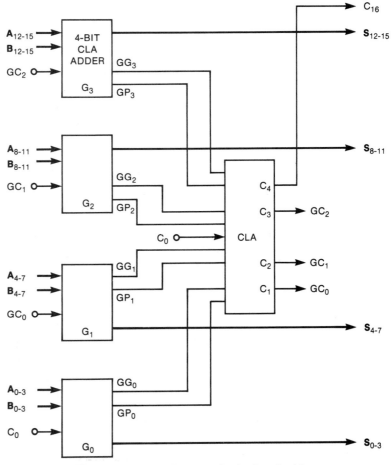

Figure 53. A 16-bit carry-look-ahead adder.

the actual carryin C_{in} propagated from the previous stage. In typical applications the number of bits of successive stages increases to ensure that the internal carries are generated just before C_{in} arrives so that each stage contributes only one multiplexer delay.

Due to different design and layout constraints such as data bus configuration, practical fast adder designs are mostly a combination of the above schemes.

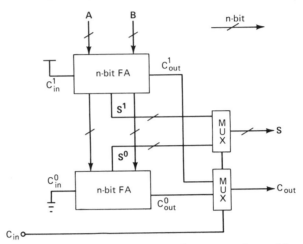

Figure 54. An *n*-bit FA stage of a carry-select adder.

EXERCISES

Consider a 32-bit adder; estimate the worst case gate delays with

1. Eight 4-bit carry bypass stages
2. Four 8-bit carry bypass stages
3. 4-bit CLA adder slices and 1 to 4-bit CLA generators
4. A chain of 4-, 4-, 5-, 6-, 7-, 6-bit carry-select stages.

3.3.2 Arithmetic/Logic Unit: ALU. An arithmetic/logic unit is the central processing unit of a processor. It performs logic and arithmetic operations according to machine instructions.

Figure 55 shows a 4-bit ALU in CMOS technology. The same ALU functions have been implemented in numerous technologies such as SN74S181 in TTL, MC10181 in ECL, and CD40181B in MOS. In Figure 55, M is the mode control. The ALU performs logic or arithmetic functions for $M = 1$ or 0, respectively. The specific functions to be carried out are controlled by four bits, S_3, S_2, S_1, and S_0, as listed in Table 3.

It is instructive to identify the function of each circuit in Figure 55. For addition operation, circuits S0, S1, S2, and S3 are *half adders generating P_i and G_i for A_i and B_i*, $i = 0, 1, 2, 3$. Circuits S4, S5, and S6 are CLA circuits generating C_1, C_2, and C_3. Circuits S7, S8, S9, and S10 generate the sum bits F_0, F_1, F_2, and F_3. Circuit S11 compares the input vectors. $P(\overline{A = B}) = 1$ if $A \neq B$ and $P(\overline{A = B}) = 0$ if $A = B$. Circuits S12 and S13 are so-called *X-Y* generators. In CLA operation the group carry C_{N+4} is given by

$$\overline{C}_{N+4} = \overline{Y} + \overline{X}\,\overline{C}_N$$

so large adders can be implemented with 4-bit ALU slices connected to CLA circuits.

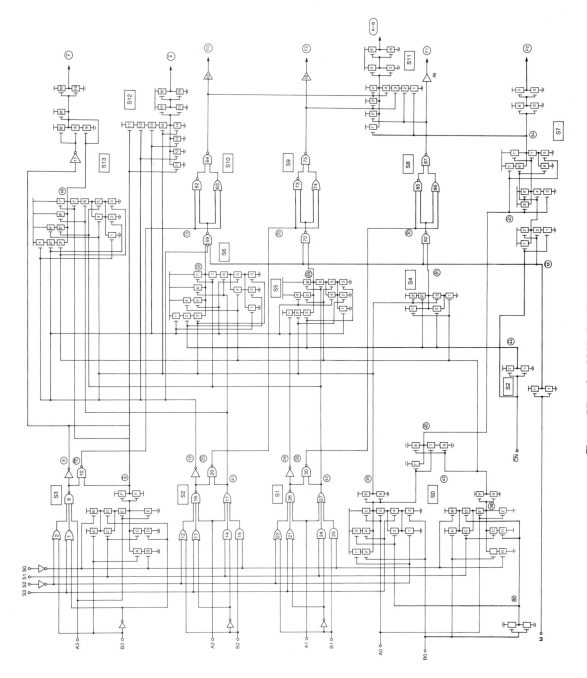

Figure 55. An ALU circuit schematic.

Table 3

S_3, S_2, S_1, S_0				$M = 1$	$M = 0, \overline{C}_n = 1$*
0	0	0	0	$\overline{A}$	A
0	0	0	1	$\overline{A + B}$	$A + B$
0	0	1	0	$\overline{A}B$	$A + \overline{B}$
0	0	1	1	Logic 0	Minus 1
0	1	0	0	$\overline{AB}$	A plus $A\overline{B}$
0	1	0	1	$\overline{B}$	$(A + B)$ plus $A\overline{B}$
0	1	1	0	$A + B$	A minus B minus 1
0	1	1	1	$A\overline{B}$	$A\overline{B}$ minus 1
1	0	0	0	$\overline{A} + B$	A plus AB
1	0	0	1	$\overline{A + B}$	A plus B
1	0	1	0	B	$(A + \overline{B})$ plus AB
1	0	1	1	AB	AB minus 1
1	1	0	0	Logic 1	A plus A
1	1	0	1	$A + \overline{B}$	$(A + B)$ plus A
1	1	1	0	$A + B$	$(A + \overline{B})$ plus A
1	1	1	1	A	A minus 1

*For $M = 0$, $\overline{C}_n = 0$, the functions are as shown plus 1.

A layout of the ALU that fits the image in Section 2 is shown inside the front cover of this book. Sharing diffusions is maximized, and most of the intercircuit wiring is confined to the circuit area. If the ALU share V_H/GD buses with many other circuits without vertical $M2$ strapping, wider V_H/GD buses are recommended to minimize noise.

3.3.3 Multipliers. A multiplier performs binary multiplication,

$$\mathbf{S} = \mathbf{X} \times \mathbf{Y}$$

where the *multiplicand* $\mathbf{X} = x_{m-1}x_{m-2}\ldots x_1x_0$ is an m-bit number, the *multiplier* $\mathbf{Y} = y_{n-1}y_{n-2}\ldots y_1y_0$ is an n-bit number. The *product* $\mathbf{S}$ is an $(m + n)$-bit number if $\mathbf{X}$ and $\mathbf{Y}$ are unsigned numbers. It is an $(m + n - 1)$-bit number if $\mathbf{X}$ and $\mathbf{Y}$ are signed numbers. The binary multiplication can be performed in the same way as hand calculation with a single-bit shift-and-add algorithm. For example, consider the following multiplication procedure:

Multiplicand:	$x_7\, x_6\, x_5\, x_4\, x_3\, x_2\, x_1\, x_0$
Multiplier:	$y_7\, y_6\, y_5\, y_4\, y_3\, y_2\, y_1\, y_0$
Partial Products	

A	$a_7\, a_6\, a_5\, a_4\, a_3\, a_2\, a_1\, a_0$
B	$b_7\, b_6\, b_5\, b_4\, b_3\, b_2\, b_1\, b_0$
C	$c_7\, c_6\, c_5\, c_4\, c_3\, c_2\, c_1\, c_0$
D	$d_7\, d_6\, d_5\, d_4\, d_3\, d_2\, d_1\, d_0$
E	$e_7\, e_6\, e_5\, e_4\, e_3\, e_2\, e_1\, e_0$
F	$f_7\, f_6\, f_5\, f_4\, f_3\, f_2\, f_1\, f_0$
G	$g_7\, g_6\, g_5\, g_4\, g_3\, g_2\, g_1\, g_0$
H	$h_7\, h_6\, h_5\, h_4\, h_3\, h_2\, h_1\, h_0$

$s_{15}\, s_{14}\, s_{13}\, s_{12}\, s_{11}\, s_{10}\, s_9\, s_8\, s_7\, s_6\, s_5\, s_4\, s_3\, s_2\, s_1\, s_0$

The partial product $\mathbf{A} = \mathbf{X}$ if $y_0 = 1$, and $\mathbf{A} = \mathbf{0}$ if $y_0 = 0$. Each bit of $\mathbf{A}$, $a_i = x_iy_0$ is the logic AND function, which is sometimes called a 1×1 multiplier. Similarly, $\mathbf{B} = \mathbf{X}$ or $\mathbf{0}$ depending on $y_1 = 1$ or 0, respectively. Vector $\mathbf{B}$ is shifted to the left by one bit because the order of y_1 is one bit higher than that of y_0. All other partial products are generated and placed in the same fashion. Finally, an adder adds the partial products together to produce the final result.

That the multiplier is very hardware intensive should be immediately obvious. The generation of partial products needs mn 1×1 multipliers. The addition of the partial products needs multioperand adders. Furthermore, if $\mathbf{X}$ and $\mathbf{Y}$ are signed numbers, pre-and postcomplementers may be needed to convert signed numbers to unsigned numbers, and vice versa. This subsection discusses techniques to alleviate these problems. We will first discuss the problems individually and then put all solutions together to build a model 8×8 2's complement multiplier. This approach, which uses the *modified Booth algorithm* and *Dadda's tree reduction scheme,* is one of the fastest and most popular designs among MOS multipliers today.

(1) Partial Product Reduction: In a single-bit shift-and-add scheme, a multiplier examines one bit at a time to form the partial products. For an $m \times n$ multiplier, n partial products are generated with a total of mn bits. By examining more than one bit of the multiplier at a time and allowing more complicated operations on the multiplicand the number of partial products can be reduced. One of the most celebrated schemes is the modified Booth algorithm.

The modified Booth algorithm applies to *2's complement multiplications.* Let $\mathbf{Y} = y_{n-1}y_{n-2} \ldots y_1 y_0$ be the multiplier. Let us first assume $\mathbf{Y}$ is a positive number so that $y_{n-1} = 0$. By definition

$$\mathbf{Y} = y_{n-1}2^{n-1} + y_{n-2}2^{n-2} + \cdots + y_2 2^2 + y_1 2^1 + y_0 2^0$$

Each term in $\mathbf{Y}$ can be expressed as

$$y_k 2^k = \left(2y_k - \frac{1}{2} \times 2y_k\right)2^k$$

$$= y_k 2^{k+1} - 2y_k 2^{k-1}, \qquad k = 0, 1, 2, \ldots, n-1 \qquad (7)$$

Assume furthermore that n is an even number. Let us expand all $y_k 2^k$ terms in $\mathbf{Y}$ for *odd k*. We then have

$$\begin{aligned}
\mathbf{Y} = &\ (y_{n-1}2^n - 2y_{n-1}2^{n-2}) + y_{n-2}2^{n-2} \\
&+ (y_{n-3}2^{n-2} - 2y_{n-3}2^{n-4}) + y_{n-4}2^{n-4} \\
&+ \cdots \\
&+ (y_3 2^4 - 2y_3 2^2) + y_2 2^2 \\
&+ (y_1 2^2 - 2y_1 2^0) + y_0 2^0
\end{aligned}$$

Grouping the terms of like powers of 2 yields

$$\begin{aligned}
\mathbf{Y} = &\ y_{n-1}2^n + (-2y_{n-1} + y_{n-2} + y_{n-3})2^{n-2} \\
&+ (-2y_{n-3} + y_{n-4} + y_{n-5})2^{n-4} \\
&+ \cdots \\
&+ (-2y_3 + y_2 + y_1)2^2 \\
&+ (-2y_1 + y_0 + 0)2^0 \\
= &\ \sum_{\substack{k=0 \\ k \text{ even}}}^{n-2} z_k 2^k \qquad (8)
\end{aligned}$$

where

$$z_k = -2y_{k+1} + y_k + y_{k-1}, \qquad k = 0, 2, 4, \ldots, n - 2 \qquad (9)$$

with $\qquad y_{-1} \triangleq 0$

Note that $y_{n-1} = 0$. Now the product $\mathbf{S}$ equals

$$\mathbf{S} = \mathbf{X} \times \mathbf{Y}$$

$$= \sum_{\substack{k=0 \\ k \text{ even}}}^{n-2} z_k \mathbf{X} 2^k$$

Thus, we have redefined the partial products as $z_k\mathbf{X}$ with binary weight 2^k. Since only even k's exist in the above expression, the number of partial products has been reduced to $n/2$. The reduction of partial products is achieved at the expense of more complicated operations on the multiplicand. Depending on the actual values of y_{k+1}, y_k, and y_{k-1}, the following operations on $\mathbf{X}$ are performed by Equation (9):

TABLE 4

y_{k+1}	y_k	y_{k-1}	z_k	$z_k\mathbf{X}$
0	0	0	0	**0**
0	0	1	1	$+1\mathbf{X}$
0	1	0	1	$+1\mathbf{X}$
0	1	1	2	$+2\mathbf{X}$
1	0	0	-2	$-2\mathbf{X}$
1	0	1	-1	$-1\mathbf{X}$
1	1	0	-1	$-1\mathbf{X}$
1	1	1	0	**0**

In the last column, **0** replaces $\mathbf{X}$ by $\mathbf{0}$, $+1\mathbf{X}$ copies $\mathbf{X}$, $+2\mathbf{X}$ shifts $\mathbf{X}$ left by one bit, $-1\mathbf{X}$ replaces $\mathbf{X}$ by its 2's complement, $-2\mathbf{X}$ shifts the 2's complement of $\mathbf{X}$ left by one bit. All these operations can be accomplished by simple circuits so that hardware impact is minimal.

Table 4 also applies to negative multipliers without modification. In this case, the multiplier

$$\mathbf{Y} = y_{n-1}2^{n-1} + y_{n-2}2^{n-2} + \cdots + y_2 2^2 + y_1 2^1 + y_0 2^0$$

with $y_{n-1} = 1$. Taking the 2's complement of $\mathbf{Y}$, we have

$$-\mathbf{Y} = 2^n - \mathbf{Y}$$

$$= (2^n - 1) - \mathbf{Y} + 1$$

$$= \sum_{k=0}^{n-1}(1 - y_k)2^k + 1$$

$$= \sum_{k=0}^{n-1} \bar{y}_k 2^k + 1$$

Expanding the odd-numbered terms in $-\mathbf{Y}$ by Equation (7) yields

$$-\mathbf{Y} = \bar{y}_{n-1}2^{n-1} + (-2\bar{y}_{n-1} + \bar{y}_{n-2} + \bar{y}_{n-3})2^{n-2}$$

$$+ (-2\bar{y}_{n-3} + \bar{y}_{n-4} + \bar{y}_{n-5})2^{n-4}$$

$$+ \cdots$$

$$+ (-2\bar{y}_3 + \bar{y}_2 + \bar{y}_1)2^2$$

$$+ (-2\bar{y}_1 + \bar{y}_0 + 1)2^0$$

$$= \sum_{\substack{k=0 \\ k \text{ even}}}^{n-2} -z_k 2^k \tag{10}$$

$$\Rightarrow \qquad \mathbf{Y} = \sum_{\substack{k=0 \\ k \text{ even}}}^{n-2} z_k 2^k \tag{10(a)}$$

where z_k is defined in Equation (9). Note that $\bar{y}_{n-1} = 0$.

Thus Equations (8) and (10(a)) have exactly the same form regardless of the sign of $\mathbf{Y}$. Hence the modified Booth algorithm applies to both positive and negative multipliers.

Decoders for the multiplier are shown in Figure 56. Both true and complement forms are needed for each bit. The outputs of the decoder, $+2$, -2, $+1$, -1, represent the operations in Table 4 that are applied to the partial product bit generators PPM and PPL as shown in Figure 57. In PPM the inputs 1 and $\bar{1}$ are connected to x_k and $\bar{x}_k$, respectively. Since $+/-2$ means shifting left one bit, inputs 2 and $\bar{2}$ are connected to x_{k-1} and $\bar{x}_{k-1}$, respectively. For the PPL generating the lowest-order bit of the partial product, the inputs are 1 and $\bar{1}$. (See Equation (8)).

EXERCISES

1. Prove Equation (10) by showing

$$-2\bar{y}_k + \bar{y}_{k-1} + \bar{y}_{k-2} = -(2y_k + y_{k-1} + y_{k-2})$$

2. Verify the logic of PPM and PPL in Figure 57.

3. Devise a scheme that inspects four consecutive bits of the multiplier each time. Construct a table like Table 4 for the scheme. What kinds of hardware operations are needed?

The 2's complement of $\mathbf{X}$ can be obtained by first complementing all bits in $\mathbf{X}$ and then adding a 1 (the round bit) to the least significant bit. According to Table 4, the logic for the round bit for $-2\mathbf{X}$ and $-1\mathbf{X}$ is

$$r_k = y_{k+1}(\bar{y}_{k-1} + \bar{y}_k)$$

The generation of partial products for the proposed 8×8 multiplier is shown below. Because of the $+/-2$ operation, partial products are 9 bits long. Each partial product is displaced 2 bits to the left relative to its predecessor. There are, of course, $8/2 = 4$ partial products in total.

Multiplicand $x_7 x_6 x_5 x_4 x_3 x_2 x_1 x_0$	Multiplier decode
$a_8\ a_8\ a_8\ a_8\ a_8\ a_8\ a_8\ a_7\ a_6\ a_5\ a_4\ a_3\ a_2\ a_1\ a_0$	$y_1\ y_0\ 0$
$b_8\ b_8\ b_8\ b_8\ b_8\ b_7\ b_6\ b_5\ b_4\ b_3\ b_2\ b_1\ b_0 \qquad r_0$	$y_3\ y_2\ y_1$
$c_8\ c_8\ c_8\ c_7\ c_6\ c_5\ c_4\ c_3\ c_2\ c_1\ c_0 \qquad r_1$	$y_5\ y_4\ y_3$
$d_8\ d_7\ d_6\ d_5\ d_4\ d_3\ d_2\ d_1\ d_0 \qquad r_2$	$y_7\ y_6\ y_5$
$\qquad\qquad\qquad\qquad\qquad r_3$	

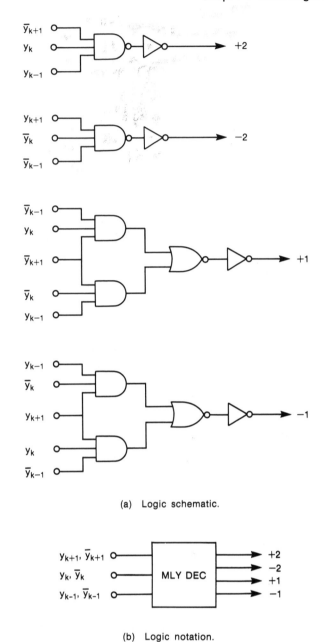

(a) Logic schematic.

(b) Logic notation.

Figure 56.   Modified Booth algorithm multiplier decode.

One problem associated with the 2's complement is sign bit extension. Since the partial products must be added together to form a 15-bit product, the sign bit of each partial product must be extended to the left most position to form a 15-bit number. A large number of sign bits greatly increases the number of adders required to sum all partial products in a later stage.

To alleviate this problem, let us sum the sign bits in the above separately with their proper weights:

$$T = a_8(2^8 + 2^9 + \cdots + 2^{14}) + b_8(2^{10} + 2^{11} + \cdots + 2^{14}) + c_8(2^{12} + 2^{13} + 2^{14})$$

$$+ d_8 2^{14}$$

$$= a_8(2^{15} - 2^8) + b_8(2^{15} - 2^{10}) + c_8(2^{15} - 2^{12}) + d_8(2^{15} - 2^{14})$$

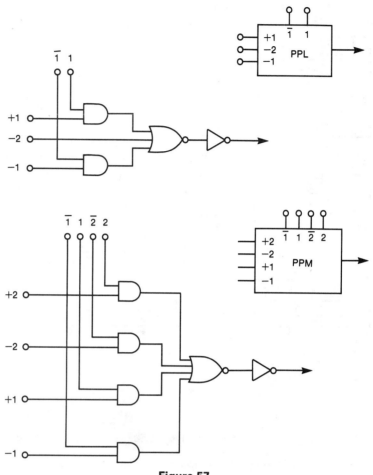

Figure 57.

$$= -a_8 2^8 - b_8 2^{10} - c_8 2^{12} - d_8 2^{14} \qquad\qquad \mathrm{mod}\ 2^{15}$$

$$= (2^{15} - 1) - a_8 2^8 - b_8 2^{10} - c_8 2^{12} - d_8 2^{14} + 1 \qquad \mathrm{mod}\ 2^{15}$$

Since $2^{15} - 1 = 11 \cdots 1$, we have

$$T = \bar{d}_8 2^{14} + 2^{13} + \bar{c}_8 2^{12} + 2^{11} + \bar{b}_8 2^{10} + 2^9 + \bar{a}_8 2^8 + \sum_{k=0}^{7} 2^k + 1$$

$$= \bar{d}_8 2^{14} + (2^{13} + \bar{c}_8 2^{12}) + (2^{11} + \bar{b}_8 2^{10}) + (2^9 + \bar{a}_8 2^8) + 2^8$$

The above complicated expression has a simpler hardware implementation: complement d_8 (the first term), add a 1 on top of a_8 (the last term), complement c_8, b_8, and a_8, and add a hardwired 1 to their higher-order bit positions. The complete partial product generation is shown in Figure 58.

EXERCISE

Derive the logic expression of PPH in Figure 58 for the sign bits $\bar{a}_8 \cdots \bar{d}_8$.

(2) Partial Products Addition: The partial products can now be added row by row to form the final product. It will take more than $8 \times 8/2 = 32$ adders and a very

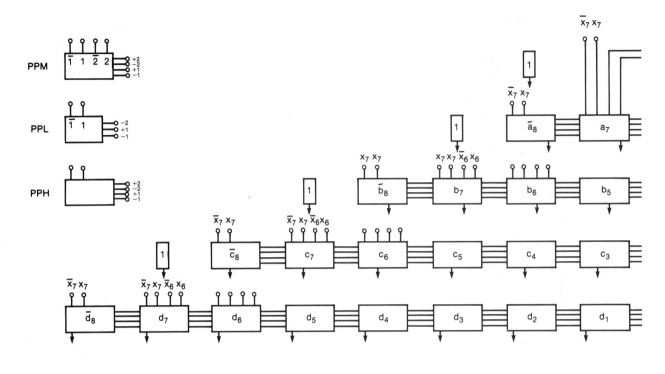

COLUMN   14   13   12   11   10   9   8   7

Figure 58. Partial products of an 8×8 multiplier.

long delay to accumulate the sum and carry bits. An alternative method is to add the bits in each column together through a full adder tree. Each adder, called *a carry-save adder or CSA,* reduces three input bits to two bits. A tree consists of several levels of CSAs. The number of bits to be added decreases down the tree. In the bottom of the tree the remaining two bits (a sum and a carry) will then be added horizontally by a carry propagation adder (CPA) in the usual fashion. This is called *Dadda's reduction scheme,* and the trees are called *Wallace trees.*

To be specific, let us consider column 6 of Figure 58 where 5 bits a_6, b_4, c_2, d_0, and r_3 are to be added. There are two carry bits generated in column 5 that must also be added. The CSA configuration of column 6 is shown in Figure 59 with only two levels of CSAs. The S output of the bottom CSA goes to the CPA in column 6 whereas the C output goes to the CPA in column 7. The reader should complete the CSA adder trees for all partial products shown in Figure 58 as an exercise.

Since the final carries propagate through the CPA, faster adders such as CLA adders must be used to achieve high speed.

Fast multipliers are essential for high-speed digital-signal processing. The design of onchip multipliers has been, and still is, an active area of research and development. New algorithms and layout styles are invented constantly. Interested readers are encouraged to seek out the appropriate literature.

3.3.4 Programmable Logic Arrays: PLAs. A programmable logic array, or PLA, is an orderly structure that can be personalized to implement logic functions. It implements the basic two-level logic in AO forms. Cascaded PLAs can be used to generate more complex functions if necessary.

The physical design of a PLA consists of two NOR networks with inverting I/O buffers. As shown in Figure 60, inputs to the PLA are inverted before they are

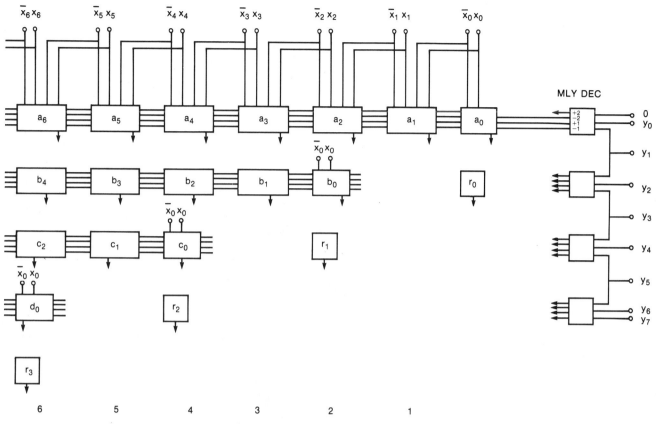

Figure 58. *(cont.)*

applied to the T/C generators that control the first NOR network. Outputs of the second NOR network are inverted before they emerge from the PLA. As a result, the first array performs the AND function on the PLA inputs, and the second array performs the OR function for the PLA outputs. From a logic viewpoint, therefore, the PLA consists of two arrays: an input AND array and an output OR array.

The true and complement forms of each input are generated by T/C GEN in the AND array to produce product terms such as AB, $\bar{A}B$, and $A\bar{B}$. Each output line of the AND array carries a product term and is called a *product line* or *wordline*. Follow the wordline of $\bar{A}B$. Since there are two pulldowns controlled by A and $\bar{B}$, the NOR output is given by

$$\overline{A + \bar{B}} = \bar{A}B$$

Each intersection of a decode line of the T/C GENs and a wordline is a possible site for a pulldown device. Since a PLA is a uniform structure, all pulldown devices are the same. The layout configuration of a pulldown is called a PLA cell. The design of PLA cells depends on the technology. For example, in a single-poly, single-metal process, two cell layouts are possible: (1) poly decode line with metal wordline and (2) metal decode line with poly wordline. In choosing the particular cell configuration, bear in mind that metal lines are much faster than poly lines for the same drivers.

Wordlines are inputs to the OR array. With an inverting buffer, each NOR circuit in the OR array performs a simple OR function on the product terms whose wordlines are tied to the pulldowns of the NOR circuit. The output of each NOR circuit is called a *dataline*. Again, different cell layout schemes can be used for the OR array.

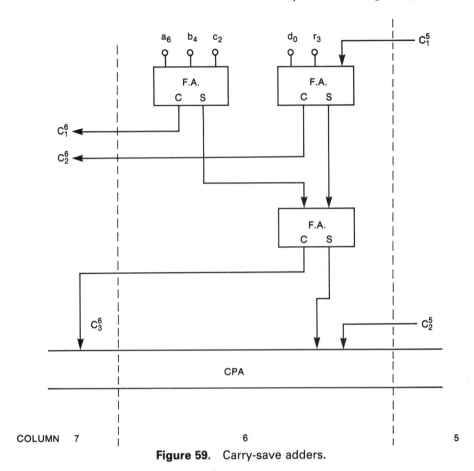

Figure 59. Carry-save adders.

Performance of a PLA is basically limited by the risetime of the long word- and datalines. Density of a PLA depends of course on the number of product terms and outputs but is generally inversely proportional to the number of input variables. Thus, large PLAs tend to be sparse and slow. Three approaches are commonly used to improve PLA performance: (1) *2-bit partitioning*, (2) *folding*, and (3) *dynamic precharge*.

1. *2-Bit Partitioning*. The T/C GENs in Figure 60 generate the true and complement form of each input variable independently. This is called 1-bit partitioning. Sometimes it is more efficient to consider two bits at a time. Instead of generating A, $\bar{A}$, B, and $\bar{B}$, the same four lines can be used to generate AB, $\bar{A}B$, $A\bar{B}$, and $\bar{A}\bar{B}$ as inputs to the AND array. This is called 2-bit partitioning or *decoded* PLA. Since decoding circuits requires little of the total area in large PLAs, 2-bit partitioning offers more flexibility in optimization of PLAs.

2. *PLA Folding*. Not all input variables appear in all product terms in practice. This is particularly true of large PLAs where variables tend to lump together forming separate groups. Figure 61 shows a PLA containing two separate groups. Straightforward implementation as in Figure 61(a) produces an AND array with two active (marked as I and II) and two wasteful empty (filled by dashed lines) areas. By folding the second group to the other side of the array and breaking the decode lines in the middle, both groups are accommodated in half the total area. This technique, called PLA folding, also applies to the OR array.

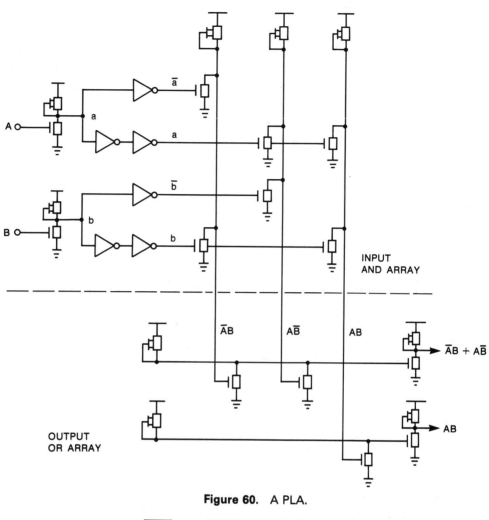

Figure 60.   A PLA.

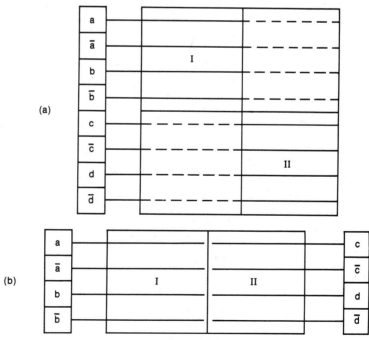

Figure 61.

In the example two separate PLAs can be used for each group. A more sophisticated case is shown in Figure 62. Here the decode lines break along the diagonal and more area is saved. In a more general form, the AND array can further be split into two parts. The OR array, with its own folding, locates in the middle of the AND arrays as shown in Figure 63. Such a structure is called an *AND-OR-AND* PLA.

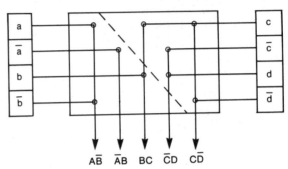

Figure 62. Diagonal folding. Pulldown connections are represented by open dots.

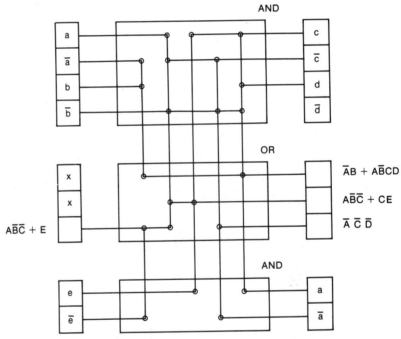

Figure 63. An AND-OR-AND PLA.

Success of PLA folding depends largely on the logic functions to be implemented. Some AND-OR-AND PLAs allow product terms that contain variables from both AND arrays. Three such feedthrough lines are shown in Figure 63. If feedthrough lines are not allowed, however, duplication of certain input variables in both AND arrays may be necessary. The duplication of input A in Figure 63 allows better folding in the OR array. Because of its great complexity, there are automatic PLA generators that perform the bit partitioning and folding based on the logic design, thus relieving human designers from the drudgery of PLA optimization.

3. *Dynamic Precharge.* As mentioned in Section 3, dynamic circuits can offer better performance and density for large logic circuits. Using dynamic T/C generators for array inputs and dynamic decoders for array outputs,

wordlines and datalines can be precharged and discharged quickly with clock control.

PLA represents a particular design style. The simplicity of its logic structure and ease of its physical layout make PLA particularly attractive to some system applications. As a matter of fact, PLA was once considered *the* VLSI solution to all logic problems by some system designers. Many automatic design tools were developed. Aside from the obvious advantages, however, large PLAs do suffer from low density and slow performance compared with ordinary random logic. The recent advances in simulation and checking tools for random logic designs finally have brought PLA down to its proper perspective: it is just another style of logic design. While two-level logic implementation remains an important conceptual tool for CAD tool developers, the actual physical realization of PLAs is quite limited in practice. Some libraries offer small standard-sized PLAs as macros so the chip designer has the option to use them when frequent ECs are expected or TAT is of primary concern.

3.3.5 Read Only Memory: ROM/ROS. The onchip ROM, or ROS (read-only storage) is mostly used for microcode storage. An $m \times n$ ROM contains m words. Each word is n-bits long. The basic architecture of an onchip ROM is similar to an SRAM and its memory array is very much like a PLA. A typical ROM array is shown in Figure 64. Since the design of all supporting circuitry such as decoders

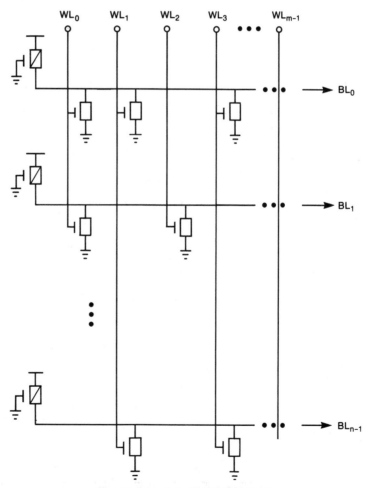

Figure 64. A typical ROM array.

and output drivers are much the same as those in SRAMs, the reader is referred to Chapter 6 for details.

Onchip ROMs of several kilowords are available in most libraries.

3.3.6 General-Purpose Register: GPR. A general-purpose register or GPR is a small, fast RAM used by the CPU as a scratch pad memory. It is also called a *register file*.

A GPR is organized in words. For example, a 64 × 32 GPR contains 64 words, each of which is 32-bits long. A GPR is accessed through "ports." A port is a register bank that holds the word address and data. For example, a port for the 64 × 32 GPR has 6 bits for the address and 32 bits for data.

Most GPRs are multiport RAMs so they can be accessed from different buses and read and written simultaneously. In high-performance GPRs, each port is dedicated either to READ or to WRITE with its own clock. The cell of a typical GPR with two READ and two WRITE ports is shown in Figure 65.

Ports can be operated either asynchronously or synchronously, depending on the design. In asynchronous operation there are obvious timing restrictions between competing operations. Simultaneous READ and WRITE of the same word from different ports result in uncertain data. In synchronous operation, READ and WRITE are controlled by clocks. For example, the CPU may read the GPR when the clock is high and write the GPR when the clock is low. In this way internal subclocks can be generated to restore the bitlines between clock transitions. In either case simultaneous WRITE to the same word from different ports must be avoided.

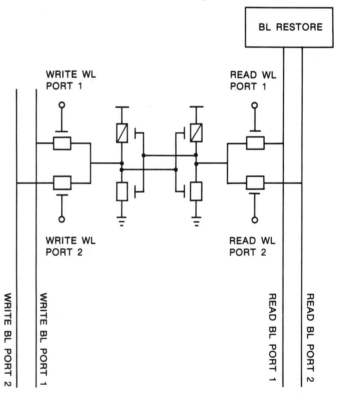

Figure 65. A GPR cell with two READ and two WRITE ports.

4. DESIGN FOR TESTABILITY

Any product must be tested before shipment. The design of the product, therefore, must be testable. All defects that cause malfunction must be detectable from chip I/Os. Since the number of I/Os does not increase with the number of circuits as the chip density increases, testability must be built in with circuit and logic design. In fact, testability is one of the major factors that affect both circuit and logic design methodology. Depending on testing philosophy, different circuit and logic design styles, and the corresponding design rules, have been developed for practical service.

This section discusses testing problems from a circuit point of view. First, the concept of stuck-at fault models and its relationship with physical defects will be introduced. Typical NMOS and CMOS faults and peculiarities will be studied. After a brief introduction of test generation the level-sensitive scan design (LSSD) will be discussed. Finally, the aspects of selftest with pseudorandom test generation are briefly mentioned.

4.1   Stuck-At Fault Models

Physical defects manifest themselves by affecting logic functions. Technology-independent models can be constructed to simulate physical defects. With a model established for each circuit, technology-specific statistics can be used to relate all possible physical defects to a particular fault in the model. These data, applied to all circuits in the chip, are used both for yield and reliability projection and as a guideline for process improvement.

One of the most popular technology-independent fault models is the *stuck-at fault model*. It assumes a single fault has occurred at one of the I/O pins of a primitive logic block that causes the value of the pin stuck at 1 (denoted s-a-1) or stuck at 0 (denoted s-a-0). For example, consider the m-input NAND shown in Figure 66. The output, node (10), can be either stuck at 1 or 0. If P10 = s-a-1, a test vector $(PA1, PA2, \ldots, PAm) = (1, 1, \ldots, 1) = 11\ldots1$ will detect the fault because P10 = 0 in a good circuit. Similarly, a test vector with at least one 0 component will detect the fault P10 = s-a-0.

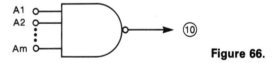

Figure 66.

As to the inputs, consider first PA1 = s-a-1. To produce different responses at node (10) for different values of $A1$, all other inputs must be 1; that is, $PA2 = PA3 = \ldots = PAm = 1$ so that they do not affect the output. *Thus, the noncontrolling value of the inputs to an AND or NAND circuit is equal to 1. Similarly, it is easy to show that the noncontrolling value of the inputs to an OR or NOR circuit is equal to 0.* Now refer back to the case PA1 = s-a-1. The input sequence $011 \cdots 1$ constitutes a test vector. On the other hand, it is easy to see that PA1 = s-a-0 implies P10 = s-a-1, so the two cases merge and no additional test is needed.

By the same argument it can be shown that the input sequence $11 \cdots 0 \cdots 11$ with a unique 0 in the kth position tests PAk = s-a-1, and PAk = s-a-0 implies P10 = s-a-1 for all k. There are, therefore, $m + 1$ test vectors (or simply tests) needed to test an m-input NAND gate.

EXERCISE

Generate the tests for an m-input NOR circuit.

The stuck-at fault model can be directed to physical defects through circuit layout. The layout of a 2-input NAND in NMOS technology is shown in Figure 67.

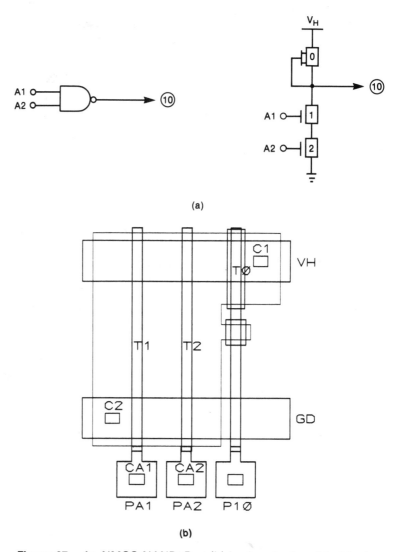

(a)

(b)

Figure 67. An NMOS NAND. Part (b) is also displayed inside the back cover of this book.

Correspondence between various physical defects and the logic faults are shown below.

Physical Defects	Logic Faults		
	PA1	PA2	P10
C1 missing			s-a-0
C2 missing			s-a-1
T0 open			s-a-0
T0 short			s-a-1
T1 open			s-a-1
T2 open			s-a-1
T1 short	s-a-1		
T2 short		s-a-1	
CA1 open	s-a-0		s-a-1
CA1 short to V_H	s-a-1		
CA2 open		s-a-0	s-a-1
CA2 short to V_H		s-a-1	

With statistical data obtained from device characterization such as shown in the following table, an assessment of a particular untested fault can be made.

Physical Defects	Percentage
M1–M1 shorts	40%
M1 line and contact opens	15%
Diffusion-diffusion and device shorts	15%
Diffusion and device opens	15%
Obvious and unknown	15%

4.2 CMOS Faults

In the previous example all defects have been adequately modeled by stuck-at faults. The "always-on" pullup of an NMOS circuit eliminates the ambiguities at the output when a pulldown is open. In CMOS circuits, however, not all defects can be modeled as stuck-at faults. Consider the simple CMOS NAND shown in Figure 68. Assume the contact C2 is missing so device T_1' loses control of node ⑩. To detect the defect, P10 must be set to 0 first by applying an input sequence 11, followed by the sequence 01. If P10 = 1, the circuit is good. If P10 = 0, the fault will be detected. Without presetting P10 = 0, the defect will leave node ⑩ floating when 01 is applied, producing ambiguous results. Designating such faults by a "logic change," physical defects and the affected logic in the circuit are shown in the following table:

Physical Defects	Logic Faults		
	PA1	PA2	P10
C1 missing			s-a-0
C2 missing			logic change
C3 missing			logic change
C4 missing			s-a-1
C5 missing			s-a-1
T1 open	s-a-0		s-a-1
T1 short			logic change
T1' open			logic change
T1' short	s-a-0		s-a-1
T2 open		s-a-0	s-a-1
T2 short			logic change
T2' open			logic change
T2' short		s-a-0	s-a-1
CA1 open	s-a-0		s-a-1
CA1 short to V_H	s-a-1		
CA2 open		s-a-0	s-a-1
CA2 short to V_H		s-a-1	

Defects that can not be modeled by stuck-at faults must be tested with ordered sequences. A defective CMOS circuit behaves like a sequential machine. Since generation of such tests greatly increases the complexity of automatic test generation, most CMOS fault models ignore such CMOS open faults. Experiments show that the simple stuck-at fault model is adequate for most CMOS circuits.

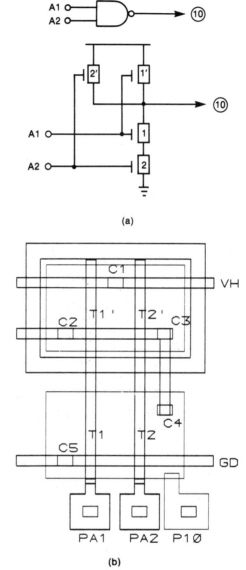

(a)

(b)

Figure 68. A CMOS NAND. Part (b) is also displayed inside the back cover of this book.

4.3 Generation of Test Patterns

The stuck-at fault model, applied to the primitive logic blocks, generates test vectors or patterns by direct reasoning. By "primitive" logic blocks we mean inverters, OR/NOR and AND/NAND gates. A more complex network must first be broken down into primitive blocks so each one can be considered separately. Input sequences will then be applied to the network to drive the faults of each block to the outputs. In a network, inputs that are directly assessable to test patterns are called *primary inputs* (PIs). Outputs that are directly assessable for measurement are called *primary outputs* (POs). Thus, from a test point of view, PIs are inputs directly controllable and POs are outputs directly observable. A test pattern then is an input sequence applied to the PIs producing a different output sequence at the POs manifesting a particular fault.

Each fault of a block will be driven forward to at least one PO by assigning *noncontrolling value to all other inputs* of the primitive blocks along its path. This is called *forward drive*. After the forward drive, a *backtrace* is carried out to arrive at a consistent input sequence at the PIs. If such a sequence exists it constitutes a test pattern, otherwise the fault is nontestable. A good example of a nontestable fault is redundant logic. A block is redundant if the outputs at POs do not change for all inputs after removing the block. By definition, then, a network that contains redundant logic cannot be fully testable.

EXAMPLE 1

Consider the network shown in Figure 69. Generate a test pattern for P4 = s-a-1.

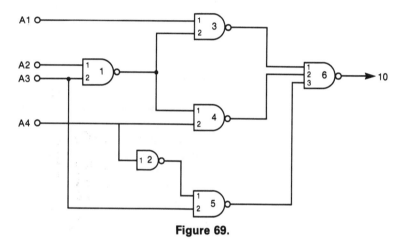

Figure 69.

Solution

To simplify the notation we shall use the following conventions. The PIs are denoted by Ai where $i = 1, 2, \ldots, m$. The POs are denoted by numbers 10, 20, \ldots, and so on. Since each primitive block has only one output, both the block and its output node (pin) are denoted by the same number. The inputs to a block are ordered as $1, 2, \ldots, n$. The ith input of the kth block is denoted by P$i.k$ where P stands for pins.

   Forward Drive: There is only one path, ④–⑥. To drive the fault to P10, P1.6 = P3.6 = 1. *Backtrace:* To test P4 = s-a-1, the inputs to block 4 must be such that P4 = 0. This implies P1.4 = 1 and P2.4 = A4 = 1. For P1.6 = 1 at least one input of block 3 must be 0. Since P2.3 = P1.4 = 1, A1 = 0. Now P3.6 = 1 ⇒ at least one input of block 5 must be 0. Since A4 = 1 ⇒ P1.5 = 0, P2.5 = X (don't care).

   A2 and A3 are determined by block 1. Since P1 = 1, at least one input of block 1 must be 0. Let A2 = 0 and A3 = 1. The sequence 0011 is therefore a test for P4 = s-a-1.

EXAMPLE 2

Generate a test pattern for P1 = s-a-1.

Solution

Forward Drive: To drive the fault from P1 to P10, let P1.3 = P2.4 = 1. Therefore, A1 = A4 = 1. The next block is block 6. To let the fault pass through, P3.6 = 1. *Backtrace:* To test P1 = s-a-1, the input of block 1 must be such that $P1 = 0$. This implies $A2 = A3 = 1$. Since $A4 = 1$, $P2.5 = 0$ which then ensures $P5 = 1$. The sequence 1111 is therefore a valid test.

In this example the fault is driven through two paths ①–③–⑥ and ①–④–⑥ *simultaneously* by letting $A1 = A4 = 1$. This is because driving through a single path with $A1 = \overline{A4}$ in this case will lead to conflicts at the PIs, and no test can be generated. The signal at P1, fanned out to gates 3 and 4, *reconverges* at gate 6. The reconvergent fanout relates the values of P1.6 and P2.6. A single-path backtrace assuming P1.6 and P2.6 being independent of each other may therefore lead to conflict at the PIs.

Formal algorithms that search all forward paths simultaneously to out drive a fault have been developed. The most popular is the D-algorithm and its variations. Other approaches, such as the Boolean difference and graph coloring method are also available. Since test generation is a vast subject, however, the reader is encouraged to consult with the references.

4.4 Level-Sensitive Scan Design

The stuck-at fault model is best suited for combinational logic circuits. For sequential machines, because of the limited number of primary I/Os, testing for a single fault may take several machine cycles. First, a homing sequence is applied to drive the machine to a given state. The test pattern for the fault follows. Several input sequences are then applied to drive the fault to a PO. Since the fault may very well affect machine states, additional errors can be generated during the drive. These errors can mask out the original fault so new test patterns must be generated. Test pattern generation for sequential machines is very complicated in general.

Level-sensitive scan design, often referred to as LSSD, is a design style that avoids generating complex test patterns for sequential machines. A logic system is said to be level sensitive if the steady state-response of the network is a function of input levels only, independent of circuit delays and insensitive to skews among inputs. In an LSSD system, the memory of the sequential machine shown in Figure 34 is implemented with SRLs. The SRLs break the feedback loop so logic signals are transferred from latches to latches controlled by clocks. The logic between the latches is combinational in nature and level sensitive as well, *if the clock periods are long enough to allow the signals to settle*.

A typical SRL used in LSSD systems is shown in Figure 70. Three clocks are required. Clock C, *the system clock,* gates system data D into L1. Clock A, *the test clock,* gates test patterns into the shift register chain. Clock B, *the shift clock,* transfers data from L1 to L2.

EXERCISE

Implement the SRL in Figure 70 with (1) NMOS E/D latches and (2) CMOS latches.

Two approaches can be used for normal logic operation. Figure 71 shows a two-latch design. Clock C receives data from a previous logic stage, and clock B shifts it from L1 to L2 driving the combinational logic. As mentioned in Section 3, the separation from C to B should be short, whereas the period from B to C must be

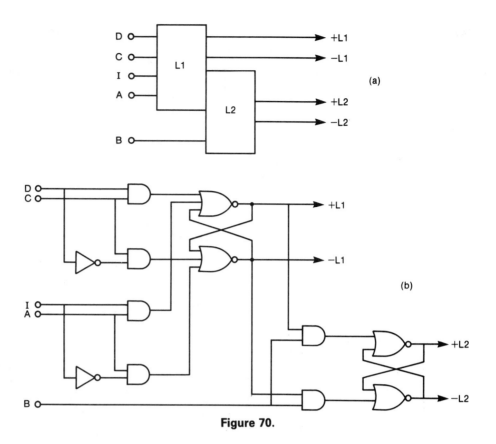

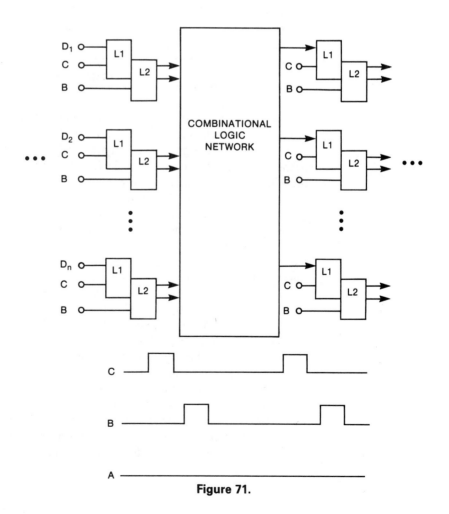

Figure 70.

Figure 71.

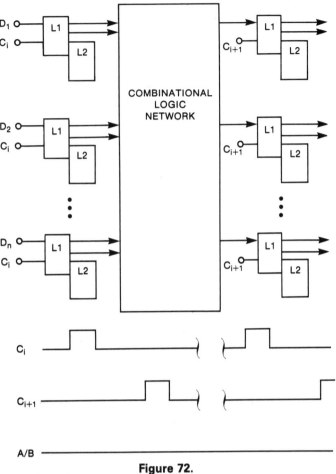

Figure 72.

long enough to allow signals to pass through the network. Since both L1 and L2 are used, this approach is called a *two-latch* or *latch/trigger design*.

A two-latch design is inherently level sensitive. The drawback of this approach is the delay between data settled in L1 and data available from L2. Another approach, which uses L1 latches only, eliminates the delay at the expense of more complex system timing. As shown in Figure 72, C_i and C_{i+1} are two system clocks that propagate signals through the combinational logic between the latches. Since L2 latches are not used in normal operations, it is called a *one-latch design*.

One-latch designs are not necessarily level sensitive. To ensure design integrity, various LSSD rules are issued by the design system. Some of the general rules are listed at the end of this subsection.

The scan input port *I* controlled by clock *A* has a parallel relationship with the data port *D* controlled by clock *C* in the SRL. As shown in Figure 73, the SRLs are chained together via the scan inputs to form a shift register during testing. Test patterns are shifted into the SRLs at *SI* (scanin) by clocks *A* and *B*. After scanin, clock *C* turns on, gating the outputs of the combinational logic into the SRLs. These data are then shifted out (scanout) by clock *A* and *B* at shift register output *SO*. The output sequence will subsequently be compared with results obtained from simulation or outputs from a "good machine" to detect and diagnose all faults.

Since the state of the machine is set by scanin and observed by scanout, combinational logic can be thoroughly tested with test patterns derived from the stuck-at fault model. Since the inputs to the combinational logic change constantly during

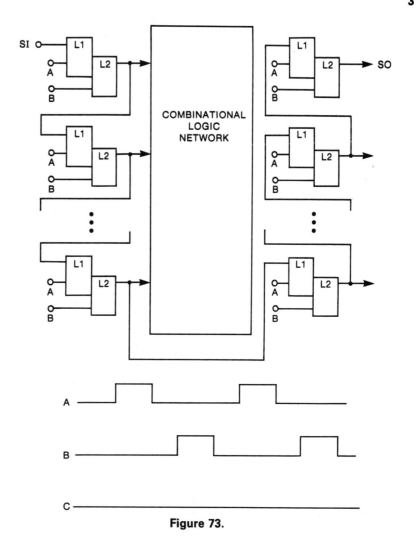

Figure 73.

scanin, however, the LSSD test is inherently slow. Thus, the LSSD test is mainly used to detect *solid faults* due to physical defects. (A solid fault is not affected by timing. Faults that vary with time are called *intermittent faults*.) Machine performance must be tested or guaranteed by other means such as functional tests where the machine is tested under normal operating conditions.

As mentioned earlier, logic design rules must be generated for both level-sensitive design and automatic test generation (ATG). Since an LSSD system always uses synchronous logic, most rules are related to clocks.

1. All internal storage except memory arrays must be implemented with SRLs.
2. All clock inputs to the SRLs must be directly controllable at chip I/Os.
3. All SRLs must be contained in a shift register chain whose input (*SI*) must be directly controllable and whose output (*SO*) must be directly observable at chip I/Os.
4. All latches can be isolated by turning clocks off. This ensures proper operation of scanin and scanout.
5. To avoid the race condition, a clock PI cannot feed the data input of a latch or the data/address inputs of a memory array.

6. A latch that drives the data input of another latch cannot be set by the same clock.

7. To avoid oscillation, no instantaneous feedback (versus through SRLs) in the combinational logic network is allowed. All bidirectional buses must be controlled by exclusive OR gating as shown in Figure 74.

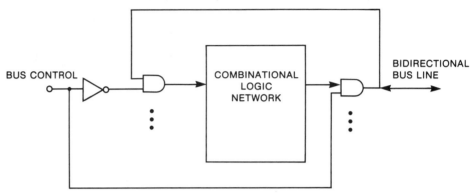

Figure 74.

8. It is recommended that all clocks be nonoverlapping. If overlap is unavoidable, the minimum delay through the combinational logic network $(t_D)_{min}$ must be greater than the clock overlap t_O as shown in Figure 75.

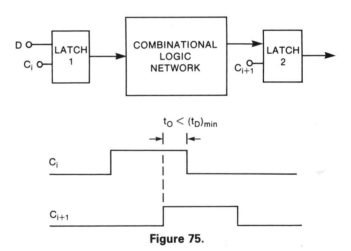

Figure 75.

9. If LSSD and non-LSSD designs are merged together, logic isolation must be provided at the interface. Since a non-LSSD design may contain asynchronous sequential logic, single-change test patterns may be required to avoid the race condition. In this case an L3 latch controlled by another clock as shown in Figure 76 can be used. Clock P is off during initial scanin. After all inputs have been shifted in, it turns on, applying the test patterns to the sequential logic inputs simultaneously.

In addition to these major rules there are many others that may pertain to a particular technology. Since the purpose of these rules is to ensure a testable design, most of them should be followed even by non-LSSD designs.

LSSD methodology, of course, is not without its shortcomings. The complicated SRL design has a direct impact on chip density and performance. In NMOS technology, since a latch always dissipates DC power, chip size is also limited by the

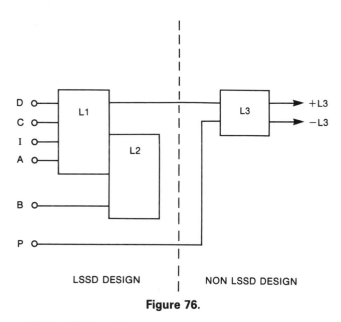

Figure 76.

extensive use of SRLs. This problem has been alleviated to a large extent in CMOS circuits.

4.5 Selftest and Pseudorandom Test Pattern Generation

Two major problems in testing large chips are test time and the large volume of test data. In high-volume production, the tester time for each chip is very much limited. Loading and unloading of huge amounts of test data including both test patterns and expected results, however, severely slow down the tester throughput. Selftest, which generates test patterns by hardware either onchip or by the tester, and checks test results by comparing some short sequences, offers an easy to implement hardware testing system for LSSD designs.

There are many selftest structures. A typical onchip selftest configuration is shown in Figure 77. Test patterns are automatically generated by a shift register chain. The output of the shift register is fed back to selective stages through XOR gates. The shift register, called *linear feedback shift register* or *LFSR* (pronounced as liftzer), changes its state continuously as outputs. These output patterns, called *pseudorandom patterns* because they are generated deterministically, are then used as test patterns for the combinational logic network.

With proper feedback, an n-bit LFSR can generate all $2^n - 1$ distinct states before repeating. Such LSFRs are called maximum-length LFSRs. Since it is a direct application of finite field theory (and will take us far off track), the basic theory of LSFRs is discussed in the appendix.

Maximum-length LSFRs are also used for compressing output data. As shown in Figure 77, outputs of the combinational logic network are XORed into a LFSR, modifying its state and thereby imprinting "signatures" into the LFSR. This device is called a *multi-input shift (signature) register* or *MISR*. At the end of the test, contents of the MISR will be compared with expected results. If the signatures agree, the chip passes the test; otherwise, it is faulty. Since only the signatures are needed for comparison, the data storage requirement is minimal.

Since the MISR compresses a huge amount of data into a short pattern, faulty signatures may be compressed into the final signature of a good machine. This is called *aliasing*. To calculate the probability of aliasing in an *n*-bit MISR, let the data stream be of length k. There are then $2^k - 1$ possible data streams and 2^n possible

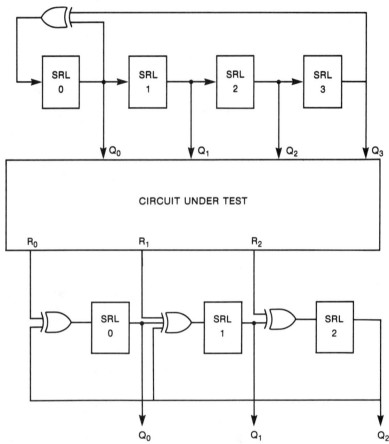

Figure 77. A selftest configuration with a 4-bit LFSR and a 3-bit MISR. The primitive polynomials associated with the LFSR and the MISR are $x^4 + x^3 + 1$ and $x^3 + x + 1$, respectively. (See appendix.)

signatures. If the data streams are evenly compressed into all possible signatures, each signature corresponds to 2^{k-n} data streams. Thus, for a good signature at the end of testing there is one correct data stream and $2^{k-n} - 1$ faulty ones. The probability of mapping a faulty data stream into a correct signature is therefore equal to $(2^{k-n} - 1)/(2^k - 1) \cong 2^{-n}$. For a 16-bit MISR, for example, the probability is about 0.000015, which is acceptable in most applications.

While reducing the tester time and test data volume dramatically, selftest does have some drawbacks. Since the test patterns are generated "aimlessly," it is very difficult to calculate the test coverage (test coverage = number of faults tested/number of faults modeled). The quality of selftest depends on many factors. For example, a 22-input AND gate takes only 23 patterns to test all stuck-at faults, whereas its probability of being tested by random patterns, $(1/2)^{22}$, is extremely small. Such "selftest–resistant" circuits must be avoided in logic design if the selftest is to be effective. Another problem is the regulating effects of logic functions. The random patterns generated by the LFSR are "biased" by logic functions as they pass through each logic stage, preventing certain input patterns from reaching subsequent logic circuits.

From a logic design viewpoint, perhaps the most severe drawback of selftesting is its inability to reduce the amount of logic simulation. Since the number of test patterns in a selftest is orders of magnitude larger than that in a deterministic test, both test coverage calculation and signature generation require extensive logic simulation.

Many algorithms are suggested to estimate the test coverage without fault simulation, but the final signature still needs to be supplied by the logic designer.

5. TIMING VERIFICATION

5.1 Delay Calculation

Logic designs must be simulated to verify their functions and performances. Most logic simulators offer delay calculations. If performance is not critical, a unit delay simulation that assumes uniform delay for all primitive logic blocks may be adequate for logic function verification. In most cases, however, performance *is* a design criterion and delays of critical paths must be carefully evaluated.

The delay calculators discussed in Chapters 4 and 5 are incorporated into the simulation models of all primitive logic blocks. Large macros can be treated either as a network of primitive blocks or as a single-delay block. The former is more convenient for statistical analysis. The logic simulator then invokes the delay calculators to determine the path delay from one net (the logic terminology of a node) to another under actual loading conditions extracted from the layout.

Since delays are statistical in nature, it is recommended to run delay calculations for several hundred cases with device parameters in each model as random variables. The values of the same parameter are correlated by a tracking window specified by the technology and the designer's confidence level. Tracking is specified as a uniform distribution around a nominal value. For example, consider the case of a 5% tracking of $V_{T0} \in (0.45, 0.55)$. The simulator chooses a value V_{T0} (nom) from the interval $(0.45, 0.55)$ at the beginning of each simulation. To calculate the delay of a particular inverter, a value randomly chosen from the interval $(0.95 \times V_{T0}(\text{nom}), 1.05 \times V_{T0}(\text{nom}))$ will then be passed to its delay calculator.

The delay at each net is therefore a distribution around its nominal value. The -3σ point is called the *best case (BC) delay* and the $+3\sigma$ point, *the worst case (WC) delay*. The delay time difference between two nets can also be calculated in the simulation. If the signal at net ① is earlier (later) than the signal at net ②, the $+3\sigma$ (-3σ) point of the delay-time difference, $t_D(1) - t_D(2)$, must be negative (positive).

All paths whose delays are critical to system functions must be analyzed. BC delays must be checked for possible race conditions; WC delays must be checked to ensure machine cycle times are satisfied.

In high-speed applications the pulsewidths of all clocks must be carefully examined. Since the risetime and falltime of a driver are generally different, the pulsewidth of a clock is modified by its driver. This is called the *pulse shrinkage problem*. As shown in Figure 78, the width of the positive going pulse is reduced by $\Delta t_D = t_{DLH} - t_{DHL}$ and that of the negative going pulse by $-\Delta t_D$. If the original clock pulse is shorter than Δt_D, the output of the clock driver will not be able to reach V_H. Successive "repowering" by such drivers may eventually annihilate the clock. Pulse shrinkage also affects clock separation. If cross-coupled clock drivers are not available, the clock generation tree must be laid out as symmetrically as possible so adjacent clocks drive comparable loads.

To ensure proper operation of the latches, the distribution of the setup time and hold time of each latch must be compared with those allowed by the WC design. Poor margins must be either corrected or proved adequate by detailed circuit simulation.

5.2 Timing Analysis

Delay simulation determines the delay along a particular path. Since the number of paths increases exponentially with the number of gates, sensitizing paths for delay

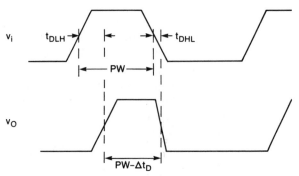

Figure 78. Pulsewidth shrinkage.

calculation is extremely time consuming. Timing analysis (TA), which estimates the delays for register-register transfers, can therefore be used for early screening.

The TA procedure is best illustrated by an example. Consider the network shown in Figure 79 where each logic circuit is represented by a delay block. *The delay block indicates the polarity of the output while ignoring all other logic properties.* Inversion of the output is represented by a wedge at the block output.

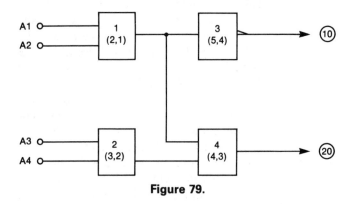

Figure 79.

Each block is represented by a vector of delay numbers. For example, the delays of block 1 are given by $DT(1) = (t_{DLH}, t_{DHL}) = (2, 1)$. The vectors can be added and subtracted componentwise:

$$(x_1, y_1) + (x_2, y_2) = (x_1 + x_2, y_1 + y_2)$$
$$(x_1, y_1) - (x_2, y_2) = (x_1 - x_2, y_1 - y_2)$$

and the maximum and minimum operators are defined by

$$\max\{(x_1, y_1), (x_2, y_2)\} = (\max(x_1, x_2), \max(y_1, y_2))$$
$$\min\{(x_1, y_1), (x_2, y_2)\} = (\min(x_1, x_2), \min(y_1, y_2))$$

For inversion blocks, vector transposition is necessary to match signal levels. Transposition is denoted by an upper bar on the net name.

For example,

$$DT(1) = (2, 1) \quad \Rightarrow \quad DT(\bar{1}) = (1, 2)$$

There are two stages in timing analysis: *forward delay computation* and *backward slack evaluation*.

1. *Forward delay computation.* Assuming all PIs are synchronized by the same clock, cumulative delays are calculated from PIs to POs by adding block delays *level by level*. There are two blocks in level 1: block 1 and 2. The arrival time (AT) of the signal at P1 is given by $AT(P1) = DT(1) = (2, 1)$. Similarly, $AT(P2) = DT(2) = (3, 2)$. For level 2, $AT(P10) = DT(3) + AT(\bar{P1}) = (5, 4) + (1, 2) = (6, 6)$. Notice that block 3 is an inversion block. For block 4, $AT(P20) = \max\{AT(P1), AT(P2)\} + DT(4) = \max\{(2, 1), (3, 2)\} + (4, 3) = (7, 5)$. Since P10 and P20 are POs, the forward delay computation is completed.

2. *Backward slack evaluation.* To meet the performance requirement each PO has a required delay time (RT). At each PO the AT is compared with the RT and the difference $\Delta T = RT - AT$ is computed. This is called the *slack* (S) at the PO. Obviously, negative slacks means possible inadequate performance.

For example, let $RT(P10) = RT(P20) = (6, 6)$. The slack $S(P10) = (6, 6) - (6, 6) = (0, 0)$, which meets the requirement. On the other hand, $S(P20) = (6, 6) - (7, 5) = (-1, 1)$, which indicates a timing problem. The same procedure is then applied to all internal nets level by level from POs toward PIs. The required arrival time at net ①, $RT(P1) = \min\{RT(P10) - DT(\bar{3}), RT(P20) - DT(4)\} = \min\{(6, 6) - (4, 5), (6, 6) - (4, 3)\} = \min\{(2, 1), (2, 3)\} = (2, 1)$. Since $AT(P1) = (2, 1)$, $S(P1) = (0, 0)$. For net ②, $RT(P2) = RT(P20) - DT(4) = (6, 6) - (4, 3) = (2, 3)$. Since $AT(P2) = (3, 2)$, $S(P2) = RT(P2) - AT(P2) = (2, 3) - (3, 2) = (-1, 1)$. Thus the path $A3$ (or A4)–②–⑳ may be too slow.

Since each block is processed only twice, timing analysis can be applied to very large networks. Since the TA ignores the actual logic functions, however, negative slacks at a net indicate only potential timing problems. Not all delay paths "flagged" by the TA are logically possible. On the other hand, valid logic paths with negative slacks must be reevaluated by more accurate methods such as delay calculators or detailed circuit simulation.

6. CONCLUSION

In this chapter we have discussed MOS logic chip designs with equal emphasis on circuits and methodology. In practice there are two different design philosophies. One approach adopts a rather restrictive logic and circuit design style while leaving the test generation completely to the design system. The advantage of such a design is the uniform quality obtained from different designers. The drawback of such an approach is that many circuits that take advantage of MOS device characteristics such as dynamic circuits and pass gates are not allowed simply because of difficulties in modeling or generating test patterns based on a *particular fault model*. As a result, the design tends to be less optimal. The other approach allows the designer to implement logic functions in whichever way that optimizes performance or density. The advantage here is that the design is highly optimized. The problem is that the designer now has to take full responsibility for specifying the test. In either case, successful design of large logic chips depends not only on efficient design tools but also on the designer discipline.

REFERENCES

1. Burns, J. R. "Switching Response of Complementary-Symmetry MOS Transistor Logic Circuits." *RCA Review,* December 1964, pp. 627–661.

2. Roth, J. P. "Diagnosis of Automata Failures: A Calculus and a Method." *IBM J. Res. Development,* Vol. 10, July 1966, pp. 278–291.

3. Peterson, W. W., and E. J. Weldon. *Error-Correcting Codes.* Cambridge, MA: The MIT Press, 1972.

4. Eichelberger, E. B., and T. W. Williams. "A Logic Design Structure for LSI Testability." *Proc. 14th Design Automation Conference,* 1977, pp. 462–468.

5. Bottorff, P. S., et al. "Test Generation for Large Logic Networks." *Proc. 14th Automation Conference,* 1977, pp. 479–485.

6. Frohwerk, R. A. "Signature Analysis: A New Digital Field Service Method." *Hewlett-Packard J.,* 28, 1977, pp. 2–8.

7. Hwang, K. *Computer Arithmetic: Principles, Architecture, and Design.* New York: John Wiley, 1979.

8. Goldstein, L. H. "Controllability/Observability Analysis of Digital Circuits." *IEEE Trans. Circuits Systems,* CAS-26, 1979, pp. 685–693.

9. Mead, C., and L. Conway. *Introduction to VLSI Systems.* Reading, MA: Addison-Wesley, 1980.

10. Veendrick, H. J. M. "The Behavior of Flip-Flops Used as Synchronizers and Prediction of Their Failure Rate." *IEEE J. Solid-State Circuits,* Vol. SC-15, no. 2, April 1980, pp. 169–176.

11. Hill, F. J., and G. R. Peterson. *Introduction to Switching Theory and Logic Design.* New York: John Wiley, 1981.

12. Uehara, T., and W. M. van Cleemput. "Optimal Layout of CMOS Functional Arrays." *IEEE Trans. Computers,* Vol. C-30, no. 5, May 1981, pp. 305–312.

13. Goel, P. "An Implicit Enumeration Algorithm to Generate Tests for Combinational Logic Circuits." *IEEE Trans. Computer,* Vol. C-30, March 1981, pp. 215–222.

14. Waser, S., and M. J. Flynn. *Introduction to Arithmetic for Digital Systems Designers.* New York: Holt, Rinehart and Winston, 1982.

15. Pless, V. *Introduction to the Theory of Error-Correcting Codes.* New York: John Wiley, 1982.

16. Hitchcock, R. B., G. L. Smith, and D. D. Cheng. "Timing Analysis of Computer Hardware." *IBM J. Res. Development,* Vol. 26, no. 1, January 1982.

17. Donze, R. L., and G. Sporzynski. "Masterimage Approach to VLSI Design." *IEEE Computer,* December 1983.

18. Breuer, M. A., and A. D. Friedman. *Diagnosis & Reliable Design of Digital Systems.* Rockville, Maryland: Computer Science Press, 1984.

19. Heller, L. G., W. R. Griffin, J. W. Davis, and N. G. Thoma. "Cascode Voltage Switch Logic: A Differential CMOS Logic Family." *ISSCC Dig. of Tech. Papers,* February 1984, pp. 16–17.

20. Savir, J., G. S. Ditlow, and P. H. Bardell. "Random Pattern Testability." *IEEE Trans. Computer,* Vol. C-33, no. 1, January 1984.

21. Glasser, L. A., and D. W. Dobberpuhl. *The Design and Analysis of VLSI Circuits*. Reading, MA: Addison-Wesley, 1985.

22. Weste, N., and K. Eshraghian. *Principles of CMOS VLSI Design—A Systems Perspective*. Reading, MA: Addison-Wesley, 1985.

23. Annaratone, M. *Digital CMOS Circuit Design*. Hingham, MA: Klumer, 1986.

24. Dillinger, T. E. *VLSI Engineering*. Englewood Cliffs, NJ: Prentice Hall, 1988.

25. Chung, P. W., and N. N. Wang. "An Edge-Triggered Latch and Production of an Edge-Triggered Signal Therefor." IBM Corporation, Patent Pending.

APPENDIX

LINEAR FEEDBACK SHIFT REGISTERS

In Section 4 we used linear feedback shift registers to generate random test patterns for selftest. With proper feedback the n-bit LFSR can generate $2^n - 1$ distinct states which approximate true random patterns very well. In this appendix we shall relate the operation of LFSRs to the algebraic structure of a finite field. In particular we shall show that the states of a LFSR represent residue classes of polynomials generated by a primitive polynomial which is implemented by the feedback connections.

1. OPERATIONS ON POLYNOMIALS

A polynomial over $GF(2)$ is an expression

$$f(x) = a_n x^n + a_{n-1} x^{n-1} + \cdots + a_1 x + a_0, \qquad n > 0$$

with $a_k \in GF(2)$. The set $GF(2)$, to be explained shortly, consists of only two elements, 0 and 1, with two operations. The addition, $+$, and the multiplication, $*$, are defined in the following tables

+	0	1		*	0	1
0	0	1		0	0	0
1	1	0		1	0	1

Thus, from logic viewpoint, $a + b = a \oplus b$ (exclusive OR) and $a * b = ab$ (AND) for any two elements a and b of $GF(2)$. As usual, $a * b$ is also written as ab.

The degree of $f(x)$ is the largest n such that $a_n \neq 0$. Since the coefficients $a_k = 0$ or 1, a polynomial of degree n can be represented by an $(n + 1)$-tuple of 0's and 1's of its coefficients. For example, the polynomial $f(x) = x^4 + x^3 + 1$ can be represented by 11001, where the coefficients are ordered from left to right with descending power of x. Addition and multiplication of polynomials over $GF(2)$ are defined in the same fashion as with ordinary polynomials. For example, letting $g(x) = x^3 + x + 1$, the sum $f(x) + g(x)$ is then given by

$$
\begin{array}{r}
1\ 1\ 0\ 0\ 1 \\
1\ 0\ 1\ 1 \\
\hline
1\ 0\ 0\ 1\ 0
\end{array}
$$

Thus $f(x) + g(x) = x^4 + x$. Similarly, the product $f(x)g(x)$ is given by

$$
\begin{array}{r}
1\ 1\ 0\ 0\ 1 \\
1\ 0\ 1\ 1 \\
\hline
1\ 1\ 0\ 0\ 1 \\
1\ 1\ 0\ 0\ 1 \\
0\ 0\ 0\ 0\ 0 \\
1\ 1\ 0\ 0\ 1 \\
\hline
1\ 1\ 1\ 0\ 0\ 0\ 1\ 1
\end{array}
$$

Therefore, $f(x)g(x) = x^7 + x^6 + x^5 + x + 1$.

Divisions and subtractions can also be carried out in the usual fashion. For example, letting $h(x) = x^6 + x^4 + x^3 + x + 1$ be the dividend and $f(x) = x^4 + x^3 + 1$ be the divisor, we then have

$$
\begin{array}{r}
1\ 1 \\
1\ 1\ 0\ 0\ 1\ )\overline{1\ 0\ 1\ 1\ 0\ 1\ 1} \\
1\ 1\ 0\ 0\ 1 \\
\hline
1\ 1\ 1\ 1\ 1\ 1 \\
1\ 1\ 0\ 0\ 1 \\
\hline
1\ 1\ 0\ 1
\end{array}
$$

Thus $x^6 + x^4 + x^3 + x + 1 = (x + 1)(x^4 + x^3 + 1) + x^3 + x^2 + 1$. Notice that $0 - 1 = 1$ since $1 + 1 = 0$.

2. LINEAR SHIFT REGISTERS

Multiplication and division of polynomials in $GF(2)$ can be accomplished with linear shift registers. A linear shift register consists of delay elements, adders, and multipliers. The delay elements, being SRLs in most cases, perform the shift function. The adders, being XORs in $GF(2)$, have two inputs and one output. The binary multiplier has a simple implementation: a connection if the multiplier equals 1, no connection otherwise. The symbols and two typical linear shift registers are shown in Figure 80.

The linear shift register in Figure 80(b) is a multiplier which multiplies a polynomial $f(x) = a_n x^n + a_{n-1} x^{n-1} + \cdots + a_1 x + a_0$ by a *fixed* polynomial $g(x) = b_k x^k + b_{k-1} x^{k-1} + \cdots + b_1 x + b_0$. Coefficients of $f(x)$ enter the multiplier in descending order, that is, with a_n first, followed by a_{n-1}, then a_{n-2}, and so on. Let T be the shift register period. Perior to $t = 0$, all registers have been cleared to 0. Now at $t = 0$, a_n enters so the output is equal to $a_n b_k$, which is the desired coefficient of x^{n+k}. Then at $t = T$, the shift registers are updated to $a_n x g(x)$, with x^{k+1} term truncated during shifting. The output of SR_k is therefore equal to $a_n b_{k-1}$. At the same time, a_{n-1} enters and the output of the multiplier is then equal to $a_n b_{k-1} + a_{n-1} b_k$, which is the desired coefficient of x^{n+k-1}. Again, at $t = 2T$ content of the shift registers is updated to $a_n x^2 g(x) + a_{n-1} x g(x)$, with x^{k+2} and x^{k+1} terms truncated. The output of SR_k is therefore $a_n b_{k-2} + a_{n-1} b_{k-1}$. Since a_{n-2} enters at $t = 2T$, the output of the multiplier is equal to $a_n b_{k-2} + a_{n-1} b_{k-1} + a_{n-2} b_k$, which is the coefficient of x^{n+k-2}. It is now easy to see that after $n + k + 1$ shifts all

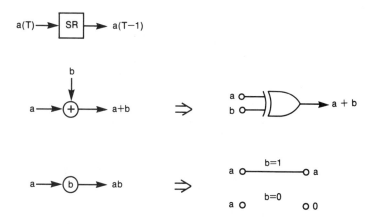

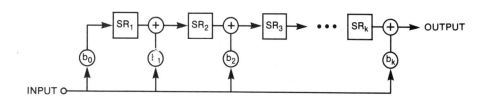

(b)   A multiplier.

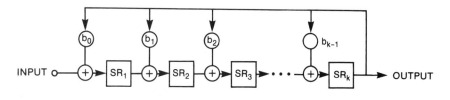

(c)   A divider.

Figure 80.   Linear shift registers.

coefficients of the product $f(x)g(x)$ have appeared at the output in descending order of powers of x.

Division by a fixed polynomial can also be implemented by linear shift registers *with feedback*. A divider is shown in Figure 80(c) which divides an arbitrary polynomial $f(x)$ by $g(x) = x^k + b_{k-1}x^{k-1} + \cdots + b_1 x + b_0$.

For convenience, let us also define $g(x) = x^k + g_1(x)$, where $g_1(x) = b_{k-1}x^{k-1} + b_{k-2}x^{k-2} + \cdots + b_1 x + b_0$.

The registers must be cleared before $t = 0$. At $t = 0$, a_n enters the divider. The output however remains 0 as the coefficients of $f(x)$ marches in until $t = kT$ when it is equal to a_n. This is, of course, the coefficient of the highest power of x in the quotient. It can be shown (see exercise below) that the remainder after the first comparison and subtraction is equal to $a_n g_1(x) + [f(x) - x^n]$, which is exactly the content of the registers at $t = T$. The compare and subtract process continues as the linear shift register progresses. At $t = nT$, the quotient have all appeared at the output and the remainder stays in the shift registers.

EXERCISE

Let $f(x) = a_n x^n + a_{n-1} x^{n-1} + \cdots + a_1 x + a_0 = a_n x^n + f_1(x)$ and $g(x) = x^n + b_{n-1} x^{n-1} + \cdots + b_1 x + b_0 = x^n + g_1(x)$. Prove that the remainder of $f(x)/g(x)$ is given by

$$r(x) = f_1(x) + a_n g_1(x)$$

Hint: See Section 5 of this appendix.

Use the above fact to justify the connection of the divider in Figure 80(c).

3. RESIDUE CLASSES AND THE EUCLIDEAN ALGORITHM

Let $g(x)$ be a polynomial of degree n. Let $f(x)$ be an arbitrary polynomial. Dividing $f(x)$ by $g(x)$, we obtain

$$f(x) = q(x)g(x) + r(x)$$

where $q(x)$ is the quotient and $r(x)$ is the remainder. By definition, the degree of $r(x)$ is less than n. For an arbitrary $f(x)$, there are 2^n possible $r(x)$ consisting of all possible polynomials of degree $n - 1$ as its remainder. By grouping all polynomials with the same remainder in the same class we have partitioned the set of all polynomials into 2^n equivalence classes. These are called *residue classes modulo $g(x)$*. Letting $\{r(x)\}$ be the residue class containing $r(x)$, it is easy to show that the addition and multiplication operations defined as

$$\{r(x)\} + \{s(x)\} = \{r(x) + s(x)\}$$
$$\{r(x)\} * \{s(x)\} = \{r(x) * s(x)\}$$

are valid binary operations among the residue classes. It can also be shown that for any three polynomials r, s, and t

$$\{t\}(\{r\} + \{s\}) = \{t\}\{r\} + \{t\}\{s\}$$

and that

$$(\{r\} + \{s\})\{t\} = \{r\}\{t\} + \{s\}\{t\}$$

so that the distributive laws hold. From now on we shall omit the variable x in the polynomials and the $*$ symbol for multiplications.

A polynomial f is said to be *congruent* to another polynomial h modulo g, denoted by $f \equiv h \mod g$, if they belong to the same residue class. In particular, $g \equiv 0 \mod g$.

Let f, s, and t be polynomials. If $f = st$, f is said to be divisible by s, and s is said to divide f. s is also called a divisor of f. Given two polynomials f and h, *the greatest common divisor of f and h*, denoted by $g.c.d.(f, h)$, is a polynomial that divides both f and h and that any common divisor of f and h also divides $g.c.d.(f, h)$. To find $g.c.d.(f, h)$, proceed with the following operations

$$f = q_1 h + r_1 \tag{A1}$$

$$h = q_2 r_1 + r_2 \tag{A2}$$

$$r_1 = q_3 r_2 + r_3 \qquad (A3)$$

$$\vdots$$

$$r_{k-2} = q_k r_{k-1} + r_k \qquad (A4)$$

$$r_{k-1} = q_{k+1} r_k \qquad (A5)$$

The process terminates in finite steps because the degrees of the polynomials r_1, r_2, . . ., and so on are monotonically decreasing. From Equation (A1), any common divisor of f and h must also divide r_1. From Equation (A2), it divides r_2, and from Equation (A3) it divides r_3. Conversely, since r_k divides r_{k-1} which divides r_{k-2} which divides r_{k-3}, . . ., r_k divides both f and h. By construction, then, $r_k = $ g.c.d. (f, h).

The process is the constructive proof of an extremely important mathematical algorithm. Starting with Equation (A4) and working backwards by substituting r_i into its previous expression involving r_{i-1} for all i, we arrive at an equation of the form

$$sf + th = \text{g.c.d.}(f, h) \qquad (A6)$$

This is the celebrated *Euclidean algorithm*. The euclidean algorithm asserts that given any two polynomials f and h, there exist polynomials s and t, such that $sf + th = $ g.c.d. (f, h). In particular, if f and h are relatively prime so that g.c.d. $(f, h) = 1$, there exist polynomials s and t such that $sf + th = 1$.

A polynomial g is said to be *irreducible* if the only divisors of g are 1 and g itself. Let f be another polynomial. Then the g.c.d. (f, g) is either 1 or g. In case g.c.d. $(f, g) = g$, f is divisible by g and hence $f \equiv 0 \mod g$.

4. THE FINITE FIELD OF RESIDUE CLASSES

A field F is a set of elements with two binary operations, addition $+$ and multiplication $*$ satisfying the following conditions:

For addition, let a, b, and c be any three elements of F:

(A-1) Associativity:

$$(a + b) + c = a + (b + c)$$

(A-2) Commutativity:

$$a + b = b + a$$

(A-3) There exists an identity element, 0, such that:

$$a + 0 = 0 + a = a$$

(A-4) For any element a, there exists an inverse of a, denoted $-a$ such that:

$$a + (-a) = (-a) + a = 0$$

For multiplication, let a, b, and c be any three elements of F:

(M-1) Associativity:

$$(a * b) * c = a * (b * c)$$

(M-2) Commutativity:

$$a * b = b * a$$

(M-3) There exists an identity element, 1, such that:

$$a * 1 = 1 * a = a$$

(M-4) For any element a, a ≠ 0, there exists an inverse of a, denoted a^{-1} such that:

$$a * a^{-1} = a^{-1} * a = 1$$

The two operations are related by the distributive law

(D-1) $$(a + b) * c = (a * c) + (b * c)$$

(D-2) $$c * (a + b) = (c * a) + (c * b)$$

Condition (A-1)–(A-4) are the definitions of an *abelian group*. A group is said to be abelian if the binary operation is commutative. Conditions (M-1)–(M-4) imply that the nonzero elements of F also constitutes an abelian group. Thus a field is a set with two group structures associated with two operations related by the distributive law. The group associated with "+" is called the additive group and that associated with "*" is called the multiplicative group.

A field is said to be finite if it has a finite number of elements. Finite fields are also called *Galois fields*. A finite field of n elements is denoted by $GF(n)$. The set of coefficients of the polynomials we have considered, $GF(2)$, is therefore a finite field of two elements. Another good example of finite fields is provided by the following lemma:

> **Lemma:** The residue classes created by an irreducible polynomial form a finite field.
>
> *Proof.* Let g be an irreducible polynomial of degree n. We prove that the 2^n residue classes created by g form a field. It is straightforward to show that conditions (A-1)–(A-4) with $0 = \{0\}$ and (M-1)–(M-3) with $1 = \{1\}$ are satisfied. To prove (M-4), let $f \not\equiv 0$ be an arbitrary polynomial. By Euclidean algorithm, there exist polynomials s and t such that
>
> $$sf + gt = 1$$
> $\Rightarrow$ $$sf \equiv 1$$

The residue class $\{s\}$ is therefore the multiplicative inverse of $\{f\}$. Verification of the distributive laws have (presumably) been done by the reader in the previous section.

Up to now we have been able to derive all necessary theorems in the development of finite field of residue classes of polynomials. The theory of finite fields, however, is much too involved to be discussed here. The following theorems, which are the foundation of the linear switching circuit, will thus be stated without a proof.

> **Theorem 1:** For each n there is an irreducible polynominal g of degree n such that the elements of the multiplicative group of the the residue classes

modulo g are congruent to x^m, $m = 1, 2, \ldots, (2^n - 1)$. That is, each $x^m \equiv$ a polynomial of degree $n - 1$ or less modulo g for $m = 1, 2, \ldots, (2^n - 1)$. Since there are $2^n - 1$ elements in the group, obviously $x^{2^n} = 1$ and the sequence of x^m, $m = 1, 2, \ldots,$ is cyclic with a period equal to $2^n - 1$.

Polynomial g is called a *primitive polynomial*. All irreducible polynomials are not primitive polynomials. If g is a primitive polynomial the residue classes x^m, $m = 1, 2, \ldots, (2^n - 1)$ are all distinct constituting the whole multiplicative group. There are no systematic ways other than trial and error to detect all primitive polynomials. Tables that list all irreducible and primitive polynomials up to degree 34, however, are available in the literature. [3]

5. LINEAR FEEDBACK SHIFT REGISTERS

We are now ready to relate the linear feedback shift register to the generation of residue classes of polynomials. To explain the implementation of linear feedback shift registers, however, the following fact is helpful. Let f be a polynomials of degree n or less; that is,

$$f(x) = a_n x^n + a_{n-1} x^{n-1} + \cdots + a_1 x + a_0$$
$$= a_n x^n + f_1(x) \tag{A7}$$

and g be a polynomial of degree n

$$g(x) = x^n + b_{n-1} x^{n-1} + \cdots + b_1 x + b_0$$
$$= x^n + g_1(x)$$

dividing f by g, we obtain

$$f(x) = a_n g(x) + r(x)$$
$$= a_n x^n + a_n g_1(x) + r(x) \tag{A8}$$

comparing Equation (A7) with (A8), we conclude that

$$r(x) = f_1(x) + a_n g_1(x) \tag{A9}$$

Thus, the residue class of $f(x)$ modulo $g(x)$, represented by $r(x)$, is given by $f_1(x) + a_n g_1(x)$. This concept is essential in implementing linear feedback shift registers.

Consider the linear feedback shift register in Figure 81(a). As shown, the n-bits stored in the shift register represent a polynomial of degree $n - 1$: $f_1(x) = a_{n-1} x^{n-1} + a_{n-2} x^{n-2} + \cdots + a_1 x + a_0$. The feedback connection adds $a_{n-1} g_1(x)$ to the shift register where $g_1(x) = b_{n-1} x^{n-1} + b_{n-2} x^{n-2} + \cdots + b_1 x + b_0$. Without feedback, after one shift cycle the shift register would represent $a_{n-2} x^{n-1} + a_{n-3} x^{n-2} + \cdots + a_1 x^2 + a_0 x$, that is, $xf(x)$ with $a_{n-1} x^n$ rounded off. With feedback, however, $a_{n-1} g_1(x)$ is added to the shift register during the same cycle. According to Equation (A9), then, the value left in the shift register after one complete cycle is exactly the remainder of $xf(x)$ divided by $g(x) = x^n + g_1(x)$.

The construction of an n-bit maximum length LFSR is now obvious. Choose a feedback $g_1(x)$ so that $g(x) = x^n + g_1(x)$ is a primitive polynomial. Start the shift

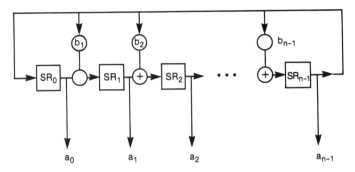

(a) General configuration.

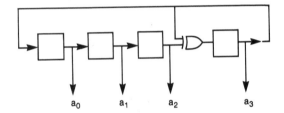

(b) A LFSR with primitive polynomial $x^4 + x^3 + 1$.

Figure 81. Linear feedback shift registers.

register with an arbitrary initial state not equal to **0**. Since $g(x)$ is primitive, the state in the shift register is congruent to x^m for some $m < 2^n - 1$. The next state after one cycle will be congruent to x^{m+1}, and the next, x^{m+2}, and so on. Since $g(x)$ is primitive, the period of the sequence is $2^n - 1$ states, which is, of course, the maximum length. A LFSR that generates 15 distinct states based on $g(x) = x^4 + x^3 + 1$ is shown in Figure 81(b) and a LFSR based on $g(x) = x^3 + x + 1$ is shown in Figure 77.

Another method of constructing maximum length LFSR is based on the following theorem.

Theorem 2:   Consider the linear feedback shift register shown in Figure 82. Let

$$g(x) = x^n + b_{n-1}x^{n-1} + \cdots + b_1x + 1$$

If $g(x)$ is a primitive polynomial, then the shift register is of maximum length. The period of the output sequence is $2^n - 1$.

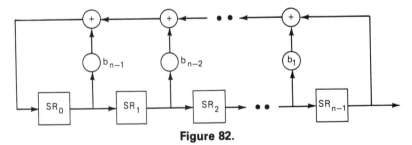

Figure 82.

A LFSR that generates 15 distinct states based on $g(x) = x^4 + x^3 + 1$ in this configuration is shown in Figure 77.

The generation of pseudo random test patterns is a straightforward application of the finite field theory. More sophisticated applications can be found in algebraic coding and information theory in electrical and communication engineering. Interested readers should consult with the references.

INDEX

Active pullup, 262–63
Adders:
　carry bypass, 317–18
　carry look-ahead (CLA), 318–19
　carry save, 328
　carry select, 319–20
　full (FA), 314–16
　half, 320
　Manchester carry chain, 316–17
Address transition detection (ATD),
　229–31
AI, 279
Aliasing, 345
ALU (arithmetic/logic unit), 320–22
AOI, 280–84
Aspect ratio, 246
Availability, 246

Bandwidth:
　circuits, 26
　memory, 268
Barrel shifter, 288
Best case (BC):
　delay, 347
　performance, 112
Bidi, 279
Bipolar (parasitic), 46
Bird's beak, 37–38, 41
Bitlines:
　DRAM, 243–44
　　droop, 261–62
　　folded, 267–68
　SRAM, 231–34
　　dynamic, 231
　　precharge, 212

restore, 225
　static, 231–34
Body effect, 41
Book (logic), 274
Booth algorithm (modified), 323–25
Bootstrap efficiency, 101
Breakdown:
　S/D junction, 41
　snapback, 46
Buried channel, 43
Buried contact, 38
Byte mode, 268

Capacitance (device), 19–22
CAS (column address selection/strobe),
　238
CCC (corrugated capacitor cell), 55
Cells:
　1-device, 237, 245–49
　4-device dynamic, 234–37
　4-device static, 231–34
　logic, 278
　PLA, 329
Channel:
　buried-channel device, 43
　length, 10
　length dilation, 14
　length modulation, 46
　narrow-channel effect, 37, 42
　short-channel effect, 41
　stopping, 36
　tailoring, 42
　width, 10
Characteristic impedance, 182, 199
Charge pump, 207

Chip image, 274–79
Chip modes, 268–69
Clock feedthrough, 295
CMOS open faults, 337
Counters, 298–300
Crosstalk:
　backward, 193, 203–6
　forward, 194, 203–6
Cut-off frequency, 25

Dadda's reduction scheme, 323,
　328
Data flushthrough, 294
Dataline, 329
DC stability, 67
DCVS (differential cascode voltage
　switch), 311–14
Decoders:
　DRAM, 244–45
　logic, 286
　PLA, 330
　SRAM, 216–17
Delay:
　average, 115
　BC (best case), 347
　block, 115
　calculation, 115–23, 163–65
　calculators for inverters, 123–31,
　　195–97
　fall (t_{DHL}), 115
　inherent (intrinsic), 92, 95–96, 100
　rise (t_{DLH}), 115
　WC (worst case), 347
Distance (between two vectors), 210
Domino logic, 309–11

DRAM (dynamic random access memory), 54–57
Drivers:
  CMOS, 175–79
  differential, 156–57
  dynamic, 166–75
  offchip (OCD), 179–94
  short-circuit protection, 181–82
  single-ended, 157–79
  transmission line, 182–92
  tristate, 179–81
DSR (dynamic shift register), 302
Dynamic logic, 304–14
Dynamic memory, 237

Edge-triggered latch, 300–301
Electromigration, 188, 278
Electron affinity, 2
Environmental variables, 112
Epitaxial, 53
Equilibrium states, 73, 219–21, 233–34, 289

Falltime, 108
FCC (folded capacitor cell), 55
Field tailoring, 36
Finite field, 355–56
Flat band voltage, 4
Flip-flops, 297–98
Folding (of PLA), 330–32
FOX (field oxide) 36–37

Galois field, 355–56
Gate:
  polysilicon, 38
  self-aligned, 39
Gatedrive, 10
Gettering, 36
GF (2), 351
GPR (general purpose register), 334
Greatest common divisor (g.c.d.), 354

Hi-C RAM cell, 55
Holding current, 48
Holdtime, 214
Hot-carrier effect, 42
Hot electrons, 42

Inherent (intrinsic) delay:
  constant pullup, 92, 95–96
  depletion mode pullup, 100
Inversion layer, 7
Inverter:
  bootstrap, 100
  CMOS, 63
  E/D, 60
  E/E, 60
Irreducible polynomial, 355

Latches:
  general, 288–93
  shift register, 293–95
  edge-triggered, 300–301
Latchup, 46–49
LDD (lightly-doped drain), 50
LFSR (linear feedback shift register), 345–46, 351–58
LOCOS (local oxidation of silicon), 37

Logic threshold, 62
LPCVD (low-pressure chemical vapor deposition), 38
LPUL (least positive uplevel), 68
LSSD (level sensitive scan design), 340–45

Macros, 274
Masterimage, 274
Metastability, 291–92
Miller effect, 131–32
MISR (multiple input shift/signature register), 345–46
MPDL (most positive downlevel), 68
Multiplier, 322–28

NAND, 379
Narrow-channel effect, 37
Nibble mode, 268
Noise:
  margins, 70–75
  tolerance curves, 155
NOR, 279
Norm (of a vector), 210
N-well, 36

OAI, 280–84
OI, 279
Output resistance, 46

Page mode, 268
Parity, 286
Photoresist (PR), 36
Pinchoff voltage, 12
PLA (programmable logic array), 328–33
Positive logic, 210
Powerdown, 226
Primary input (PI), 338
Primary output (PO), 338
Primitive polynomial, 357
Product line, 329
Pseudorandom patterns, 345
PSG (phosphosilicate glass), 40
Pulldown, 107–9
Pullup:
  bootstrap, 100–107
  constant gate voltage, 89–97
  depletion mode, 98–100
  PMOS, 109–11
Pulse shrinkage, 347
Pulsewidth, 214
Punchthrough, 41
Pushpull drivers, 157

Random access memory (RAM):
  dynamic, 237
  static, 209
RAS (row address selection/strobe), 238
Ratioed design, 74, 302
Ratioless design, 74, 302
Ratio of ratios, 67
Receivers:
  differential, 142–49
  noise margin, 155–56
  offchip (OCR), 142–56
  single-ended, 149–55
Redundancy, 270
Reflection coefficient, 183, 201–2

Refresh (DRAM cell):
  CAS before RAS, 270
  hidden, 269
  RAS-only, 269
Register file, 334
Rent's rule, 276
Residue classes, 354
Retrograde-wells, 53
Ripple mode, 269
Risetime:
  bootstrap pullup, 102
  constant pullup, 90, 93
  depletion mode pullup, 98–99
  PMOS pullup, 110–11
ROM (ROS), 333–34
Rotator, 288
ROX (semirecessed oxide), 36–37
  etchback, 37

Salicide (self-aligned silicide), 51
Scaling theory, 49–50
Schmitt triggers, 149–55
SCR (silicon-controlled rectifier), 46
Sense amplifier:
  charge transfer, 264–65
  differential, 223–25
  dynamic, 237, 250–67
  half V_H, 266–67
Setup time, 214, 300
Short-channel effect, 41
Simultaneous switching, 194
Snapback, 46
Soft error, 271
Spacers, 50
SPT (substrate-plate trench capacitor cell), 55
Square law, 12
SRL (shift register latch), 293–95
SROX (semirecessed oxide), 36–37
Stability:
  DC network, 67
  latches, 73
  SRAM cell, 218–21
Standard cells, 274
Static column mode, 269
Stuck-at fault, 335–38
Substrate:
  bias generator, 206–8
  sensitivity, 27–28, 41
Subthreshold current, 43
Surface potential, 8
Surface states, 3–4
Sustaining voltage, 47

TC (transfer characteristic) plot, 62
T/C (true and complement), 210, 215
Temperature coefficient:
  mobility, 27
  threshold voltage, 26–27
Thermal equilibrium, 3
Threshold voltage:
  adjustment, 42
  depletion mode devices, 18
  enhancement mode devices, 10
  tracking, 76, 347
Time of flight, 182
Tracking (threshold voltage), 76, 347
Transconductance, 22–24
Transfer ratio, 248
Transit time, 24
Transition frequency, 25

Transition voltage, 62–64
Trench isolation, 52
TTC (trench transistor cell), 57
TTL compatibility, 179
Two-bit partitioning, 330

Unidi, 279
Unity gain point, 62–64

Vacuum level, 2
Valid time, 214

Velocity saturation, 44
Via:
   hole, 40
   stacked, 40

Wallace trees, 328
Wordlines:
   DRAM, 240
   PLA, 329
   SRAM, 211
Work function, 2

Worst case (WC):
   delay, 347
   performance, 112
   power, 158–63

XNOR, 284–86
XOR, 284–86

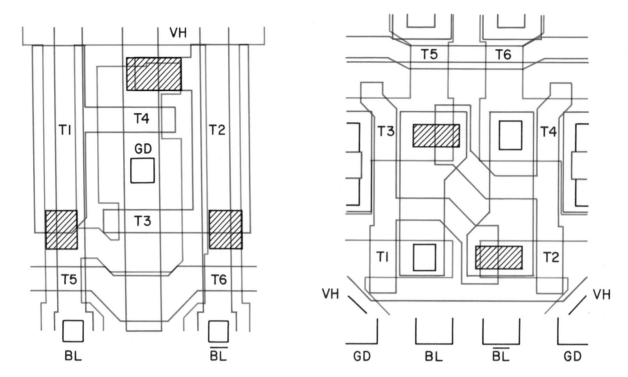

Figure 6-9. **Examples of SRAM cells.** *Color code:* Red—diffusion; Green—polysilicon; Blue—M1; Brown—M1-M2 via; Black—M1 contact or M2.

(a) **An E/D cell.** Shaded areas are buried contacts. Depletion implant mask for T1 and T2 are not shown. (Reprinted with permission from *Electronics*—August 1977. © 1977 by VNU Business Publications)

(b) **A CMOS cell.** Shaded areas are butting contacts that connect polysilicon, metal and diffusion together. The p-well mask is not shown. (Ref. 11, © 1986 by IEEE)

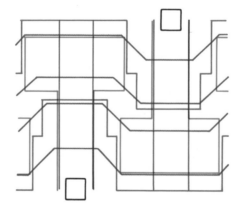

Figure 6-32. **Folded bitlines 1-device cell.** *Color code:* Red—diffusion; Green—poly 1; Blue—M1; Brown—poly 2; Black—M1 contact. Shaded areas are storage capacitors. (Ref. 12, © 1980 by IEEE)